AF718642

TECHNICAL
Sketching and Drawing

AMERICAN TECHNICAL PUBLISHERS, INC.
ORLAND PARK, ILLINOIS 60467-5703

Technical Sketching and Drawing contains procedures commonly practiced in industry and the trade. Specific procedures vary with each task and must be performed by a qualified person. For maximum safety, always refer to specific manufacturer recommendations, insurance regulations, specific job site and plant procedures, applicable federal, state, and local regulations, and any authority having jurisdiction. The material contained is intended to be an educational resource for the user. American Technical Publishers, Inc. assumes no responsibility or liability in connection with this material or its use by any individual or organization.

American Technical Publishers, Inc., Editorial Staff

Editor in Chief:
Jonathan F. Gosse
Vice President—Production:
Peter A. Zurlis
Art Manager:
James M. Clarke
Technical Editor:
Jennifer M. Hines
Technical Reviewer:
Eric T. Comiza
Copy Editor:
Diane J. Weidner
Cover Design:
Jennifer M. Hines
Illustration/Layout:
Jennifer M. Hines
Eric T. Comiza
Samuel T. Tucker
Multimedia Coordinator:
Carl R. Hansen
CD-ROM Development:
Robert E. Stickley
Daniel Kundrat
Gretje Dahl

Firefox is a registered trademark of Mozilla Corporation in the United States and other countries. Intel and Pentium are registered trademarks of Intel Corporation or its subsidiaries in the United States and other countries. Microsoft, Windows Vista, Windows XP, Windows 2000, Windows NT, and Internet Explorer are either registered trademarks or trademarks of Microsoft Corporation in the United States and/or other countries. Netscape is a registered trademark of Netscape Communications Corporation in the United States and other countries. Quick Quiz and Quick Quizzes are registered trademarks of American Technical Publishers, Inc.

© 2008 by American Technical Publishers, Inc.
All rights reserved

1 2 3 4 5 6 7 8 9 – 08 – 9 8 7 6 5 4 3 2 1

Printed in the United States of America

ISBN 978-0-8269-1164-3

This book is printed on 30% recycled paper.

made in USA

Acknowledgments

The authors and publisher are grateful for the technical information and assistance provided by the following companies, organizations, and individuals.

Boston Gear
Carrier Corporation
Kolbe & Kolbe Millwork Co., Inc.
Staedtler, Inc.
Viking Pump, Inc., A Unit of IDEX Corporation

Technical Sketching and Drawing

Contents

1 Drawing Fundamentals 1

Sketching • Materials • Sketching Basics • Composition • Figure and Ground • Sighting • Depth Illusion • Tone and Texture

Review Questions 15

2 Instrument Drawings and Techniques 21

Instrument Drawings • Standard Tools • Technical Drawing Instruments • Line Drawing • Drawing Lines • Line Types • Illustrating Technical Drawings • Inking • Shading • Rendering • Lettering Technical Drawings • Sizing • Formation • Spacing

Review Questions 39

3 Geometric Constructions 49

Geometric Shapes and Terminology • Two-Dimensional Shapes • Three-Dimensional Solids • Geometric Construction Principles • Drawing Lines • Bisecting Geometric Components • Circular Geometric Constructions • Plane Figures • Enlarging or Reducing Shapes • Complex Geometric Construction

Review Questions 75

4 Pictorial Drawings 81

Pictorial Drawings • Axonometric Drawings • Oblique Drawings • Perspective Drawings • Complex Pictorial Drawings • Interior Arrangement Drawings • Cutaway Drawings • Exploded View Drawings

Review Questions 95

5 Multiview, Auxiliary View, and Sectional View Drawings 103

Multiview Drawings • Orthographic Projection • View Relationships • View Placement • Drawing Multiviews • Auxiliary View Drawings • Primary Auxiliary Views • Secondary Auxiliary Views • Drawing Auxiliary Views • Sectional View Drawings • Cutting-Plane Lines • Section Lines • Sectional View Drawing Types

Review Questions 212

6 Dimensioning and Tolerancing 127

Dimensioning • Indicating Dimensions • Dimension Types • Units of Measure • Notes • Dimension Placement • Tolerancing • Limit Dimensioning • Plus and Minus Tolerancing • Dimensioning Methods • Geometric Dimensioning and Tolerancing • Geometric Dimensioning • Datums • Feature Control

Review Questions **141**

7 Prints 149

Prints • Print Format and Features • Abbreviations and Symbols • Title Blocks • Schedules • Zones • Notes • Specifications • Section and Detail Drawings • Architectural Prints • Architectural Plans • Architectural Elevations • Structural Prints • Mechanical Prints • Plumbing Prints • HVAC Prints • Fire Protection System Prints • Electrical Prints • Electrical Diagrams • Schematic Diagrams • Civil Prints • Manufacturing Prints

Review Questions **163**

8 Specialized Drawings 171

Information Representation • Symbols • Graphs • Topography • Contour Lines • Profiles • Manufacturing Elements • Developments • Sheet Metal Patterns • Threaded Fasteners • Nonthreaded Fasteners • Springs • Gears • Cams

Review Questions **183**

Appendix **191**

Glossary **201**

Index **209**

Interactive CD-ROM Contents

- *Using This CD-ROM*
- *Quick Quizzes®*
- *Illustrated Glossary*
- *Flash Cards*
- *Interactive Tutorials*
- *Interactive Exercises*
- *Print and Drawing Library*
- *Media Clips*
- *ATPeResources.com*

Technical Sketching and Drawing

Introduction

Technical Sketching and Drawing presents both the basic and advanced skills needed to visually communicate technical information with speed and accuracy. This textbook combines sketching principles with fundamental drawing techniques to teach a variety of visual communication methods. It addresses practical and commonly used skills as well as introducing more advanced concepts. Offering a series of progressive technical drawing techniques, the textbook provides learners with the capabilities and knowledge needed to develop their skills.

This comprehensive textbook contains eight chapters covering topics such as drawing fundamentals, geometric construction, dimensioning and tolerancing, and prints. *Technical Sketching and Drawing* includes review questions and activities that enable hands-on application of the concepts presented. Packed with step-by-step drawing instructions, this textbook illustrates advanced concepts and includes photographs and tech tips that complement the content.

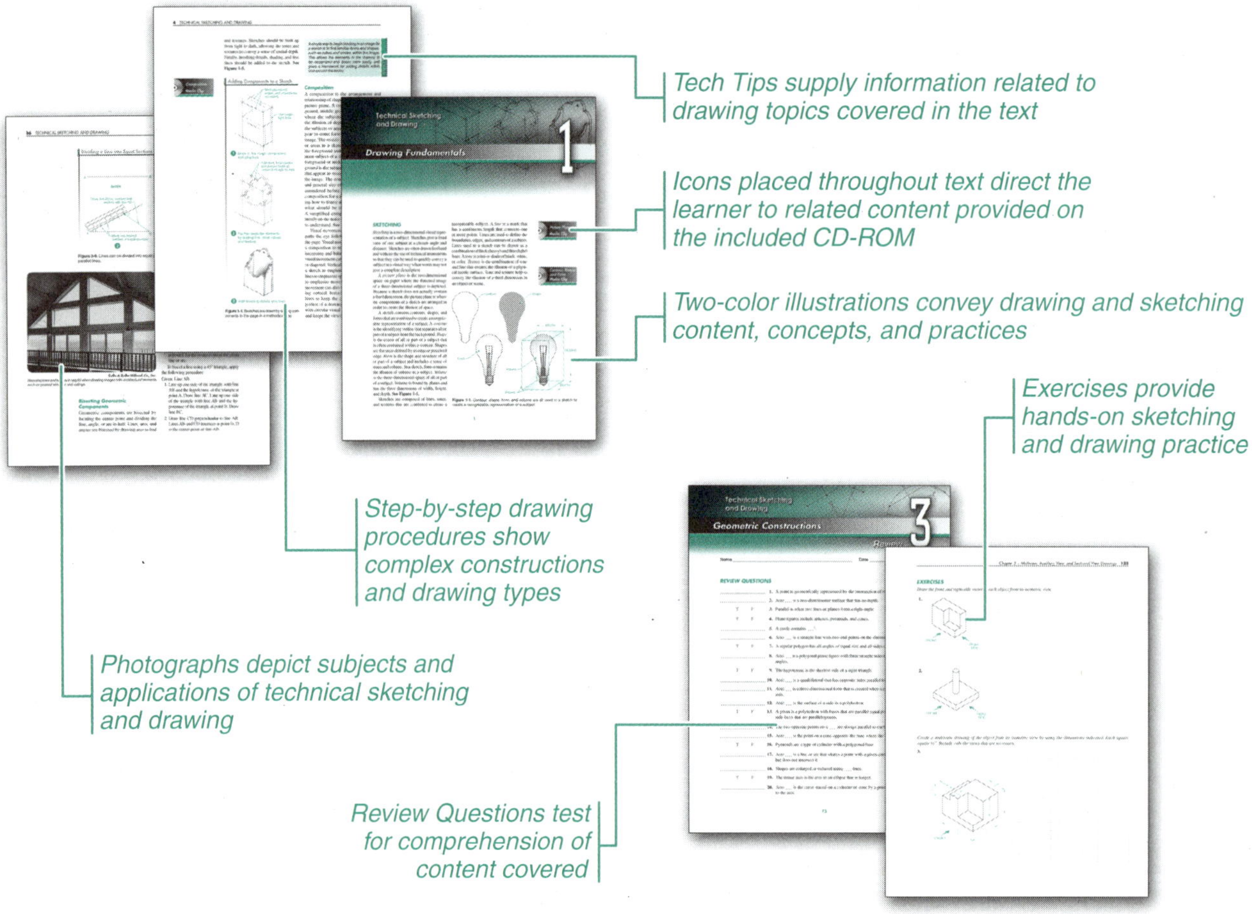

Interactive CD-ROM Contents

The CD-ROM included in the textbook is a self-study aid with the following features:

- Quick Quizzes® that provide an interactive review of topics covered in the chapters, with 10 questions for each chapter
- An Illustrated Glossary that provides a visual reference to key terms in the text
- Flash Cards that enable a self-study/review of terms and definitions, drawing tools, shapes and solids, and drawing types
- Interactive Tutorials that give step-by-step instructions for creating drawings and geometric constructions
- Interactive Exercises with components that learners drag and drop to test their comprehension
- A Print and Drawing Library that contains examples of prints and drawings in PDF format
- Media Clips that illustrate principles discussed in the book using video clips and animations
- Access to ATPeResources.com, an on-line source providing a variety of instructional resources related to sketching and drawing

To obtain information about related products from American Technical Publishers, visit our website at www.go2atp.com or log onto www.atperesources.com.

Technical Sketching and Drawing

1 Drawing Fundamentals

SKETCHING

Sketching is a two-dimensional visual representation of a subject. Sketches give a fixed view of one subject at a chosen angle and distance. Sketches are often drawn freehand and without the use of technical instruments so that they can be used to quickly convey a subject in a visual way when words may not give a complete description.

A *picture plane* is the two-dimensional space on paper where the flattened image of a three-dimensional subject is depicted. Because a sketch does not actually contain a third dimension, the picture plane is where the components of a sketch are arranged in order to create the illusion of space.

A sketch contains contours, shapes, and forms that are combined to create a recognizable representation of a subject. A *contour* is the identifying outline that separates all or part of a subject from the background. *Shape* is the extent of all or part of a subject that is often contained within a contour. Shapes are flat areas defined by an edge or perceived edge. *Form* is the shape and structure of all or part of a subject and includes a sense of mass and volume. In a sketch, form contains the illusion of volume in a subject. *Volume* is the three-dimensional space of all or part of a subject. Volume is bound by planes and has the three dimensions of width, height, and depth. **See Figure 1-1.**

Sketches are composed of lines, tones, and textures that are combined to create a recognizable subject. A *line* is a mark that has a continuous length that connects one or more points. Lines are used to define the boundaries, edges, and contours of a subject. Lines used in a sketch can be drawn as a combination of thick (heavy) and thin (light) lines. A *tone* is a tint or shade of black, white, or color. *Texture* is the combination of tone and line that creates the illusion of a physical tactile surface. Tone and texture help to convey the illusion of a third dimension in an object or scene.

Picture Plane Media Clip

Contour, Shape, and Form Media Clip

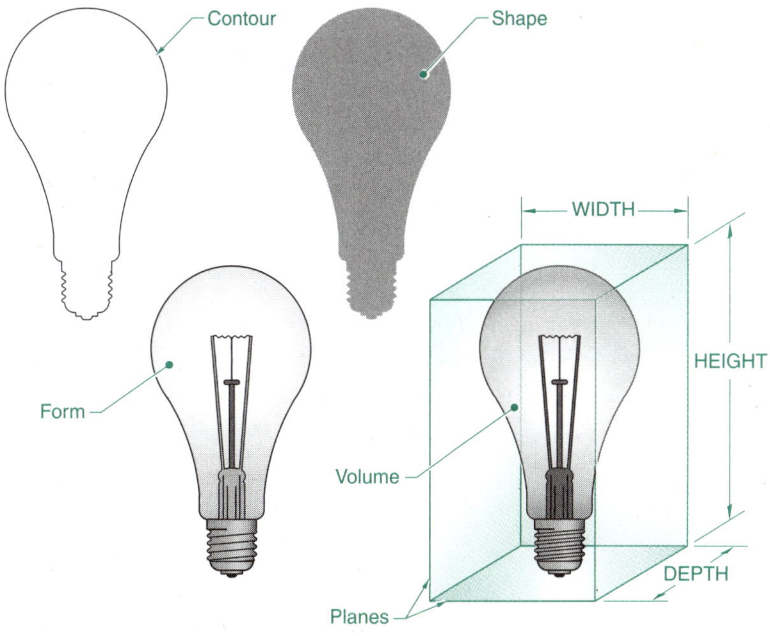

Figure 1-1. Contour, shape, form, and volume are all used in a sketch to create a recognizable representation of a subject.

Materials

When making a freehand sketch, the artist begins with a piece of paper and a pencil, pen, or other mark-making device, such as charcoal, conté crayon, or chalk. Paper that is used for freehand sketching is generally plain and unlined. Paper is characterized by its material and weight. Paper is generally made from wood pulp or cotton fibers. A paper's weight is determined by the weight of 500 sheets, or one ream, of the paper at a specific size and is given in pounds (lb). Heavier-weighted paper can withstand more rough handling and erasing. The weight and type of paper also can be a factor in what materials can be used in the sketch. Most sketching papers work well with pencil, pen, or charcoal. Ink, watercolor, chalk, or another source of color can also be used if the paper type is suited to the marking medium.

Pencils Media Clip

Wood pencils are often used in sketching. Wood pencils are designated as hard (H) or soft (B) with a number scale indicating the degree of hardness or softness. The hardness or softness of the lead is indicated with a stamp near one end of the pencil shaft. Hard leads create lighter, thinner marks on the paper, whereas soft leads create darker, thicker marks on the paper.

Soft leads are used primarily for sketching. Medium leads are used to draw general object lines. Hard leads are used to draw fine, precise lines. Grades of lead range from 6B (extremely soft) to 9H (exceptionally hard). Sketching is recommended with pencils that have a lead grade that is in the soft to middle of the range—lead that is too soft will smudge easily, and lead that is too hard will make it difficult to put smooth marks on the paper. The most commonly used lead grades for sketching include B, HB, F, and H. **See Figure 1-2.**

Pencils and other mark-making utensils should be sharpened into a point or near-point for best results. Ink pens used for lines and detail work normally have a fine or medium tip. Broad tips are generally used for more gestural lines or for larger areas of tone.

Pencil Leads

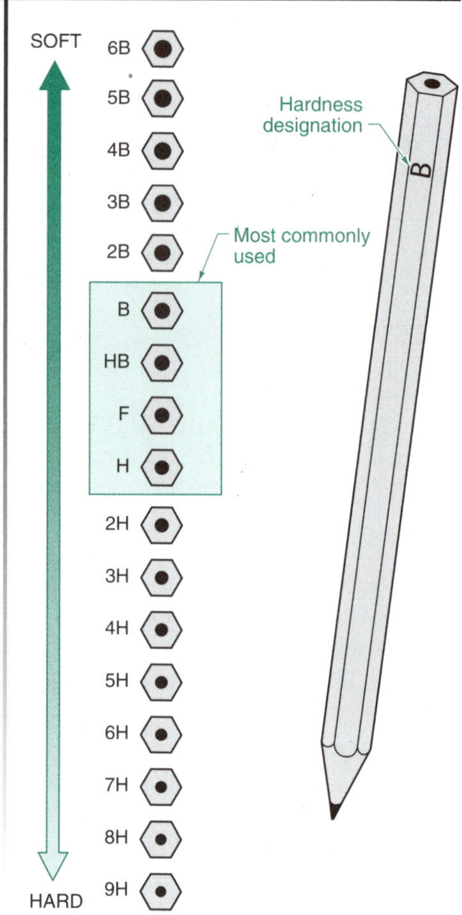

Figure 1-2. Wood pencils are designated as hard (H) or soft (B) with a number scale indicating the degree of hardness or softness.

Sketching Basics

A sketch should be drawn with confidence and without the use of technical tools or erasers. In order to be fully prepared before an initial sketch is started, it is important to warm up.

Warming Up Media Clip

Warming up before starting a sketch allows the lines in the sketch to be more natural and fluid. Drawing a series of straight horizontal and vertical lines, as well as coils, circles, and ellipses, helps to loosen the muscles of the drawing hand. Varying the direction the drawing instrument moves (from right to left or top to bottom) will allow the arm to feel natural moving in all directions. **See Figure 1-3.**

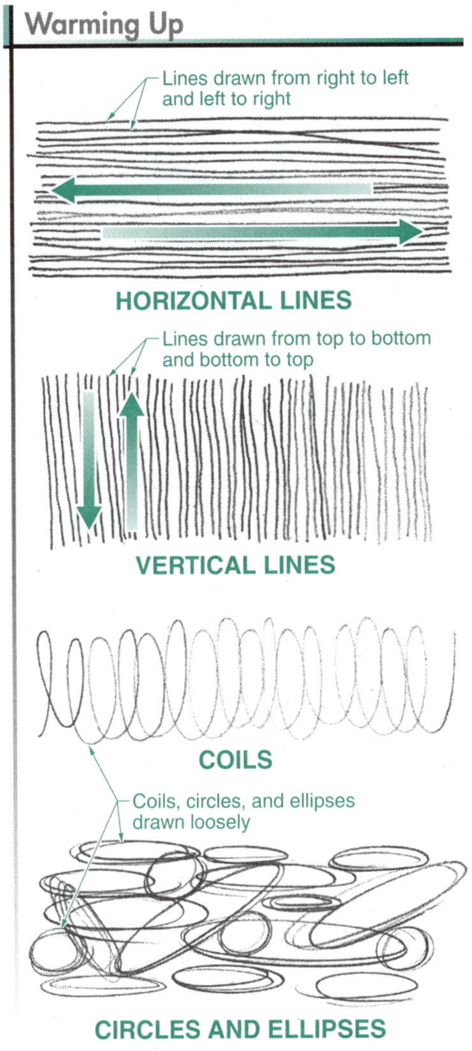

Figure 1-3. In order to be fully prepared before an initial sketch is started, it is important to warm up by drawing a series of lines, coils, circles, and ellipses.

Contour drawings can be an important exercise to strengthen eye-hand coordination before sketching. Contour drawings are an exploration of a subject's contour and shape and can be done blind or by looking at the subject. They are often completed quickly as a warm-up exercise. Blind contour drawings are drawn by looking only at the subject and not looking at the drawing once the pencil or pen touches the paper. The finished blind contour drawing is not intended to look like an accurate depiction of the subject, but is used to carefully study the contour of the subject. **See Figure 1-4.**

Blind contour drawing is an exercise that can be done occasionally or regularly, but it is not necessary to do before beginning every sketch. Contour drawing can also be done by looking at both the subject and the paper while completing the drawing. Unlike blind contour drawing, this type of contour drawing more precisely explores the contours of a subject and will give a more accurate depiction of an object.

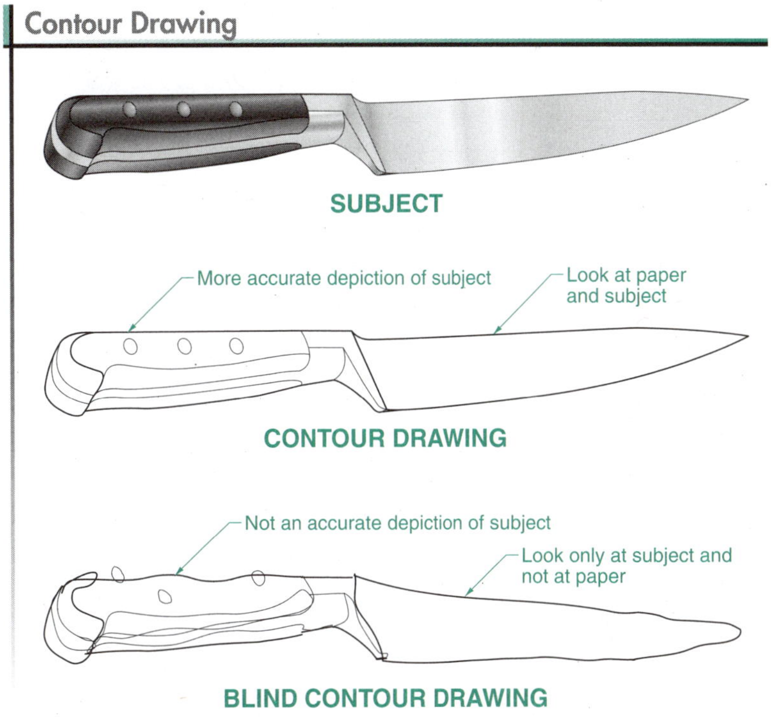

Figure 1-4. Contour drawings can be an important exercise to strengthen eye-hand coordination before sketching.

Sketches are drawn by adding components to the page in a methodical way. Addressing elements on the entire subject or page at once creates a more accurate and balanced finished sketch. To begin a sketch, block in the layout and structure using rough, light lines. *Blocking in* is the method of quickly marking the structure of a subject on a sketch by breaking the subject into common shapes and lines. The placement, shape, and proportions of the major elements in the sketch should be blocked in accurately before more detail is added. Next, objects can be further depicted by adding more lines, tonal values, and textures. Sketches should be built up from light to dark, allowing the tones and textures to convey a sense of spatial depth. Finally, finishing details, shading, and fine lines should be added to the sketch. **See Figure 1-5.**

Composition Media Clip

Adding Components to a Sketch

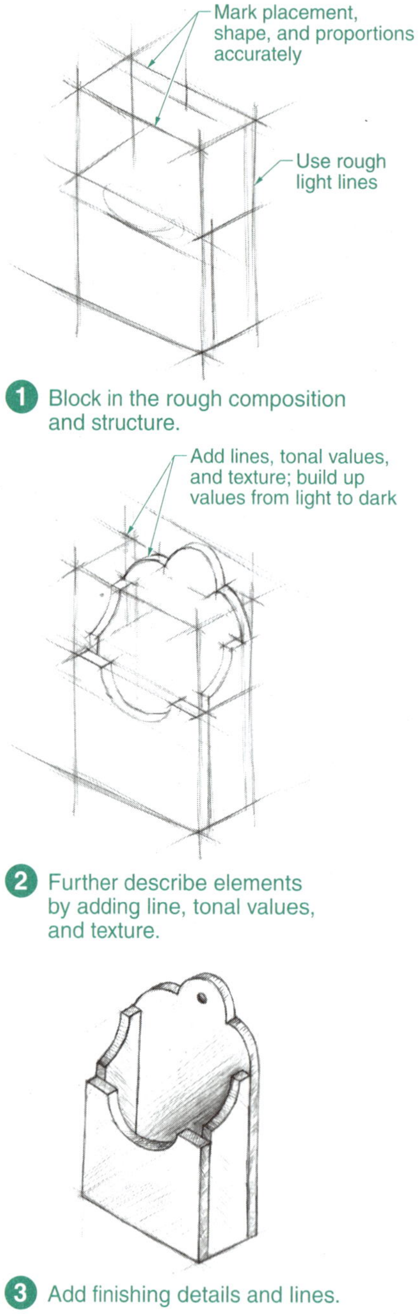

Figure 1-5. Sketches are drawn by adding components to the page in a methodical way.

> *A simple way to begin blocking in an image for a sketch is to find familiar forms and shapes, such as cubes and circles, within the image. This allows the elements in the drawing to be recognized and drawn more easily, and gives a framework for adding details within and around the forms.*
>
> TECH TIP

Composition

A *composition* is the arrangement and relationship of shapes and elements on the picture plane. A composition has a foreground, middle ground, and background where the subjects are placed to create the illusion of depth. The *foreground* is the subjects or areas in a sketch that appear to come forward to the front of an image. The *middle ground* is the subjects or areas in a sketch that exist between the foreground and the background. The main subject of a sketch is usually in the foreground or middle ground. The *background* is the subjects or areas of a sketch that appear to recede toward the back of the image. The composition, orientation, and general size of an image should be considered before starting. Choosing a composition for a sketch involves deciding how to frame an image to determine what should be included or omitted. A simplified composition that focuses mostly on the main subject is often easiest to understand. **See Figure 1-6.**

Visual movement is the visual path or paths the eye follows as it moves across the page. Visual movement is necessary in a composition to make the sketch appear interesting and balanced. The direction of visual movement can be vertical, horizontal, or diagonal. Vertical lines are often used in a sketch to emphasize height, horizontal lines to emphasize space, and diagonal lines to emphasize movement or action. Visual movement can also be circular by combining vertical, horizontal, and/or diagonal lines to keep the eye moving from one portion of a drawing to the next. A sketch with circular visual movement is dynamic and keeps the viewer interested.

Choosing a Composition

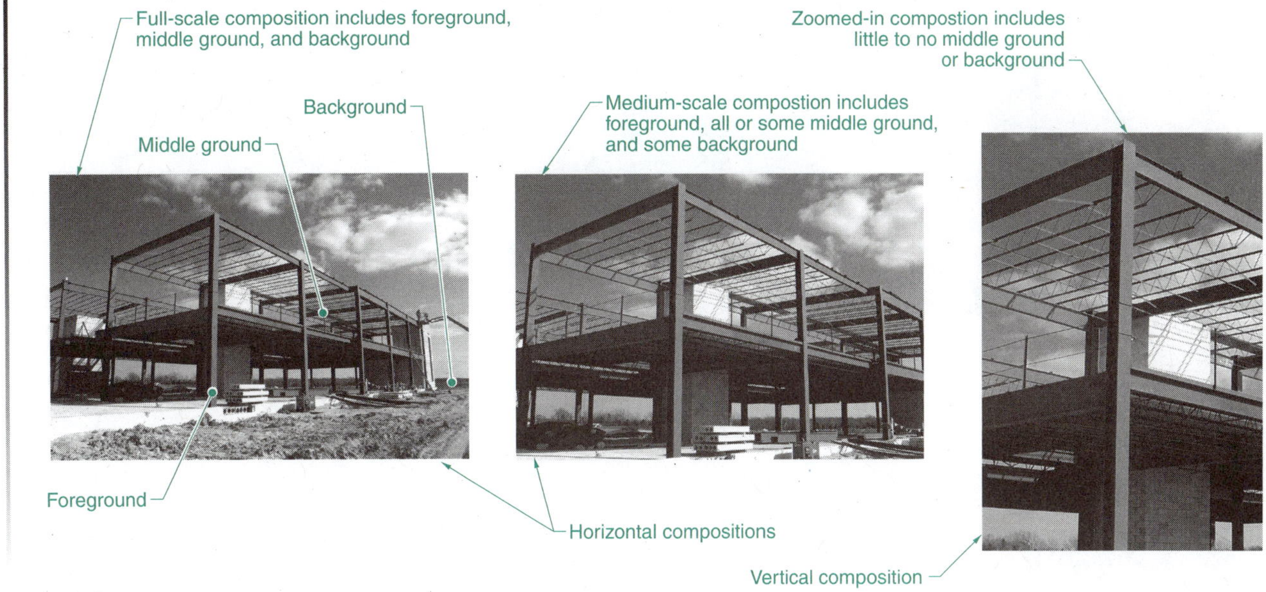

Figure 1-6. A composition is the arrangement and relationship of shapes and elements on the picture plane and contains a foreground, middle ground, and background.

Two basic ways to create visual movement in a composition include "the rule of thirds" and "the rule of odds." The rule of thirds suggests that every image can be split into three columns and three rows, thus creating nine equal segments. The main subject of the drawing should be situated near one of these lines so that the final image is visually divided into three approximate columns and/or rows. This creates a pleasing and balanced composition.

The rule of odds suggests that an odd number of subjects should be included in a sketch if the composition includes more than one subject. The rule of odds illustrates that a triangular shape, which is often formed by an odd number of objects placed in a group, creates a composition that is more balanced and pleasing to the eye. **See Figure 1-7.**

Figure and Ground

Sketches are composed using the relationship between two components, the figure and the ground. The *figure* of a sketch is the subject being drawn and emphasized. The figure or figures should appear as solid, well-defined objects. The *ground* of a sketch is the area surrounding the figure. The ground is usually depicted as a less-distinct area. Figures and grounds are shown on a two-dimensional plane using positive and negative space. **See Figure 1-8.**

The desired arrangement of objects should be emphasized when choosing a composition for a sketch.

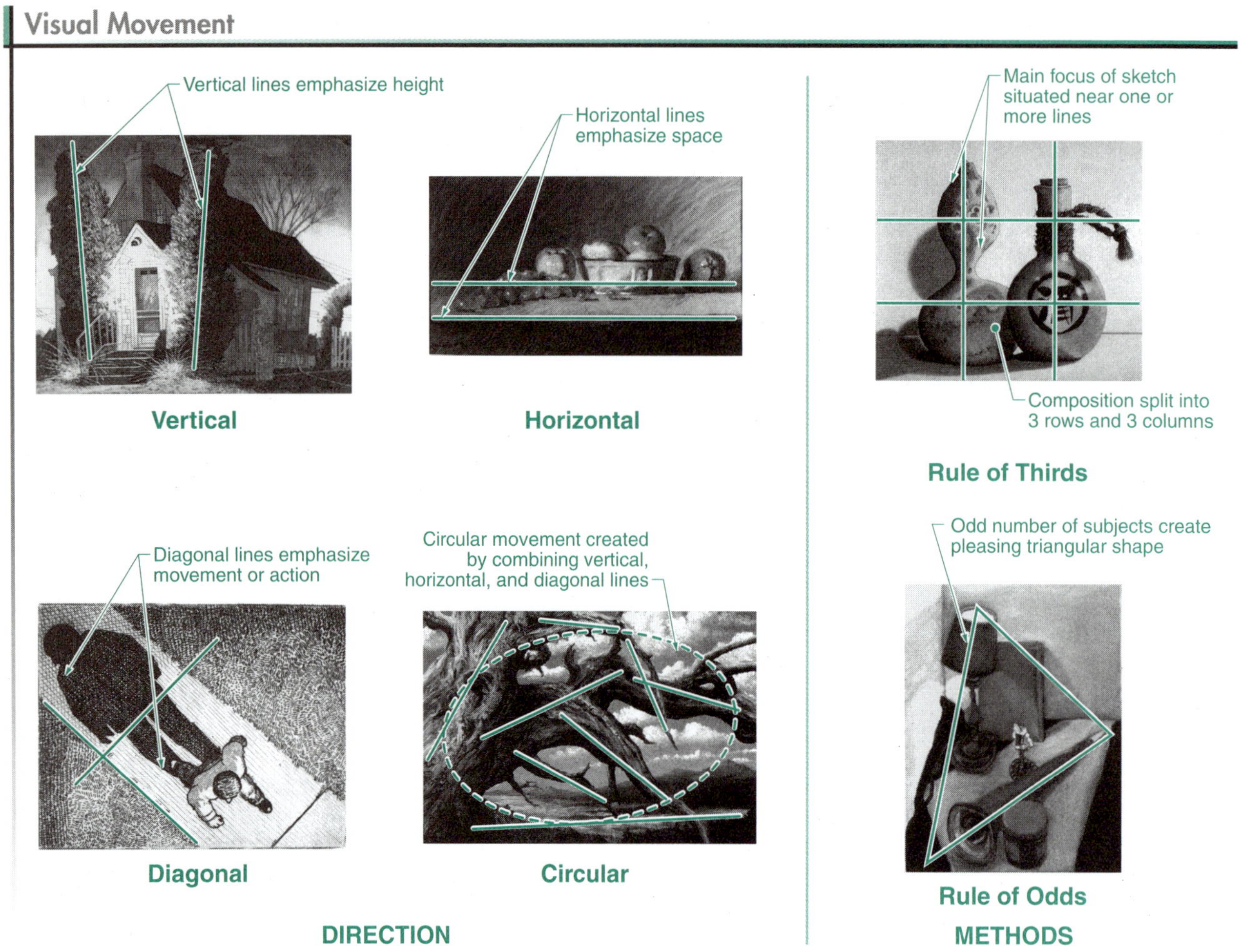

Figure 1-7. Visual movement is the visual path or paths the eye follows as it moves across the page and is necessary to give a sketch an interesting and balanced composition.

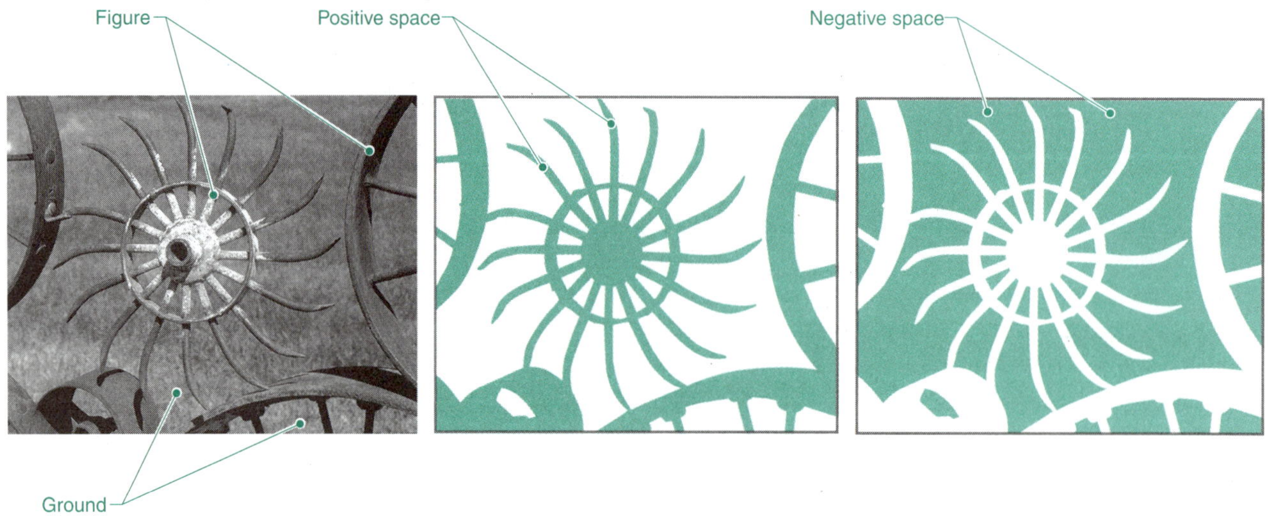

Figure 1-8. A sketch contains a figure and a ground, which are depicted using positive and negative space.

Positive space is the space that forms a shape that can be clearly seen against a background. Positive spaces are generally tangible and easily recognized, such as the object being drawn. *Negative space* is the space that both surrounds and fits between a form or part of a form. Negative space is non-tangible and has no form. The boundaries of negative spaces can be used to define an object edge. Positive and negative spaces in a drawing share the same boundaries and connect together like pieces in a jigsaw puzzle.

Sighting

Three-dimensional spaces must be sighted and transferred to the two-dimensional picture plane when creating a sketch. *Sighting* is a method of viewing and measuring a three-dimensional space in order to transfer components to their appropriate place and proportion on a sketch. *Proportion* is the comparative size relationship between one part of a sketch or object to another. In freehand sketching, sighting is generally done by eye using no technical instruments. By looking at a three-dimensional object through only one eye, the perception of depth is flattened. This is a useful method to view and transfer the accurate size and placement of aspects on a sketch. Sighting is often done with a viewfinder or the shaft of a pen or pencil.

A *viewfinder* is a viewing tool with a proportional window that can be used to sight the composition and accurately place objects in a sketch. A viewfinder is made by cutting a window into a sheet of paper or cardboard. The window should be a smaller version and have the same proportional height and width as the sketching paper size. Viewfinders can be used in a horizontal or vertical format and at different distances from the eye to visualize what elements of the subject can fit on the picture plane at various scales. Viewfinders can also be used to estimate the horizontal and vertical placement of the components of a subject on the page. Adding tick-marks or attaching thread across the window opening can further aid sighting a sketch by dividing the composition into horizontal and vertical segments, especially if the same marks are added to the margins of the drawing paper. **See Figure 1-9.**

Positive and Negative Space Media Clip

Sighting with a Viewfinder

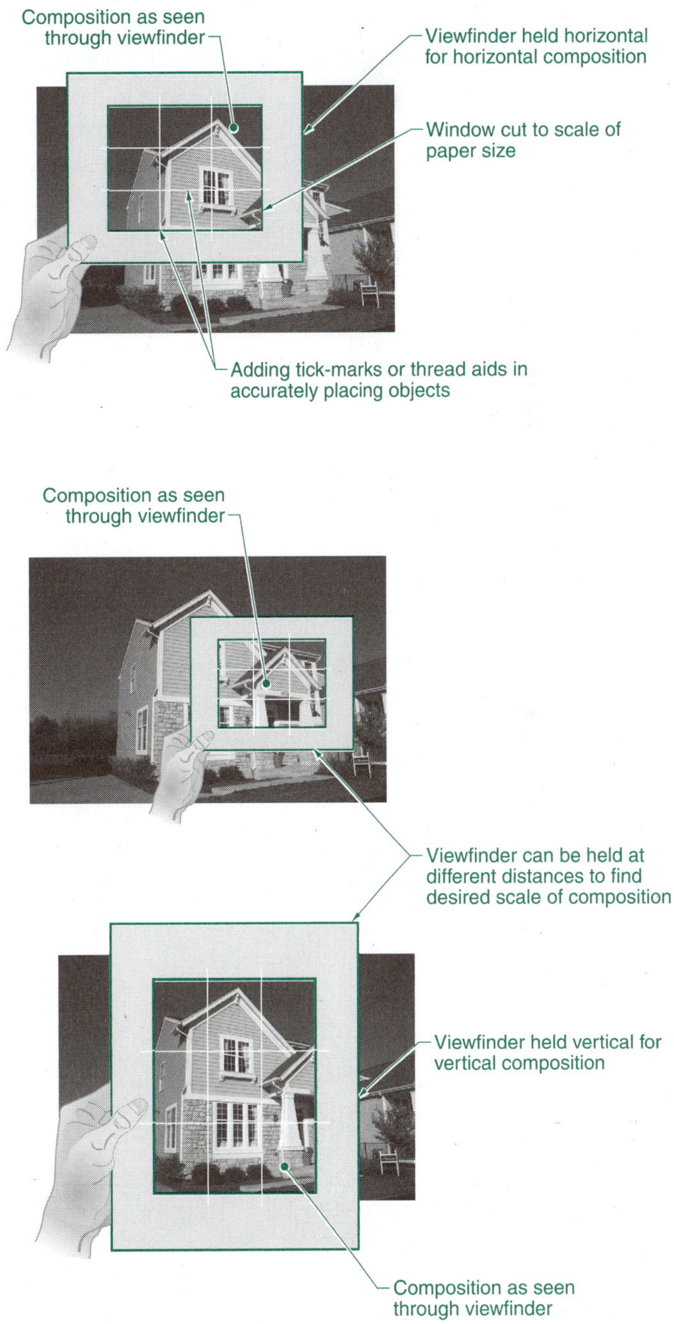

Figure 1-9. A viewfinder has a proportional window that can be used to sight the composition and accurately place objects in a sketch.

TECH TIP

Knowing the exact size of an object is not necessary for estimating proportions. When measuring is not practical, forming proportional relationships among objects to find the relative size allows for a reasonably accurate depiction of an object.

Estimating Proportions Media Clip

Lengths and angles in an object can be proportionally measured and transferred to a sketch by using the shaft of a pencil or pen. To sight the proportion of components in a sketch using a pencil or pen, hold the shaft at an arm's length parallel with the body and perpendicular to the line of sight. The relative lengths of components can be gauged by measuring a smaller length on the pencil's shaft using a thumb and/or finger and using that measurement as a "unit." The remaining measurements can be found by comparing their lengths to this unit of measurement. **See Figure 1-10.**

Angles in a composition are sighted by holding the pencil horizontally or vertically and lining up one end of the pencil to the angle. The angle degree measurement is estimated based on the difference between the angle and the horizontal or vertical line the pencil creates. Angles can also be easily sighted and marked on the paper by using a straight piece of charcoal or chalk. Because charcoal and chalk usually have no wood or paint on their sides, the entire surface, including the length, can be used to make marks on the page. To sight an angle, the charcoal or chalk can be lined up with the angle. The angle is then transferred down to the page and marked directly on the paper with the length of the charcoal or chalk. **See Figure 1-11.**

Estimating Proportions

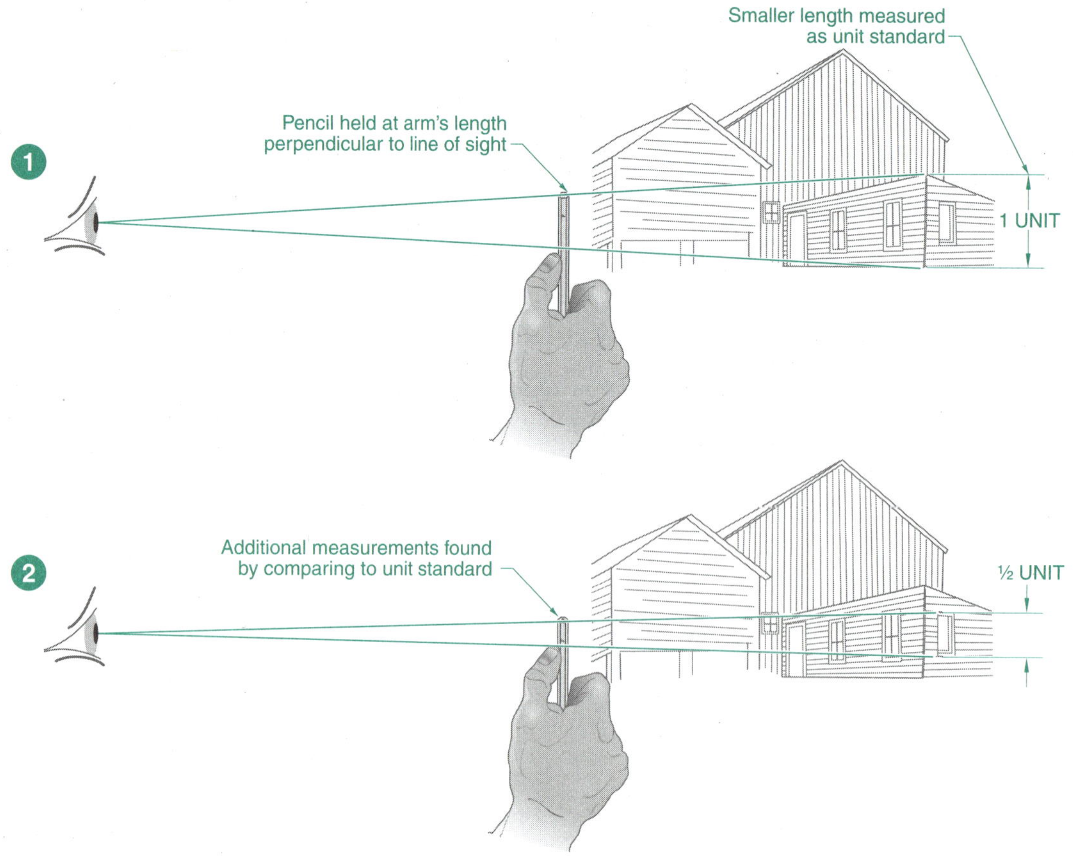

Figure 1-10. Lengths in a subject can be proportionally measured and transferred to a sketch by using the shaft of a pencil or pen.

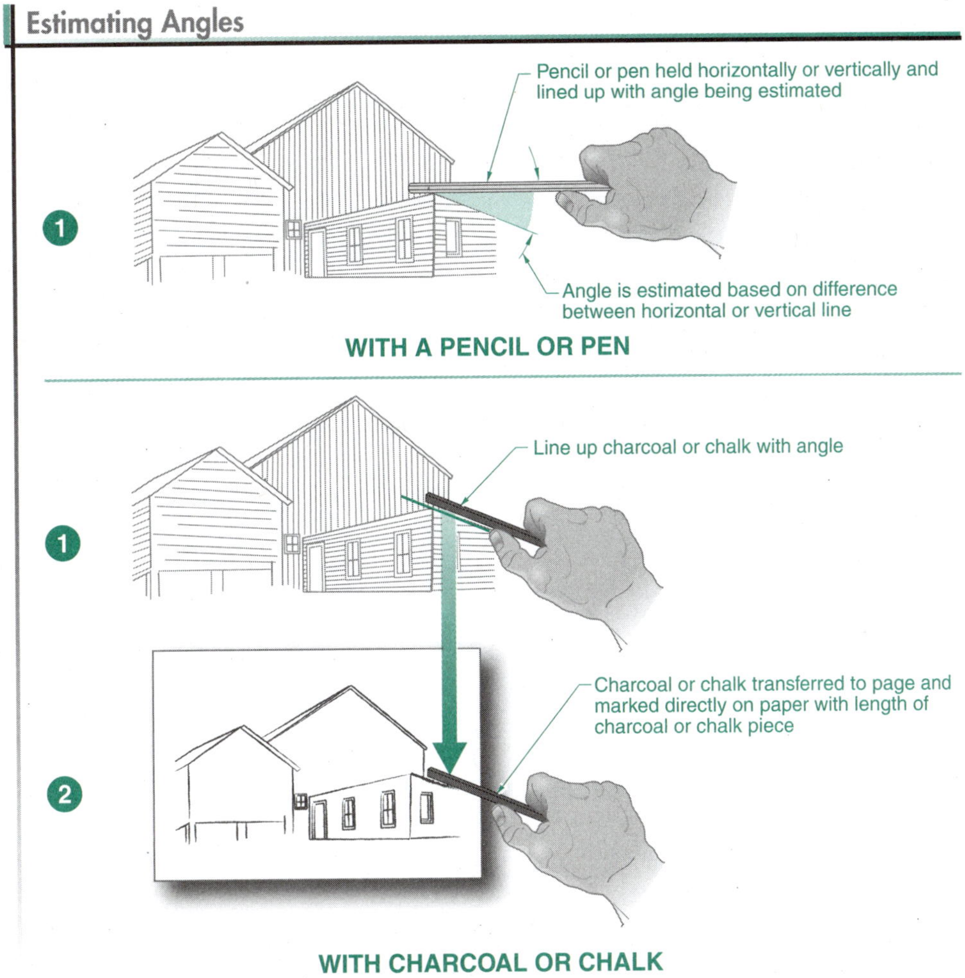

Figure 1-11. Angles in a subject can be estimated and marked using the shaft of a pencil, pen, or piece of charcoal or chalk.

Depth Illusion

Sketching is a two-dimensional representation and is therefore an interpretation of a three-dimensional subject, requiring the creation of depth illusion. Sketches use one or a combination of various methods to create the illusion of depth, including overlapping shapes, relative size, vertical positioning, and atmospheric effects, because the third dimension is lost when transferring an image to paper. The cumulative effects of multiple methods can allow for a more realistic depiction of depth on the two-dimensional plane.

Overlapping Shapes. Overlapping shapes in a sketch is a simple way to create depth. When perceiving an image, the mind understands that the objects in the foreground cover parts or all of the objects in the background. Allowing objects to overlap when sketching creates the illusion of depth by clearly showing which objects are in front of others. **See Figure 1-12.**

Relative Size. The relative size of two objects can create a sense of depth by differentiating objects in the foreground from objects in the background. Objects in front generally appear larger in scale, while objects in back generally appear smaller in scale. If the actual size of two objects are similar, the one in the background will appear smaller than the one in the foreground. **See Figure 1-13.**

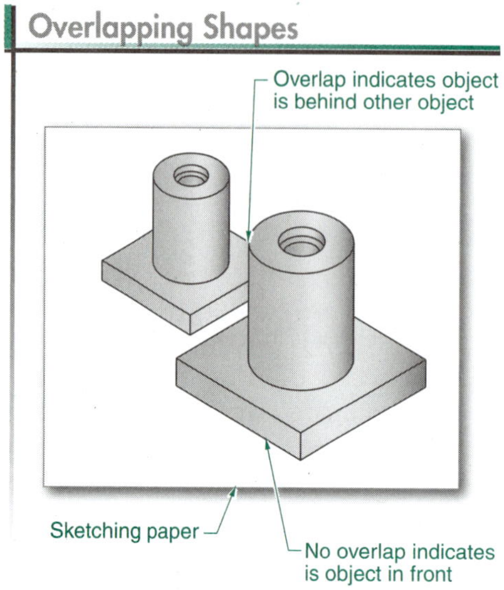

Figure 1-12. Overlapping shapes in a sketch creates the illusion of depth by separating objects in front from objects in back.

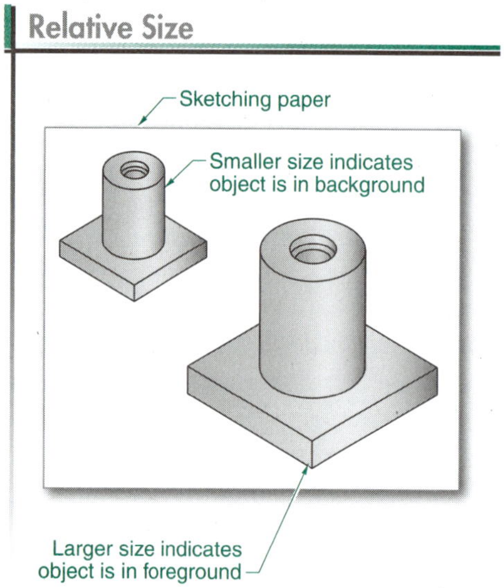

Figure 1-13. The relative size of two objects can create a sense of depth by differentiating objects in the foreground from objects in the background.

Vertical Positioning. The position of an object on the vertical picture plane can also create the illusion of depth in a sketch. Objects that are closer appear lower in the viewer's field of vision, while objects that are farther away appear higher in the field of vision. Objects can be arranged on a sketch with the foreground positioned toward the bottom half of the sheet and the background positioned toward the top half. Vertical positioning and relative size are often used together when creating an illusion of depth in a drawing. **See Figure 1-14.**

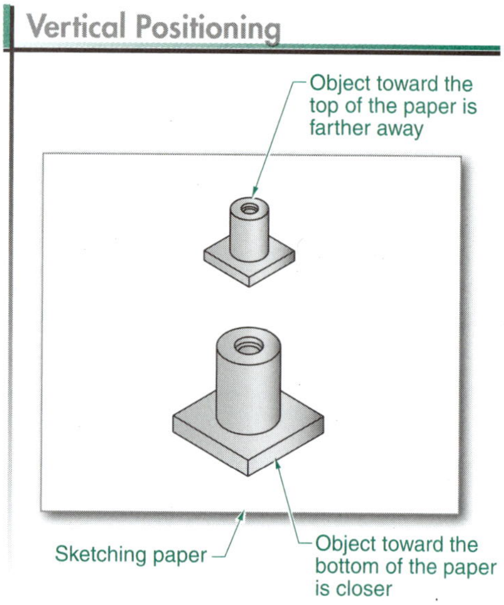

Figure 1-14. The vertical position of objects on a sketch can separate foreground objects from background objects to create the illusion of depth.

Atmospheric Effects. Adding atmospheric effects is yet another means for creating depth illusion in a sketch. Objects that are farther away often appear less defined or partially obscured due to the amount of water vapor, smoke, pollution, dust, or other particulate matter that is contained in the atmosphere. The farther away an object is from the viewer, the more it will be obscured by the contents of the atmosphere. To create the illusion of depth in a sketch using atmospheric effects, the background objects should appear fainter, less sharp, and less defined. This method is most effective when combined with other depth illusion methods. **See Figure 1-15.**

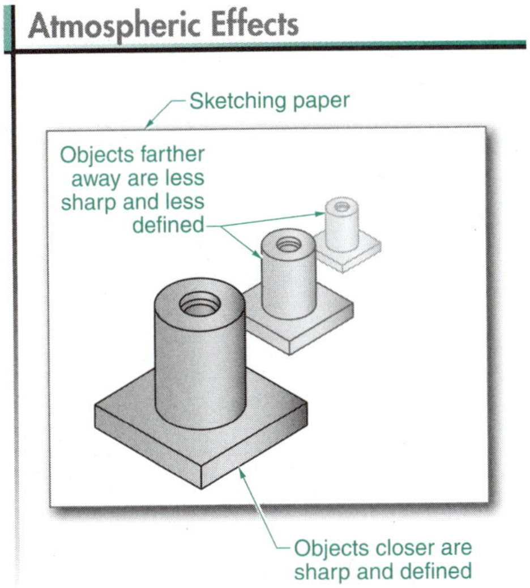

Figure 1-15. Atmospheric effects can be used to create depth illusion in a sketch by showing the background objects as fainter and less defined.

Tone and Texture

Tone and texture help to fully represent an object and to create the illusion of a third dimension. The human eye uses patterns of light and dark tones to perceive objects. Tones help to indicate light sources, volume, and space. Drawings with a variety of tones are usually the easiest to understand. Adding value, lighting effects, and surface treatment are three effective ways of creating tone and texture on a sketch.

Value. *Value* is the relative lightness or darkness of an object as perceived by the viewer. Creating values in a sketch is critical to showing objects as being three-dimensional. The values the eye sees are dependent on their context and contrast with the other components of a sketch. When a drawing is in black and white, white is the lightest value and black is the darkest value, with a range of intermediate grays in between. *Value contrast* is the relationship between two areas of value. Value contrasts are generally considered high contrast, medium contrast, or low contrast. High-contrast compositions use values at the beginning and end of the value range, but few to no values in between. Medium-contrast compositions use the full scope of the value range, from light to dark, and include a variety of intermediate grays. Low-contrast compositions use only a narrow range of values. **See Figure 1-16.** By squinting the eye when looking at the subject, the artist can more easily recognize the patterns of light and dark values, which can be transferred to the sketch.

Four common ways of indicating value on a drawing are hatching, cross-hatching, scribbling, and stippling. These methods create value through the addition of layers of lines or dots. The spacing or density of the lines or dots changes the value—as the spacing increases, the value is lighter, while as the density increases, the value is darker. **See Figure 1-17.** Value can be further manipulated through the drawing material, marking pressure, or line weight. A material that is not pure black, like pencil, will create a lighter value than a material that is very black, like ink. The pressure applied to some materials when making a mark can create a lighter or darker mark, and therefore can result in a lighter or darker value. Thicker-weighted lines or dots will make a darker tone as compared to thinner-weighted lines or dots.

In a sketch, the illusion of depth is created from a three-dimensional scene by using overlapping shapes, relative sizing, vertical positioning, and atmospheric perspective.

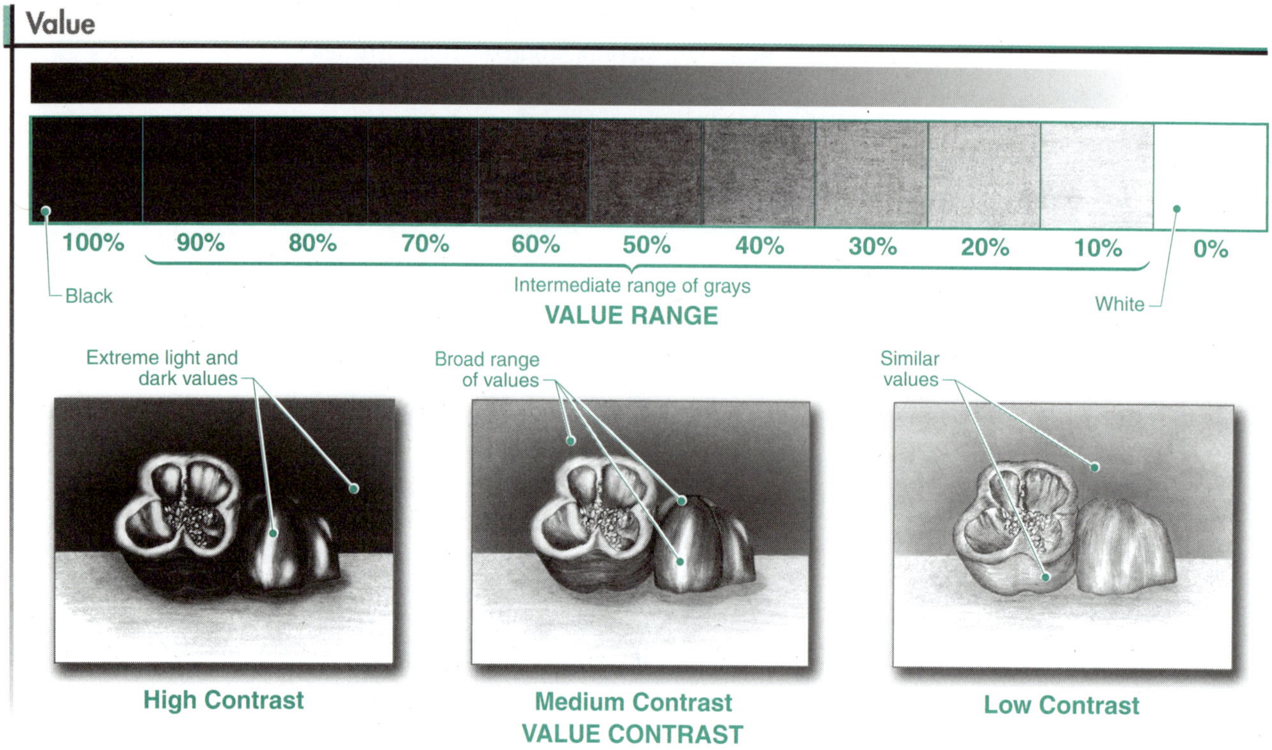

Figure 1-16. Value is the relative lightness or brightness of objects and can be used to create high-contrast, medium-contrast, or low-contrast drawings.

Figure 1-17. Creating values in a sketch is critical to showing objects as being three-dimensional.

Lighting Effects. Lighting effects help to give a sense of spatial depth and allow the eye to understand an object's form. When a light source hits an object, it creates a highlight, an illuminated surface, a shaded surface, a core shadow, and a cast shadow. **See Figure 1-18.**

A *highlight* is a bright place on the illuminated surface of an object that is directly in front of the light source. A highlight usually is the lightest value on the sketch.

An *illuminated surface* is the area of an object that is turned toward the light source. An illuminated surface is indicated on a sketch with light values. As the surface of an object moves away from a light source, the tonal values shift from light to dark.

A *shaded surface* is the area or areas of an object that are turned away from the light source. Shaded surfaces of objects are shown on sketches as intermediate gray to dark tonal values.

A *core shadow* is the area of an object that is directly opposite the light source. The core shadow is the darkest value on an object.

A *cast shadow* is the dark silhouette created by an object that blocks the light source and is cast onto another surface. Cast shadows are shown on sketches using dark tonal values. A cast shadow visually anchors an object to the surface on which it rests, giving indications of the object's relative distance to other objects. A shadow's shape can reveal the placement and direction of a light source.

Reflected light is light that shines on an object after bouncing off other objects and therefore does not come directly from the light source. It is less intense than the original light source. Reflected light is indicated on a sketch by the varying tones of a shaded or shadowed area.

Surface Treatment. The surface treatment of an object can visually evoke a recognizable physical texture, which can add believability and depth to the object being drawn. Texture is the physically tactile quality of a surface on an object or part of an object. Texture can include smoothness or roughness, or any other tangible surface characteristics or treatment.

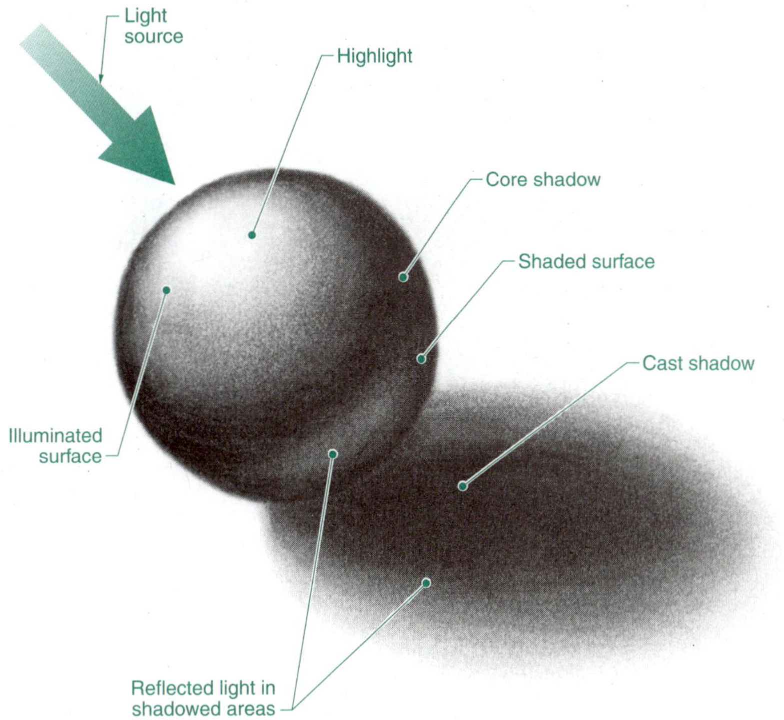

Figure 1-18. When a light source hits an object, it creates a highlight, an illuminated surface, a shaded surface, a core shadow, and a cast shadow.

Textures are made on a drawing by using line and/or tone. Textures can be actual or visual. An *actual texture* is a texture that is a physically tactile surface that can be felt or touched. *Visual texture* is the illusion of texture as shown on a sketch. All actual textures become visual textures when shown in a sketch. Nevertheless, visual texture can be a representation of an actual texture or an invented texture. Visual texture can be created based on the smoothness or roughness of the drawing surface of the paper and how the surface accepts the drawing medium. A sheet of paper with a smooth surface will produce consistent, smoother textures, whereas a rough surface will produce irregular and rougher-looking textures. **See Figure 1-19.**

Chapter 1 Quick Quiz®

Visual texture can also be created in a sketch by including elements that have different visual appearances. The eye can read these varying elements as a texture, which helps to create a pleasing composition.

Texture

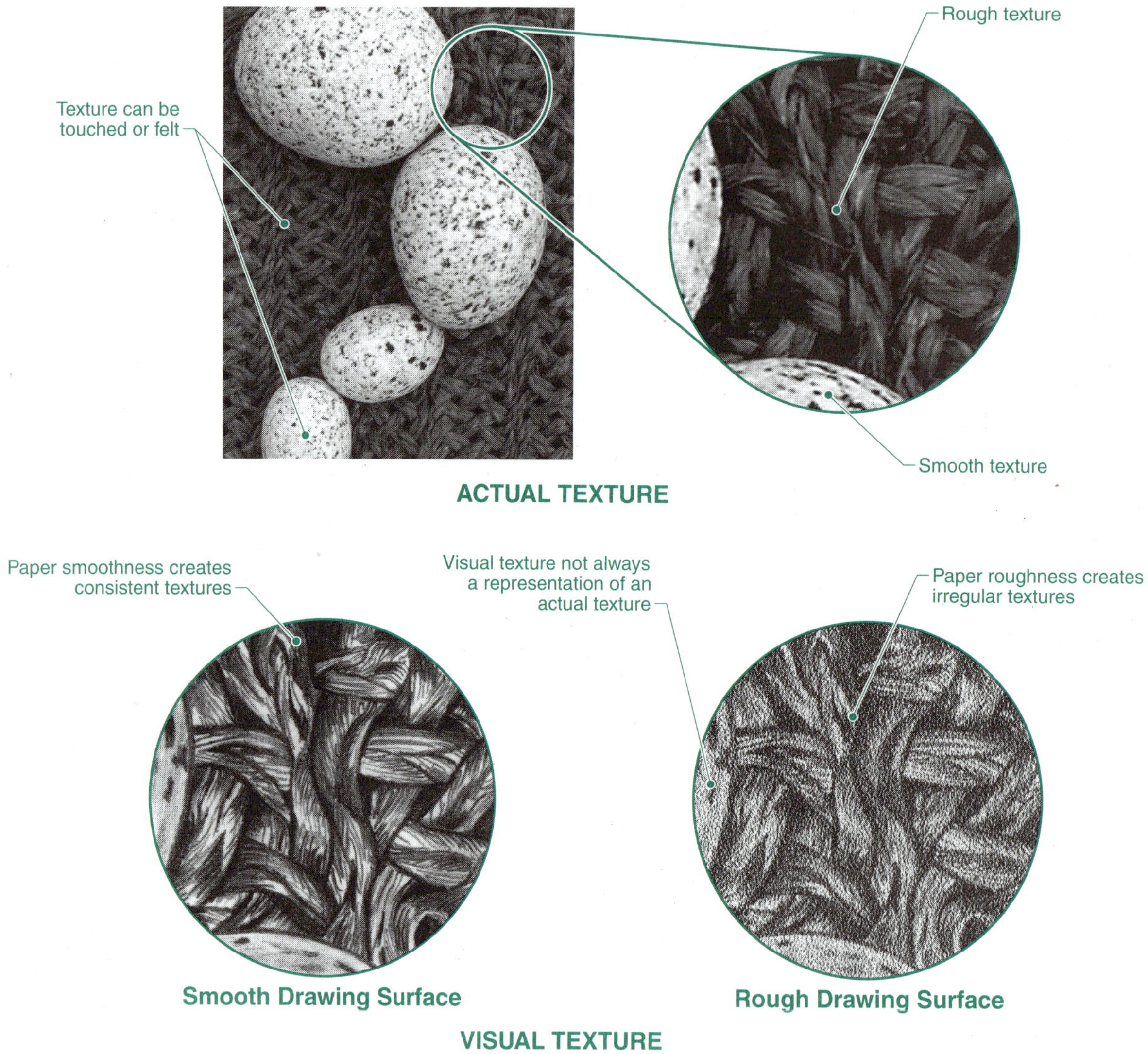

Figure 1-19. Visual texture is the illusion of texture as shown on a sketch and can be a representation of an actual or invented texture.

Technical Sketching and Drawing

Drawing Fundamentals

1

Review

Name ______________________ Date ______________

REVIEW QUESTIONS

______________ **1.** A(n) ___ is the two-dimensional space on the paper where the flattened image of a three-dimensional subject is depicted.

T F **2.** Contours are flat areas defined by an edge or perceived edge.

______________ **3.** ___ is the three-dimensional space of a subject (or part) and is bound by planes.

______________ **4.** A(n) ___ is a tint or shade of black, white, or color.

T F **5.** Lighter-weighted paper can withstand more rough handling and erasing.

______________ **6.** Grades of ___ range from 6B (extremely soft) to 9H (exceptionally hard).

T F **7.** A finished blind contour drawing is not intended to look like an accurate depiction of a subject.

T F **8.** Addressing elements on an entire object or page at once creates a more accurate and balanced finished sketch.

______________ **9.** ___ is the method of quickly marking the structure of a subject on a sketch by breaking the subject into common shapes and lines.

T F **10.** The main subject of a sketch is usually in the background.

______________ **11.** Visual movement can be ___ by combining vertical, horizontal, and/or diagonal lines to keep the eye moving from one portion of a drawing to the next.

T F **12.** The rule of odds suggests that an even number of subjects should be included in a sketch if the composition includes more than one subject.

______________ **13.** The ___ of a sketch is the subject being drawn and emphasized and appears as a solid, well-defined object.

T F **14.** Positive and negative spaces in a drawing share the same boundaries and connect together like pieces in a jigsaw puzzle.

______________ **15.** ___ is a method of viewing and measuring a three-dimensional space in order to transfer components to their appropriate place and proportion on a sketch.

______________ **16.** A(n) ___ is a viewing tool with a proportional window that can be used to sight the composition and placement of objects in a sketch.

T F **17.** In a sketch, an object in front will generally appear smaller in scale, while objects in back generally appear larger in scale.

________________ **18.** ___ is the relative lightness or brightness of an object as perceived by the viewer.

T F **19.** A highlight is a bright place on the illuminated surface of an object that is directly in front of the light source.

T F **20.** A visual texture is a physically tactile surface that can be felt or touched.

IDENTIFICATION – Subject Representation

________________ **1.** Volume

________________ **2.** Contour

________________ **3.** Shape

________________ **4.** Form

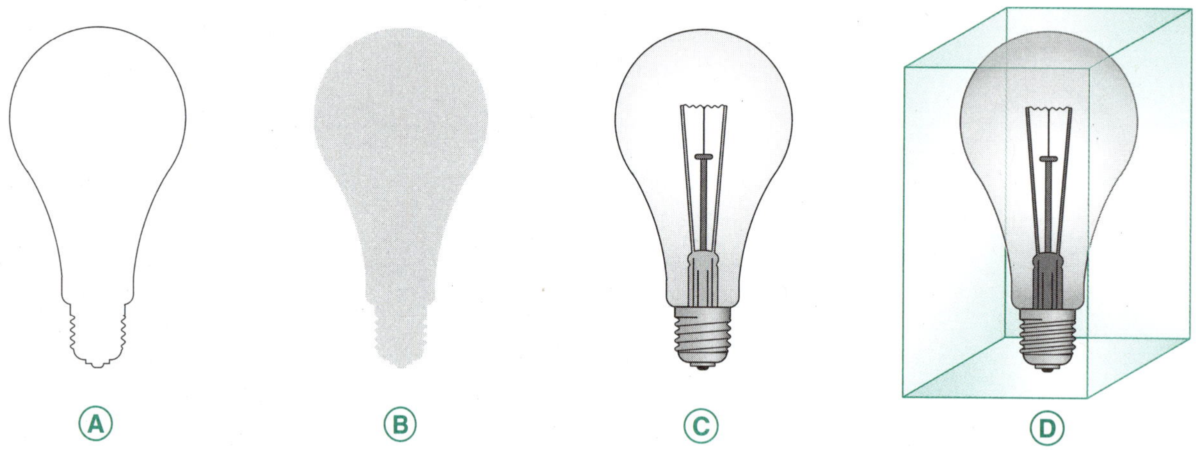

IDENTIFICATION – Lighting Effects

________________ **1.** Cast shadow

________________ **2.** Reflected light

________________ **3.** Highlight

________________ **4.** Illuminated surface

________________ **5.** Core shadow

________________ **6.** Shaded surface

________________ **7.** Light source

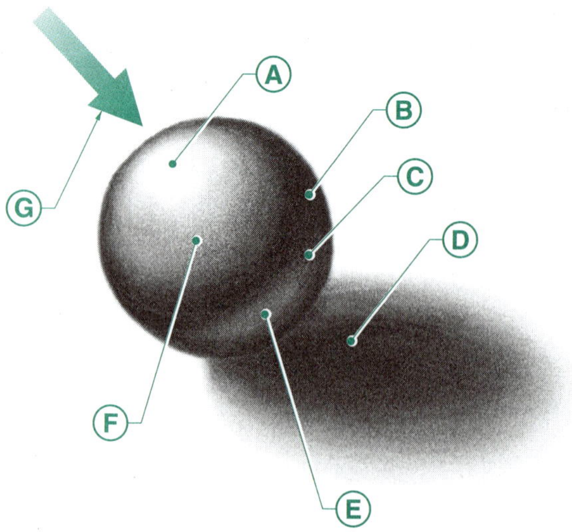

EXERCISES

1. Using the given image, create one sketch of the contour and one sketch of the shape of the object.

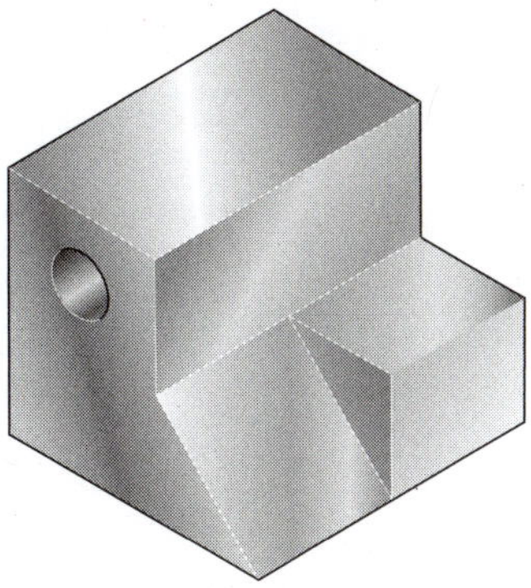

CONTOUR

SHAPE

2. Using the given image, block in the initial structure of the objects with rough, light lines.

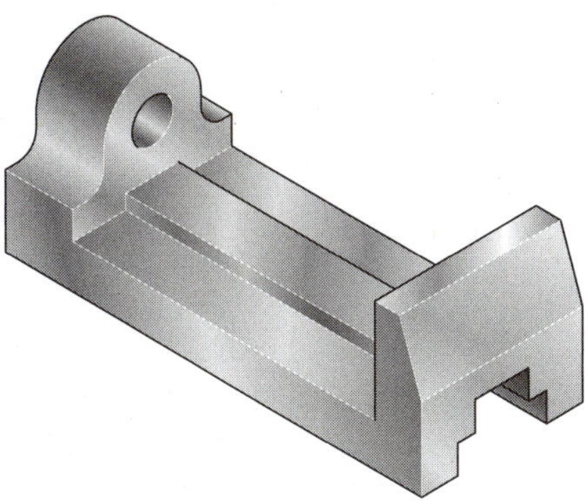

3. Scale the given image to a larger size. Use the tick-marks as guides for proper positioning and proportion of objects.

4. Duplicate and position the given object five times to create a composition that shows depth. Utilize the concepts of overlapping shapes, relative size, vertical positioning, and atmospheric effects in the composition.

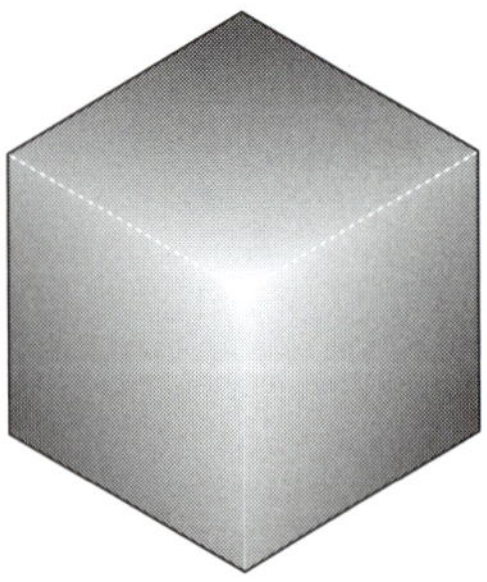

5. Draw the given image once in each square. Add value to each drawing to create a low-contrast and a high-contrast composition.

LOW-CONTRAST

HIGH-CONTRAST

6. Draw the given image once in each square. Add value to each drawing using the indicated method.

HATCHING

CROSS-HATCHING

SCRIBBLING

STIPPLING

7. Draw the given objects in the space provided. Add lighting effects to the objects based on the marked light source in the space provided. Draw all lighting effects, including a highlight, illuminated side, shaded side, core shadow, and cast shadow.

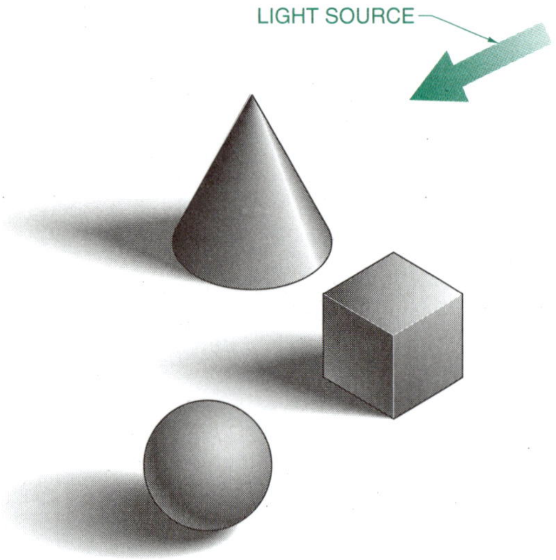

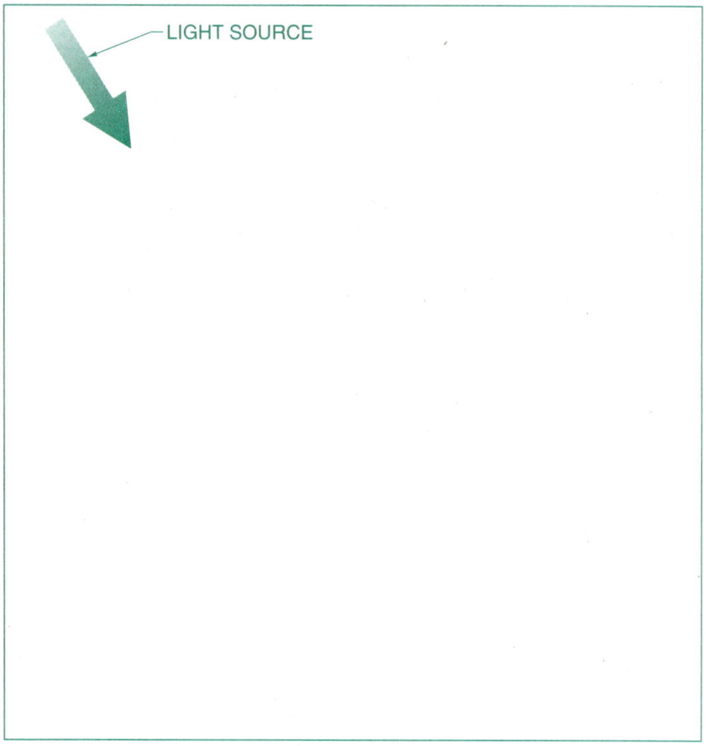

8. Use value to draw the texture shown in the given image.

Technical Sketching and Drawing

2 Instrument Drawings and Techniques

INSTRUMENT DRAWINGS

Instrument drawings are technical drawings that are made by using instruments to achieve greater accuracy than freehand sketching. Technical drawings often depict objects that can be used in various industry settings. Standard tools and technical drawing instruments are used to ensure the object is drawn properly and to the required industry standards.

Standard Tools

The standard tools required for a technical drawing are paper, a pencil or pen, and an eraser. Additional technical drawing instruments can be used as necessary for an accurate depiction of the object being drawn.

Paper. The paper size and type selected for a drawing depends upon the end use of the drawing. Paper used in technical drawing is either plain or preprinted with grids to facilitate drawing. Preprinted paper is available in a variety of grid sizes for different types of drawings. A grid size of ¼″ (quad-rule) is popular. Grids are commonly printed in light blue, nonreproducing ink. A smooth paper surface texture is preferred so that all lines and values are sharp and clear.

Paper is available in pads, sheets, or continuous rolls. Flat sheets of paper in the American system are designated by the letters A through F. Rolls are designated by the letters G, H, J, and K. Letters earlier in the alphabet indicate smaller paper sizes. International standard (ISO) paper sizes are designated by letter-number combinations. ISO paper sizes are given in metric units and are very close to the inch sizes, but have slightly different proportions. **See Figure 2-1.**

Paper Sizes

American—Flat Sheets		
Size	Width*	Length*
A	8.5	11
B	11	17
C	17	22
D	22	34
E	34	44
F	28	40

American—Rolls			
Size	Width*	Length*	
		Min	Max
G	11	22.5	90
H	28	44	143
J	34	55	176
K	40	55	143

International Standard (ISO)—Flat Sheets			
Size	Width†	Length†	Nearest USA Length
A4	210	297	A
A3	297	420	B
A2	420	594	C
A1	594	841	D
A0	841	1189	E

* in in.
† in mm

Figure 2-1. American and international paper sizes have different proportions.

Pencils Media Clip

Pencils. Pencils used in technical drawings are either wood or mechanical. Wood and mechanical pencils have lead grades ranging from hard to soft. Because hard leads do not smudge easily, they are used to draw fine, precise lines and are necessary when extreme accuracy is required. Medium leads are used to draw object lines. Soft leads are used primarily for freehand sketching, or to add tone and texture to a sketch. The most common lead grades used for technical drawings are HB, F, H, and 2H. Lead grades 3H or 4H may be used for more precise work, or to create light construction lines that will not need to be erased.

A *mechanical pencil* is a pencil that uses an internal mechanism to push a thin lead through a hole at the tip of a lead holder. Mechanical pencils produce a more consistent line width than wood pencils and most styles maintain their point without sharpening. Lead in mechanical pencils can be changed or replaced when needed. **See Figure 2-2.** Lead sizes are based on the thickness of the shaft of lead. Common lead sizes, given in millimeters, are 0.3, 0.5, 0.7, 0.9, 1.1, 1.3, and 2.0. Lead grade and size should be chosen based on the specific task required. **See Appendix.**

A pencil should be sharpened to a point or near-point for best results. Pencil sharpeners work by shaving down one end of a wooden pencil and can be powered by mechanical or electrical means. Lead pointers are a type of pencil sharpener used specifically for mechanical pencils that use thicker lead widths. Pencils may also be sharpened using sandpaper to file away the lead to create a point.

Pens. A *technical pen* is a specialized writing instrument that is used to make lines of consistent width on a technical drawing. Technical pens use black, permanent ink and can be used to darken and sharpen lines after a sketch has been drawn in pencil. Technical pens can have tapered or straight tips. Tapered tips are generally used for artwork or object drawing, while straight tips are generally used for drafting and lettering. Pen tips can vary in width from 0.005″ (0.13 mm) to 0.079″ (2.00 mm). **See Figure 2-3.** Because drawings contain both thick and thin lines, two widths of pen tips with an approximate size ratio of two-to-one should be used. Selected pen tip widths can vary from drawing to drawing, but the recommended minimum thick line width should be no less than 0.024″ (0.6 mm), and the recommended minimum thin line width should be no less than 0.012″ (0.3 mm).

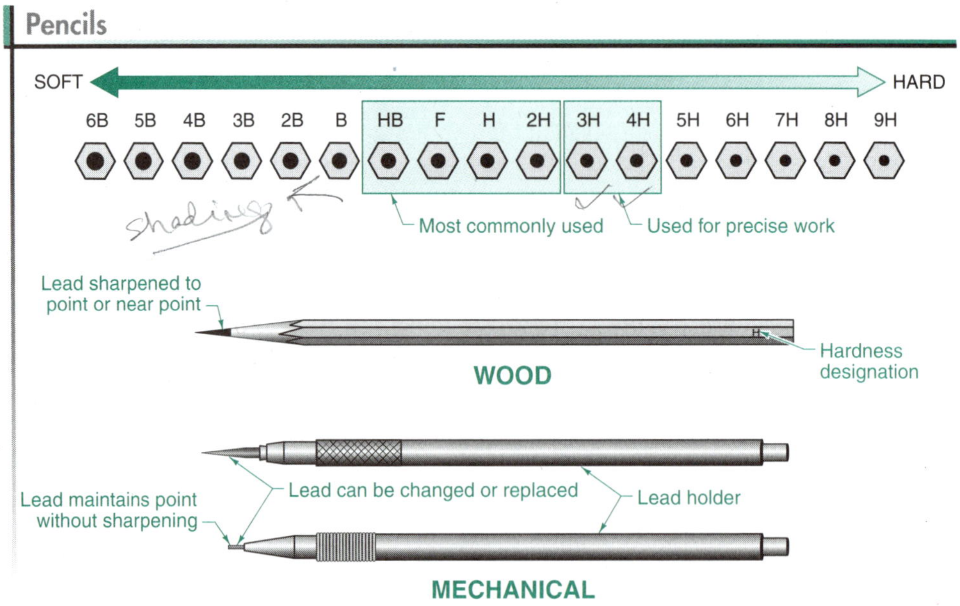

Figure 2-2. Pencils used in technical drawings are either wood or mechanical.

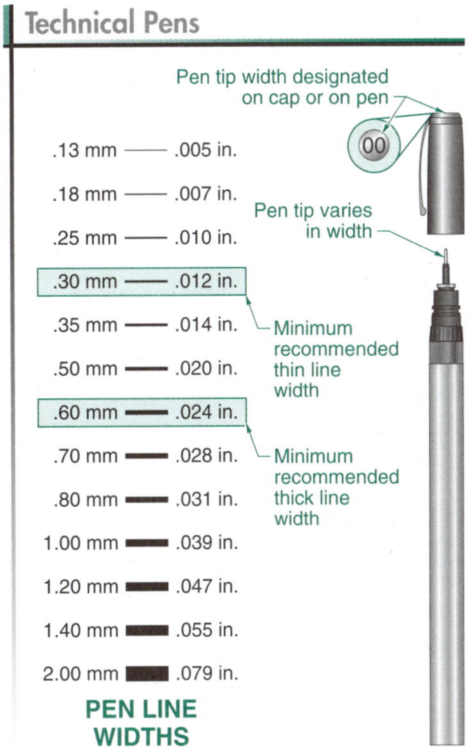

Figure 2-3. A technical pen is used to make consistent lines in a technical drawing and can accommodate a variety of pen tips with different widths.

Erasers. An *eraser* is a handheld piece of rubber or rubber-like material that removes unwanted marks from paper. An eraser should be soft enough to remove marks from a drawing without smearing marks or damaging the paper. Pink rubber, white vinyl or plastic, or gum erasers are commonly used in technical drawing. An *erasing shield* is a template that has cutouts in different sizes and shapes. These cutouts can be used to isolate a small portion of a drawing to more precisely erase an unwanted mark.

Technical Drawing Instruments

Technical drawing instruments can help to draw accurate lines, angles, curves, and proportions. T-squares, triangles, compasses, dividers, and scales are all used to measure or place components on the drawing paper. Additional tools such as protractors, templates, french curves, and flexible curves assist in the proper placement and shape of components on a drawing.

T-Squares. A *T-square* is an instrument used to draw horizontal lines and to position triangles. A T-square gets its name from its "T" shape and consists of a head and a blade. A T-square is used with the blade held horizontally on a surface and the head held firmly against one edge of the surface. T-squares are made of wood, plastic, or aluminum and are available in various lengths.

Using a T-Square Media Clip

Triangles. A *triangle* is an instrument used to draw vertical and inclined lines. Triangles are commonly made of clear plastic. Two standard triangles, 30°-60° and 45°, are available in a variety of sizes. **See Figure 2-4.** The base of the triangle is held firmly against the blade of the T-square to draw vertical or inclined lines.

Vertical lines can be drawn using either the 30°-60° or the 45° triangle. The 30°-60° triangle can create inclined lines at 30° and at 60° sloping to the left or right. The 45° triangle can create inclined lines at 45° sloping to the left or right. Triangles can be used in combination to produce inclined lines at various other angles.

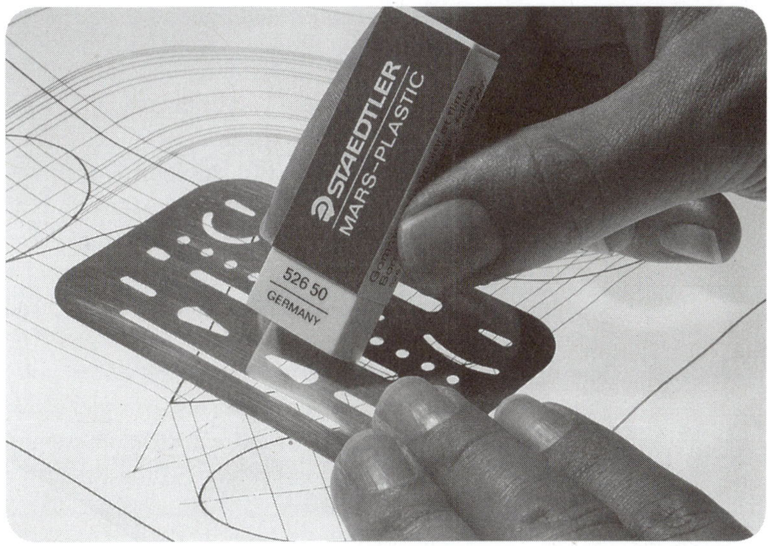

Staedtler, Inc.

Erasing shields are available with cutouts in a variety of shapes and sizes to isolate and erase only unwanted portions of lines.

T-Square and Triangles

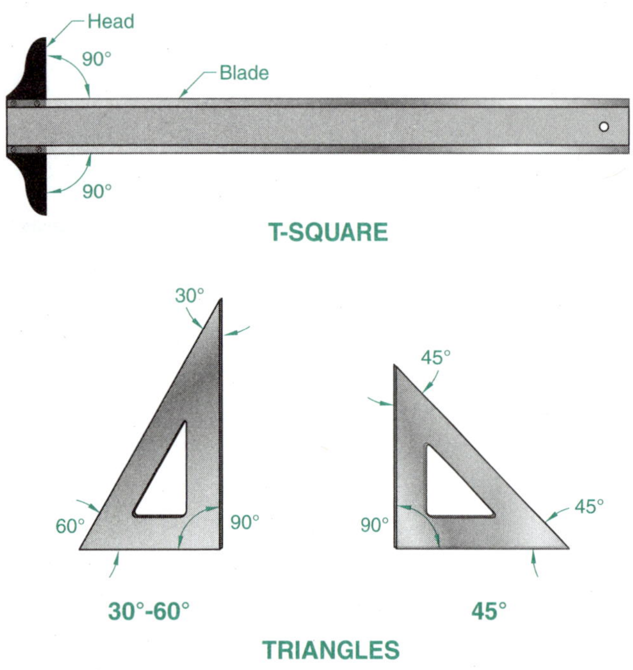

Figure 2-4. T-squares and triangles are both instruments that facilitate the drawing of accurate horizontal, vertical, and inclined lines.

> **TECH TIP**
> *A T-square head is attached to the blade at a 90° angle to allow accurate horizontal lines to be drawn when the head is held securely against the edge of a drawing surface. T-squares can also have adjustable heads that can be angled to draw inclined lines at varying angles.*

Compasses. A *compass* is an instrument used to draw arcs and circles. An *arc* is a curved line that is part of the circumference of a circle. Compasses are generally made of metal and have two legs joined by a hinge that can be adjusted to change the radius of the arc or circle. One leg of the compass has a needlepoint that is positioned on the center point of the arc or circle to be drawn. The other leg contains a pencil lead or pen. The compass is rotated around the center point and the pencil or pen marks the arc or circle on the paper. **See Figure 2-5.** Compasses are categorized as either bow or drop-bow. Bow compasses can create circles or arcs with a radius of approximately 1″ to 10″. Drop-bow compasses are generally used to create circles and arcs with a radius of less than 1″.

Dividers. A *divider*, or hinge divider, is an instrument that is used to transfer and compare measurements. Dividers look like compasses with two legs connected by a hinge, but divders have a needlepoint on both legs. To use a divider, each needlepoint is positioned at the end of the feature to be measured. The divider retains the size of the feature when moved to a scale or other feature for comparison. Dividers are also used to break a dimension into two or more equal parts. Other types of dividers include bow or proportional dividers. Bow dividers have a screw thread with which to adjust the divider legs. Proportional dividers have a sliding, adjustable pivot for enlarging or reducing dimensions to scale.

Scales. A *scale* is an instrument used to measure and proportionally enlarge or reduce lines. Three types of scales are the architect's scale, engineer's scale, and mechanical engineer's scale. **See Figure 2-6.** Each scale is available in a variety of sizes and measurement units. In addition, triangular combination scales combine the most common scales from the architect's, engineer's, and mechanical engineer's scales. A note should be placed on the drawing to indicate the scale used.

An architect's scale is used when making drawings of buildings and other structural parts. These scales are usually triangular in shape. One edge of the scale is a standard ruler divided into inches and sixteenths of an inch. The other five edges each contain two scales (one going left to right and one going right to left on the same edge) for a total of eleven scales. Scales included are 3, 1½, 1, ¾, ½, ⅜, ¼, 3⁄16, ⅛, and 3⁄32. The ¼ scale means that ¼″ = 1′-0″, and so forth.

The engineer's scale is used for making maps, survey drawings, and plot plans. The engineer's scale is triangular with six edges that are graduated in decimal units. One-inch units on the scale are divided into 10, 20, 30, 40, 50, or 60 parts. These units are used to represent the desired measuring unit, such as inches, feet, or miles. The engineer's scale is also referred to as the civil engineer's scale.

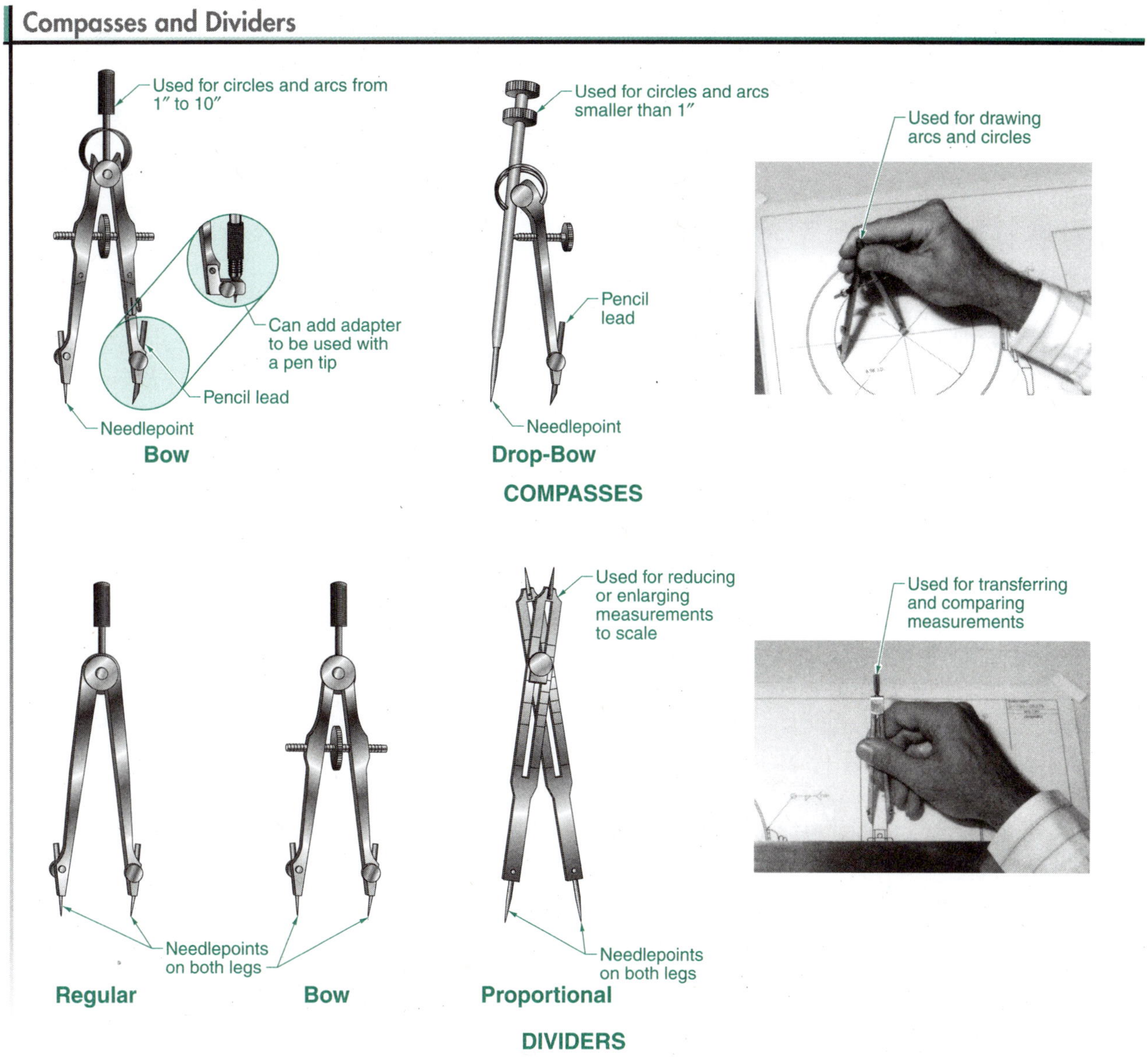

Figure 2-5. A compass is used to draw arcs and circles. A divider is used to transfer and compare dimensions.

> **TECH TIP**
> *When using pencil lead in a compass, an artist should choose a lead grade slightly softer than the lead grade used for the drawing. Since less pressure is applied when using a compass, a softer lead grade allows the compass-drawn lines to match the lines made with a drawing pencil.*

The mechanical engineer's scale is used when drawing machines and machine parts. This scale is similar to the architect's scale but is flat and has only four edges. The mechanical engineer's scale includes a full-size scale and several others, such as 1/2, 1/4, and 1/8 scales. The fractional scales have a direct size relationship between the actual measurement of the object and the scale measurement. For example, 1/2″ = 1″ in the 1/2 scale, 1/4″ = 1″ in the 1/4 scale, and so on. No conversion is needed to transfer an actual measurement to a scale measurement.

Scales
Media Clip

Scales

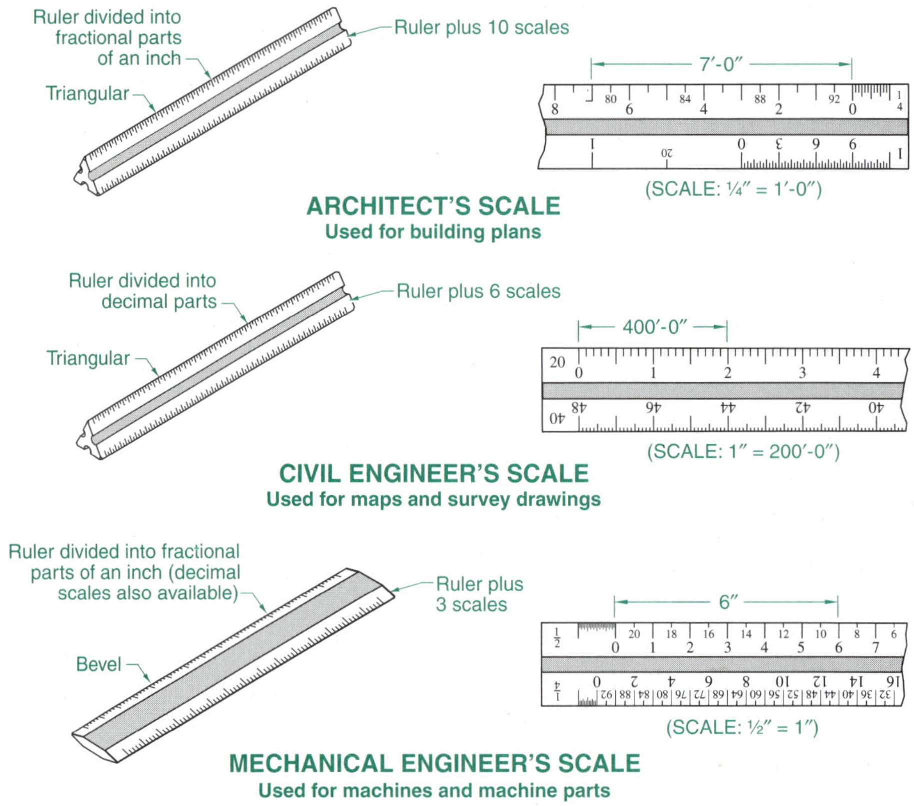

Figure 2-6. Scales are used to measure and proportionally enlarge or reduce lines and include the architect's scale, engineer's scale, and mechanical engineer's scale.

Additional Tools. There are a variety of additional tools that can be used to assist in the accurate drawing of angles, lines, and shapes in a technical drawing. These include protractors, templates, french curves, and flexible curves. **See Figure 2-7.**

A *protractor* is a flat, semicircular tool that is used to measure angles. Protractors provide angle measurements in degrees. Protractors are placed with the flat edge lined up to the horizontal blade of the T-square. The center point and desired angle are marked on the page and connected with a straight edge. Because protractors give estimated angle measurements, tools with greater precision should be used when an exact measurement is required.

A *template* is a flat piece of plastic or metal that has cutouts of commonly used shapes. A template is placed on the drawing paper and the needed shape cutout is traced using a pencil or other writing instrument. Ellipse templates are commonly used and are available in a variety of sizes. Other commonly used templates include circles, hexagons, arrowheads, lettering, and wavy lines, as well as specialized templates for industry-specific parts, such as mechanical or electrical components.

A *french curve* (irregular curve) is a template shaped for different curves and is used to draw smooth curves of varying radii. A french curve is placed on the drawing paper and the needed curve is traced with a pencil or pen. A *flexible curve* (adjustable curve) is an instrument made of flexible metal or plastic that can be bent by hand to form the necessary curve. Once bent, a flexible curve can be used as a guide to trace curves in a similar manner to french curves.

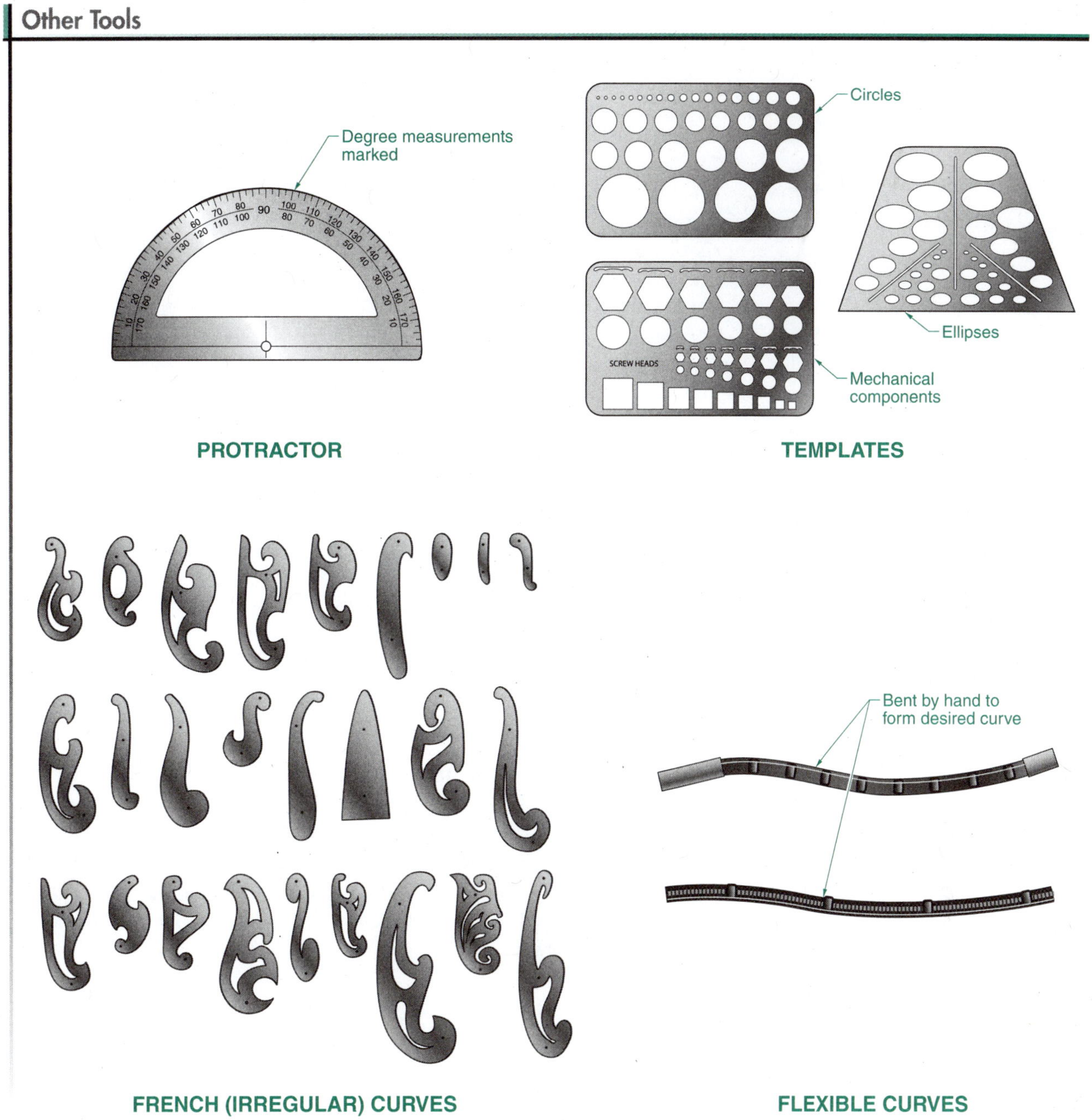

Figure 2-7. Protractors, templates, french curves, and flexible curves are used to assist in drawing accurate angles, lines, and shapes in technical drawings.

LINE DRAWING

Maintaining a consistent line-drawing technique is important in creating accurate technical drawings. To assist with accuracy, the drawing pencil or pen should be held loosely at 1½″ to 2″ from the tip. Slanting the pencil from the vertical plane when drawing helps to keep lines regular and exact. The pencil point should be pulled across the paper while sketching to avoid tearing the paper. While pulling the pencil, slowly rotate it to produce lines of consistent width.

Drawing Lines

In technical drawings, lines should be drawn sharp and dark. Line widths are classified as thick or thin. Thick lines should be a minimum of 0.024″ (0.6 mm) wide. Thin lines should be a minimum of 0.012″ (0.3 mm) wide. Line contrast in technical drawings should be shown through line width, not the line's degree of darkness. Light, trial lines can be drawn first to establish the accuracy of a line before the line is darkened. Horizontal lines, vertical lines, inclined lines, and curved lines should be drawn carefully and accurately.

Horizontal Lines. A *horizontal line* is a line that is parallel to the horizon. Horizontal lines extend from right to left or left to right. To draw horizontal lines by hand, locate the end points with dots to indicate the position and length of the line. For short lines, the end dots are connected with a smooth wrist movement from left to right (for a right-handed person). Long lines may require intermediate dots and a full arm movement to avoid making an arc. **See Figure 2-8.** The top or bottom edges of the paper may be used as a guide while sketching horizontal lines.

Horizontal lines can also be drawn using a T-square. By lining up the T-square blade with the desired placement of the line, the pencil or pen can be run across the edge of the blade from left to right (for a right-handed person). The angle of the pencil or pen should be held even for the extent of the line to avoid uneven marking.

Vertical Lines. A *vertical line* is a line that is perpendicular to the horizon. Vertical lines extend from top to bottom or bottom to top. To draw vertical lines by hand, locate the end dots and connect with a line from the top to the bottom. The side edges of the paper may be used as a guide while sketching vertical lines. Vertical lines can be drawn with a triangle and T-square by placing a triangle with the vertical edge perpendicular to the T-square. The line is then drawn upward along the triangle edge from the T-square blade.

Inclined Lines. An *inclined line* is a slanted line that is neither horizontal nor vertical. To draw inclined lines by hand, locate end dots and draw from left to right (for a right-handed person). The paper may be rotated so that the inclined lines are in either a horizontal or vertical position to facilitate sketching.

Inclined lines are also called angular lines because angles are created when two or more lines cross. An *angle* is the intersection of two lines. An *acute angle* is an angle measuring less than 90°. An *obtuse angle* is an angle measuring more than 90°. There are 360° in a full circle, 90° in a right angle, and 180° in a straight line.

TECH TIP

When drawing inclined lines using a 45° or 30°-60° triangle, aligning the long side against the T-square or straightedge gives the greatest stability and accuracy. When drawing long inclined lines, aligning a short side against the T-square or straightedge may be necessary.

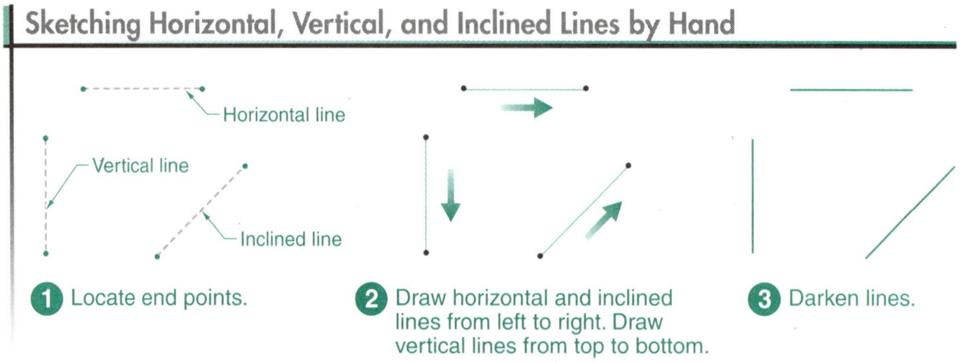

Figure 2-8. Horizontal lines, vertical lines, and inclined lines can be sketched by hand.

To draw inclined lines using a T-square and a triangle or triangles, place the triangle(s) on the horizontal T-square blade to form the desired incline and draw the line upward along the triangle edge. **See Figure 2-9.** Triangles can also be used to draw an angular line with a specific degree measurement. The 45° triangle, 30°-60° triangle, or both can be used to create inclined lines every 15°. A protractor can also be used to draw angular lines with a specific degree measurement. To draw an angular line with a protractor, line up the bottom of the protractor with the horizontal blade of the T-square and mark the center point and angle degree measurement on the paper. These marks are then connected with a line using a straightedge.

Curved Lines. A *curved line* is a line that bends and has no angles. To draw curved lines by hand, locate and mark the points of the curve on the paper. Using short, overlapping strokes, lightly connect the points with straight lines. Round the lines out, darkening them to form the desired curve. **See Figure 2-10.**

A french curve or flexible curve can be used as a template to create curved lines by tracing a pencil or pen around the desired edge. An arc or circular curved line can be drawn with a compass. The compass is positioned on the center point of the desired arc and set to reflect the radius, then rotated while applying pressure to the pencil or pen.

Drawing a Curved Line Media Clip

Inclined Lines

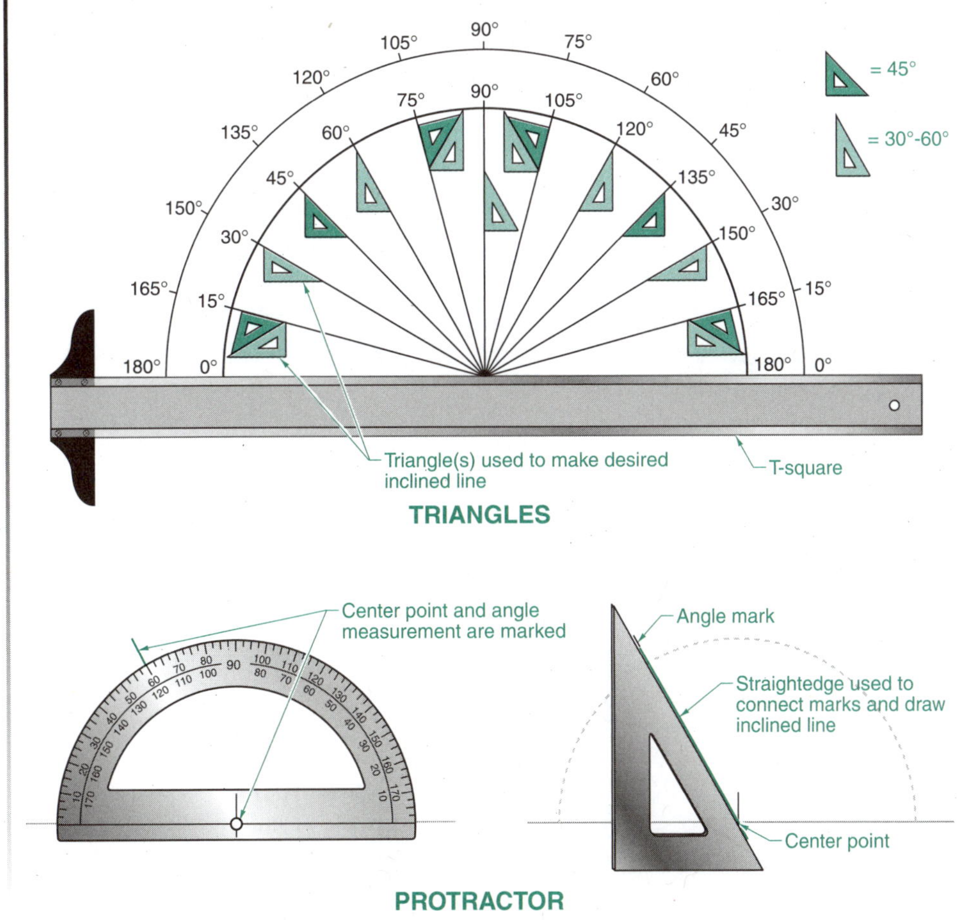

Figure 2-9. Inclined and angular lines can be sketched using a T-square with one or two triangles or a protractor to form the desired incline.

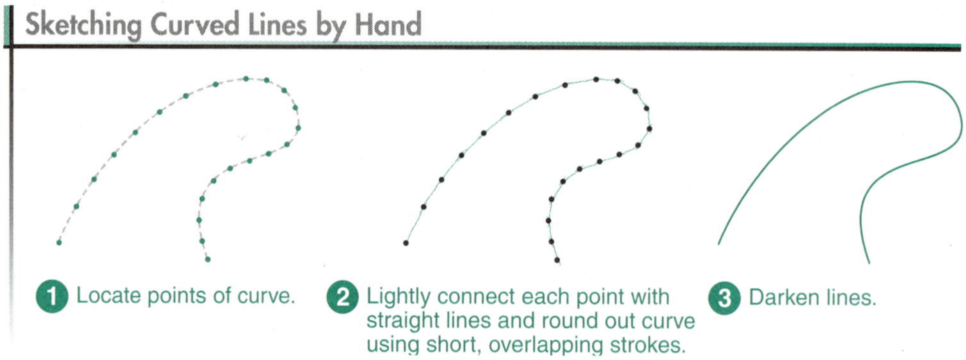

Figure 2-10. Curved lines can be drawn by hand by marking the points of the curve using dots and connecting the dots.

Line Types

Technical drawings contain different types of lines to show the shape and features of the drawn object. Common line types on technical drawings include visible, hidden, center, symmetry, cutting-plane, viewing-plane, section, break, and phantom lines. Each of these lines has a unique appearance that allows it to be easily recognized. **See Figure 2-11.**

Visible Lines. A *visible line (object line)* is a thick line that defines the visible shape of an object. Visible lines are sometimes referred to as object lines. Visible lines are continuous lines and should be drawn thick and dark. They are used wherever there is a distinct edge or contour on an object. Visible lines are the most common lines in a drawing.

Standard styles for lines, such as visible, hidden, and break lines, are used to allow drawings to be consistent and easily understood.

Hidden Lines. A *hidden line* is a thin line that represents the shape and edges of an object that cannot be seen. Hidden lines are used wherever there is a distinct edge or contour on an object that is concealed. Hidden lines are drawn with a series of short, evenly spaced dashes. The dashes are approximately 1⁄8″ (3 mm) long with 1⁄16″ (1.6 mm) spaces. Hidden lines are drawn touching the object or line and should always begin and end with a dash. When drawing an object, select a composition angle that uses the fewest number of hidden lines. A large number of hidden lines in a drawing can make the drawing difficult to read.

Center Lines. A *center line* is a thin line that locates axes or center points of arcs and circles. Center lines are thin and drawn as a series of long and short dashes separated by spaces. The long dashes are from 3⁄4″ (18 mm) to 1½″ (36 mm) long. The short dashes are approximately 1⁄16″ (3 mm) long. The spaces are approximately 1⁄16″ (1.6 mm) long. Center lines extend a short distance beyond the object or feature of the drawing unless a longer extension is required for dimensioning.

Center lines can be extended from a center point of an arc or circle. A center point is two center lines crossing. Center points are marked by two short dashes crossing one another at right angles. Center points are indicated in the centers of all circles and arcs.

Alphabet of Lines

Name and Use	Conventional Representation	Example
VISIBLE LINE Defines shape Outlines and details objects		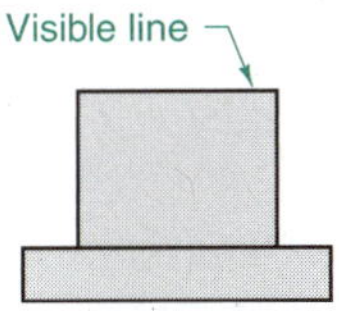 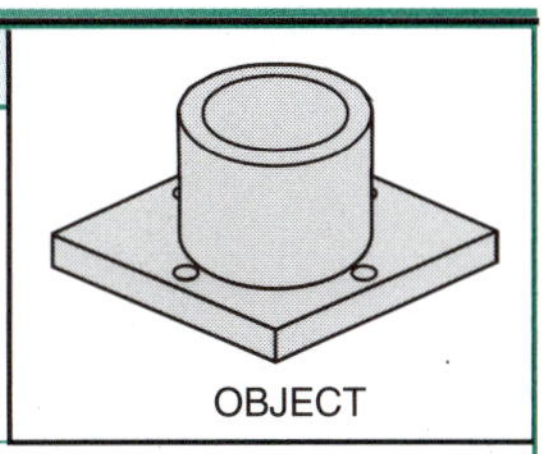
HIDDEN LINE Shows hidden features	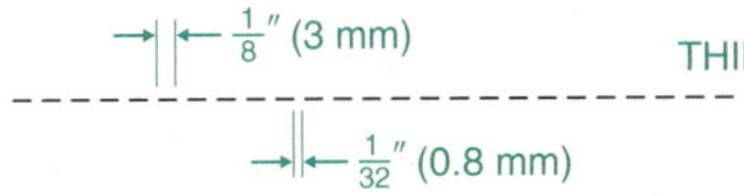	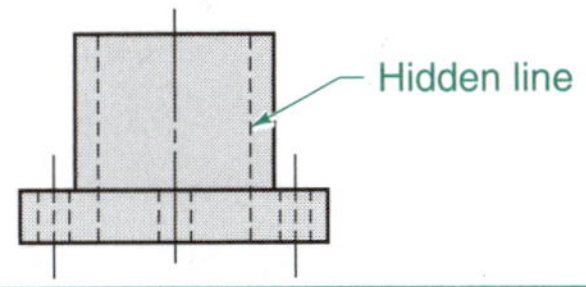
CENTER LINE Locates center points of arcs and circles **SYMMETRY LINE** Shows line of symmetry for partial view	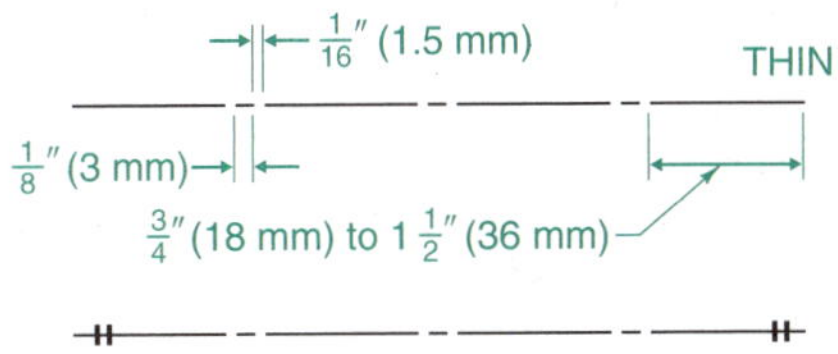	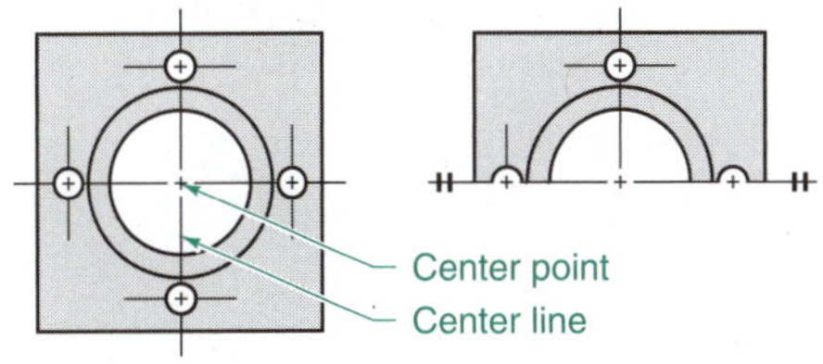
CUTTING-PLANE LINE Shows internal features **VIEWING-PLANE LINE** Shows viewpoint of removed features	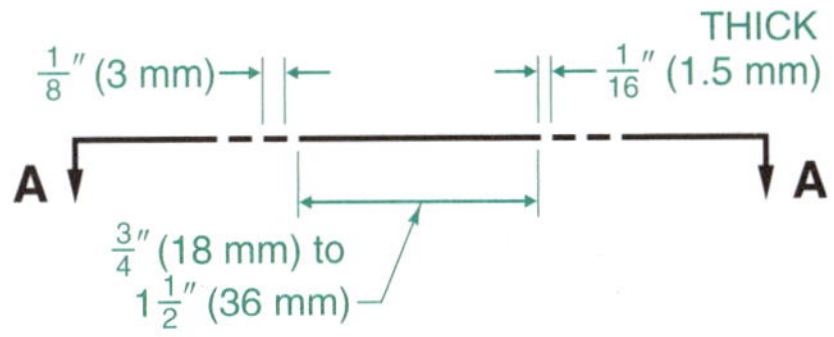	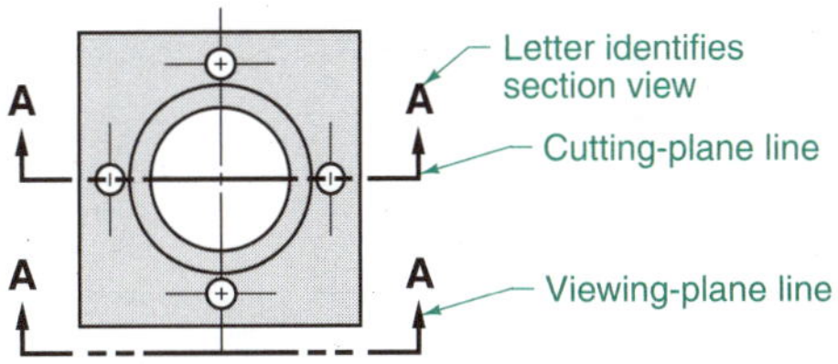
SECTION LINE Identifies internal features	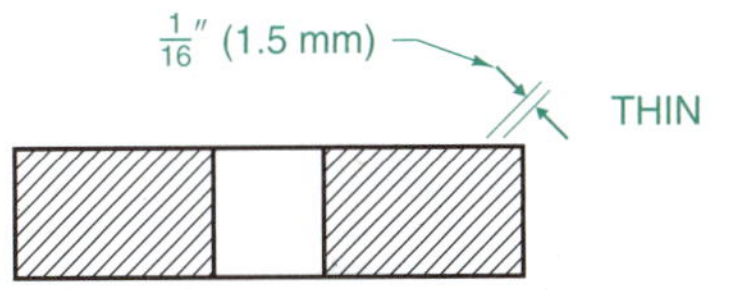	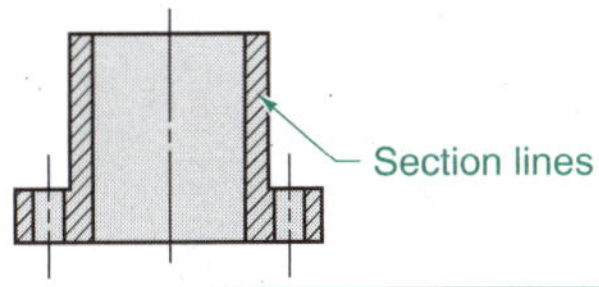
BREAK LINES **Long** Shows long breaks **Short** Shows short breaks	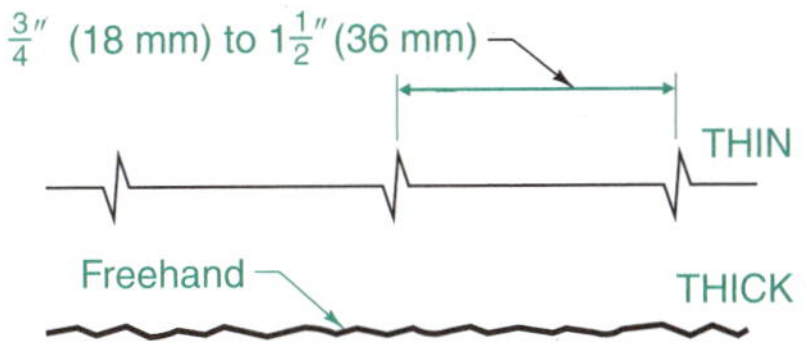	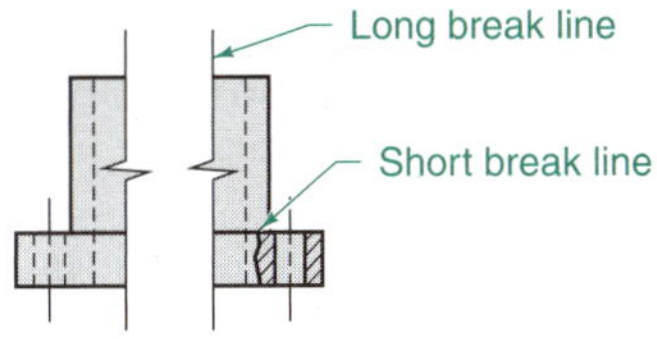
PHANTOM LINE Shows alternate position		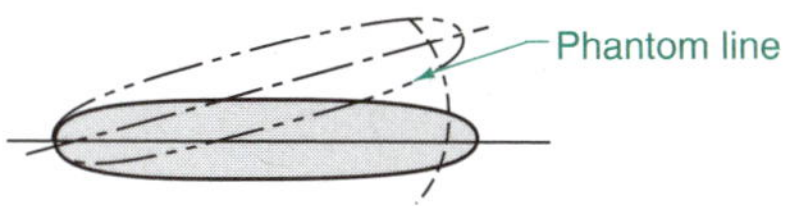

Figure 2-11. Common line types on drawings include visible, hidden, center, symmetry, cutting-plane, viewing-plane, section, break, and phantom lines.

Symmetry Lines. A *symmetry line* is a thin line that is used on a partial view of an object to indicate the remaining portion is symmetrical. Symmetry lines are helpful to save space in a drawing by showing only one half of symmetrical objects. Symmetry lines are two short parallel lines that are drawn at right angles to the center line and are located near each end to indicate the other, identical half of the object.

Cutting-Plane and Viewing-Plane Lines. A *cutting-plane line* is a thick line that shows where an object would have an imaginary cut in order to view internal features. A *viewing-plane line* is a thick line that shows the viewpoint of an additional drawing showing another view or removed portion of an object. Cutting-plane and viewing-plane lines are drawn as a series of long single and short double dashes. Long single dashes are from ¾″ (18 mm) to 1½″ (36 mm) long. Short double dashes are approximately ⅛″ (3 mm) long. Spaces are 1⁄16″ (1.6 mm) long. Cutting-plane and viewing-plane lines have arrowheads on the ends of the line to indicate the direction of sight. Letters near the arrowheads identify the section view and correspond to letters in additional views that are drawn.

Phantom lines are used when an alternate position needs to be indicated in the drawing of objects, such as when showing the movement of a cam in an engine assembly.

Section Lines. *Section lines* are a grouping of thin, parallel lines that identify the internal features of an object cut by a cutting-plane line. Section lines are drawn straight and thin. Section lines are drawn at a 45° angle and are generally equally spaced approximately 1⁄16″ (1.6 mm) apart. Alternate angles or directions can be used when there are two different parts next to each other. Specific patterns of section lines are sometimes used to identify the particular type of material being sectioned, but usually an all-purpose section line pattern is used.

Break Lines. A *break line* is a line that is used to indicate a partial view of a larger object or part. Break lines are used to show a long, continuous feature when a full view is not necessary or may take up too much space. Long break lines are drawn with a thin line containing a zigzag approximately every ¾″ (18 mm) to 1½″ (36 mm). Short break lines are drawn with a thick, dark, freehand line.

Phantom Lines. A *phantom line* is a thin line that shows an object's alternate position or a repeated detail. Phantom lines are drawn as broken lines, alternating long dashes with pairs of short dashes.

ILLUSTRATING TECHNICAL DRAWINGS

Technical drawings can be illustrated to imply a three-dimensional object. Ink can be added to sharpen and clarify lines, and line weights and breaks can be manipulated to give an object the illusion of depth. A drawing of an object can appear more three-dimensional by using a thicker line weight on the edges of the object that border open space. A thinner line weight is used on the remaining interior lines and edges. Objects can also be outlined using a thicker line to delineate the object from the background. Lines that describe parts of an object that fall behind other parts can be broken just before they meet the lines in front, allowing the broken lines to appear as if they recede behind the other lines. **See Figure 2-12.**

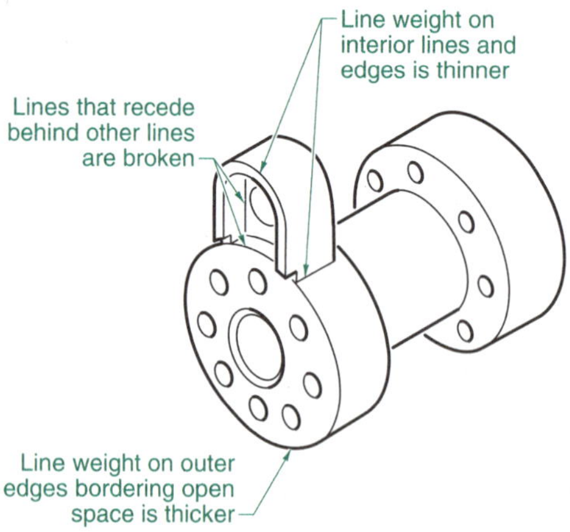

Figure 2-12. Line weights and breaks can be manipulated to give an object the illusion of depth.

Technical drawings may require further illustration or modeling for specific industry needs or to create a polished finished product. Illustration processes such as inking, shading, and rendering can be used on technical drawings to portray the object or objects being shown in a clear and professional way.

Inking

Inking is an illustration process where pencil lines are drawn over in ink. Inking a drawing adds clarity to pencil lines and makes the drawing more readable and easier to reproduce. Vellum or other translucent papers are helpful to overlay and ink the pencil drawing while keeping the original intact.

When adding ink to a drawing, choose the best pen width for the line desired. Inked lines should be centered directly over any existing pencil lines on a drawing. To draw more uniform lines, the pen should be held perpendicular to the drawing paper. Curved lines and circles should be inked first. Compasses can be used to draw circles and arcs by attaching an ink pen tip. After the curved lines are drawn, straight lines are added because they can easily be drawn to connect the curves. When inking straight lines, begin with horizontal lines, then vertical lines, and finally inclined lines. Wait for each line to dry before connecting it to another line to avoid ink running or smearing. Arrowheads, dimensions, and lettering should be inked last. **See Figure 2-13.** When inking lines along a T-square, triangle, or template edge, it is important to keep the pen vertical to avoid an uneven or misplaced line or ink seepage.

Drawing ink is permanent and does not erase easily. Any spills or pools should be blotted immediately while the ink is still wet. Unwanted marks that have dried on the paper can be erased using an electric eraser. If part of an erased line needs to be redrawn, a series of fine lines should be drawn side-by-side until the desired line width is reached. Redrawing a portion of a line with the same line width will often result in a bump, or the two parts not lining up to each other.

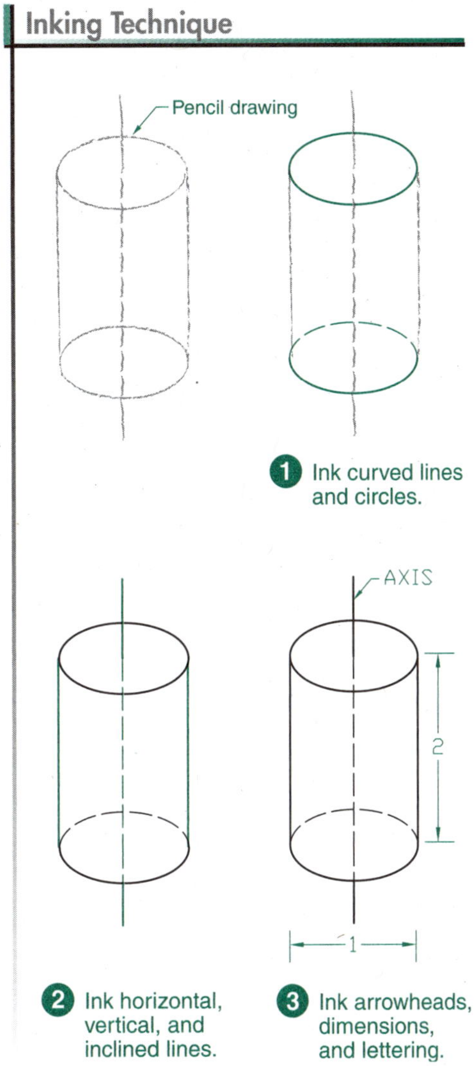

Figure 2-13. A drawing should be inked beginning with curved lines, followed by straight lines, and finishing with arrowheads, dimensions, and lettering.

Shading

Technical drawings often require shading to fully illustrate the object being drawn. A standard light source in technical drawings comes from the upper left-hand corner. Having a uniform light source helps the viewer to interpret drawings more consistently. Dark shading or cast shadows are not generally indicated on technical drawings because they can obscure aspects of the object being shown. **See Figure 2-14.**

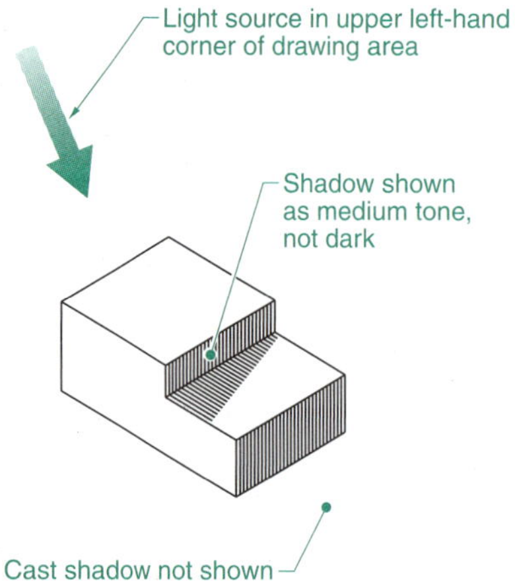

Figure 2-14. A standard upper left-hand corner light source can be adopted to help the viewer uniformly interpret drawings.

Shading on technical drawings is often created with black elements, such as shapes, lines, or dots. Patterns of repeated black elements can create the illusion of grays. Block shading and straight-line shading are two methods of shading technical drawings. **See Figure 2-15.** Cross-hatching or stippling can also be used.

TECH TIP

When adding straight-line shading to an object, the angle of the lines drawn is usually determined by the angle of the height or length axes. Vertical planes are shaded with vertical lines that align with the height axis, whereas horizontal planes are shaded with inclined lines that match the angle of the length axis. A spherical shape can be shaded with straight or curved lines that match the curve of the object, depending on preference.

Block shading is a method of shading that uses flat areas of black. Block shading is generally added to the right side and bottom of an object with a standard light source in the upper left-hand corner. Block shading can be subtly added by simply darkening the object lines on the right side and bottom of the object. It can also be added using larger areas of black. White areas can be left untouched to indicate reflected light.

Straight-line shading is a method of shading that uses a series of straight or curved parallel lines. Closely spaced lines appear as gray tones. The quantity and spacing of the lines determines how light or dark the tone appears.

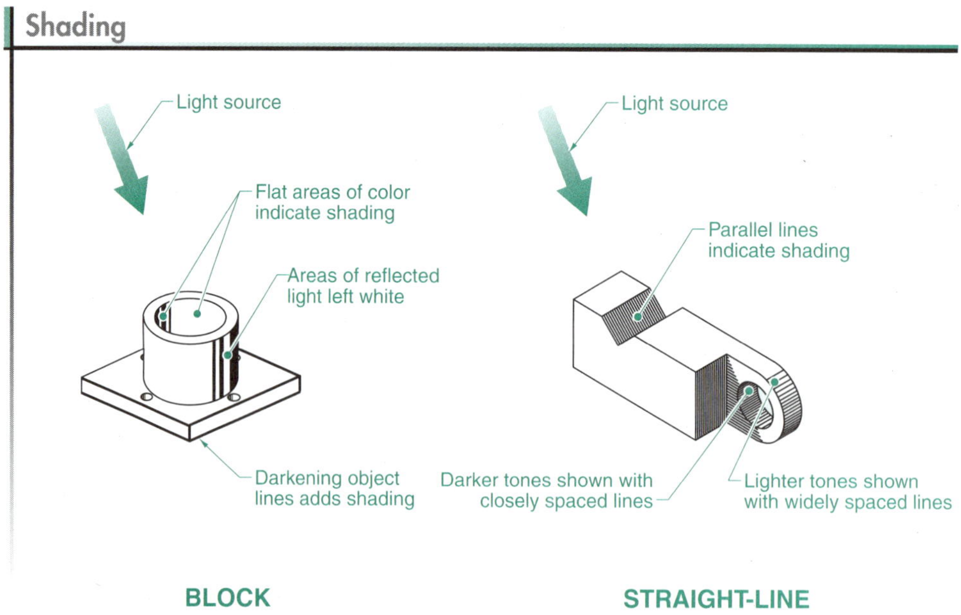

Figure 2-15. Shading is often added to technical drawings using block shading or straight-line shading.

Rendering

Rendering is an illustration technique that shades an image using blacks, whites, and intermediate grays to create a three-dimensional, photo-realistic appearance. Technical drawings are rendered when a lifelike depiction of an object is needed. Technical drawings for manufacturing or assembly uses are not usually rendered.

Rendering is done after the final line drawing is complete. Rendering can be done using various materials, such as pencil or charcoal. An airbrush can also be used to give objects smooth, realistic shading. An *airbrush* is a small spray gun that uses air to spray paint or other colored liquids onto a surface. Airbrush rendering can create softer transitions between tones than other drawing materials. Because airbrushing can be less precise for hard edges, marking the edges and lines by hand is still recommended. **See Figure 2-16.**

LETTERING TECHNICAL DRAWINGS

Technical drawings often require notes, callouts, or other lettering. Lettering size and style are based on the conventions in ANSI/ASME Standard Y14.2M, *Line Conventions and Lettering*. Uppercase, single-stroke gothic letters are used. They may be inclined or vertical. The preferred slope for inclined letters is 68° to the horizontal. Lettering is created following specific guidelines to maintain legibility and consistent sizing, formation, and spacing.

Sizing

The height of freehand lettering can vary depending on the paper size. Once a height is chosen to fit the page, all capital letters should be the same height. The widths of letters and numbers vary in freehand lettering. Most letters and numbers are approximately two-thirds as wide as they are high.

Rendering

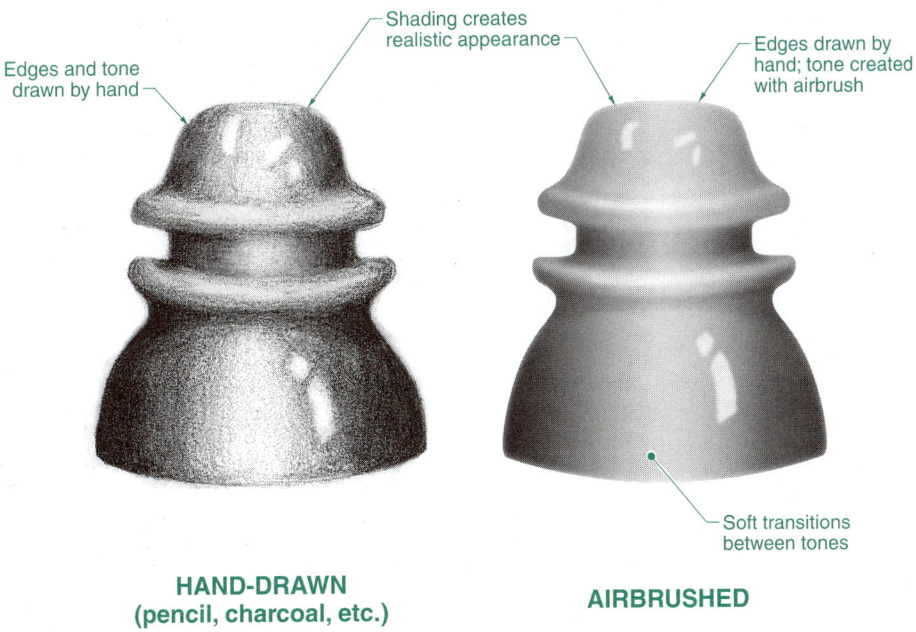

Figure 2-16. Rendering is an illustration technique that adds shading to an image to create a three-dimensional, realistic appearance.

Rendering Media Clip

Lettering Size
Media Clip

Letter width and height can be determined by placing a grid of six-by-six equally sized squares behind each number and letter. All capital letters will be the same height of 6 grid-squares high. Most letters are either 5 or 6 grid-squares wide, with the exceptions of W at 8 grid-squares wide, and I at 1 grid-square wide. Letters that are 6 grid-squares wide are A, M, O, Q, T, V, X, and Y. The remaining letters in the alphabet are 5 grid-squares wide. The width for numbers follows the same guidelines. Numbers have a general width of 5 grid-squares wide, except for the number 1. **See Figure 2-17.**

Letter Width and Form

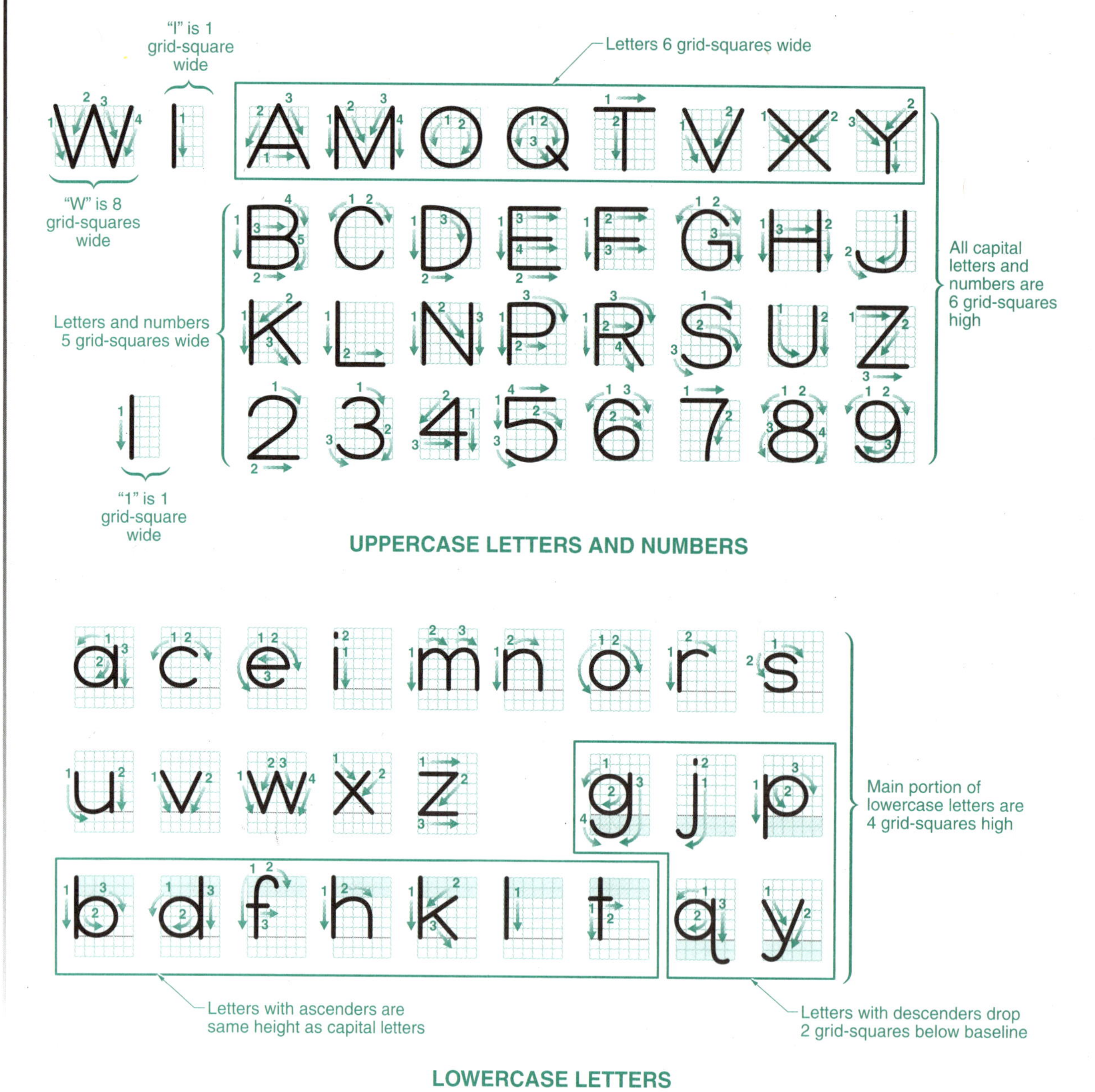

Figure 2-17. Letter width, height, spacing, and formation vary depending on the letter or number and should be consistent in a drawing.

Lowercase letters are rarely used in technical drawings. However, when they are used, lowercase letters are 4 grid-squares high according to the six-by-six grid system. The main portion of a lowercase letter should be approximately two-thirds the height of a capital letter, not counting the ascender or descender. An *ascender* is the portion of a lowercase letter that rises above the height of the lowercase letter "x". Lowercase letters with an ascender include b, d, f, h, k, l, and t. Lowercase letters with an ascender are the same height as a capital letter. A *descender* is the portion of a lowercase letter that drops below the main baseline of the line of letters. Lowercase letters with a descender include g, j, p, q, and y. The descender on these lowercase letters falls 2 grid-squares below the baseline of the capital letter. For italic lettering, the same sizing principles can be used, but with the grid-squares slanted at the same angle as the letter incline.

Formation

To form freehand letters, draw horizontal strokes from left to right (for a right-handed person) and vertical, inclined, and curved strokes from top to bottom. Light horizontal guidelines can be drawn to assist with proper placement and height of lettering.

Spacing

The shape of each letter determines the required spacing between letters in a word. This spacing should be estimated by eye based on the background areas between the letters. When letters are spaced with an equal distance between them, the letters actually appear unevenly spaced instead of balanced. Spacing between numbers follows the same guidelines. The spacing between words should be approximately equal to one letter space. The spacing between lines should be approximately ½ times to 1½ times the height of one letter space. **See Figure 2-18.**

Chapter 2 Quick Quiz®

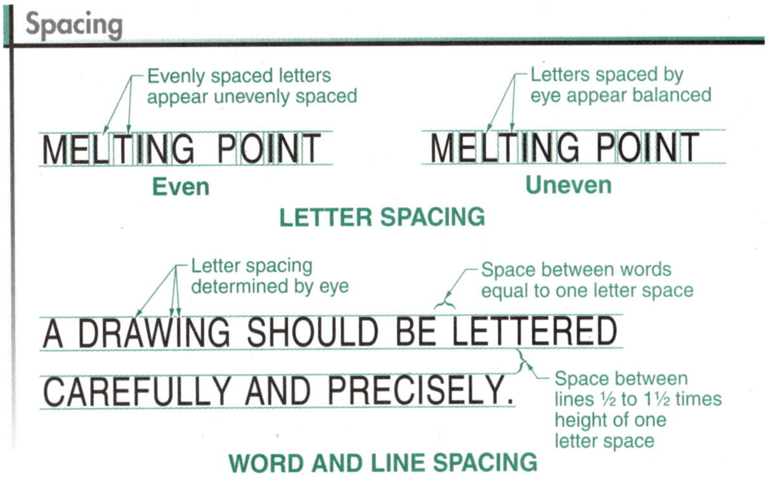

Figure 2-18. Spacing between letters and numbers should be estimated by eye. Spacing between words and lines should be sized based on one letter space.

Instrument Drawings and Techniques

Review 2

Name ______________________ Date ______________

REVIEW QUESTIONS

__________ **1.** The standard tools required for a technical drawing are ___, a pencil or pen, and an eraser.

T F **2.** Mechanical pencils produce a more consistent line width than wood pencils, and most styles maintain their point without sharpening.

__________ **3.** Two widths of pen tips with an approximate size ratio of ___-to-one should be used in drawings with both thick and thin lines.

__________ **4.** A(n) ___ is an instrument used to draw vertical and inclined lines.

T F **5.** A compass is an instrument that is used to transfer and compare measurements.

__________ **6.** An architect's scale is usually triangular in shape and has a total of ___ scales.

T F **7.** A protractor should be used when exact angle measurements are required.

__________ **8.** A(n) ___ line is parallel to the horizon and extends from right to left or left to right.

__________ **9.** Inclined lines are also called ___ lines because angles are created when two or more lines cross.

__________ **10.** The 45° triangle, 30°-60° triangle, or both can be used to create inclined lines every ___°.

T F **11.** A visible line is a thick line that defines the visible shape of an object and is sometimes referred to as an object line.

T F **12.** When drawing an object, select an orientation that uses a large number of hidden lines.

__________ **13.** Center points are indicated in the centers of all circles and arcs by two short dashes crossing one another at ___ angles.

T F **14.** Symmetry lines are helpful to save space in a drawing by showing only one half of symmetrical objects.

__________ **15.** ___ lines are a grouping of thin, parallel lines that identify the internal features of an object cut by a cutting-plane line.

T F **16.** Curved lines and circles should be inked after straight lines.

__________ **17.** A standard light source in technical drawings comes from the upper ___-hand corner.

_________________ **18.** ___ shading uses a series of straight or curved parallel lines to add tones to objects.

T F **19.** Rendering an illustration creates a two-dimensional, outlined appearance.

T F **20.** Letters should be spaced with an equal distance between them to appear balanced.

IDENTIFICATION – Line Types

f **1.** Visible line

h **2.** Hidden line

e **3.** Center line

d **4.** Symmetry line

c **5.** Cutting-plane line

b **6.** Section line

a **7.** Long break line

g **8.** Phantom line

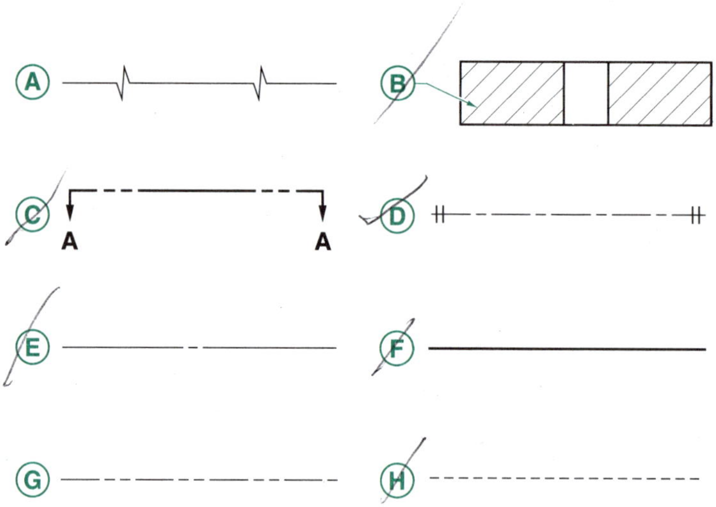

IDENTIFICATION – Measurements

_______ 1. 135°

_______ 2. 120°

_______ 3. 165°

_______ 4. 75°

_______ 5. 30°

_______ 6. 45°

_______ 7. 90°

_______ 8. 150°

_______ 9. 60°

_______ 10. 15°

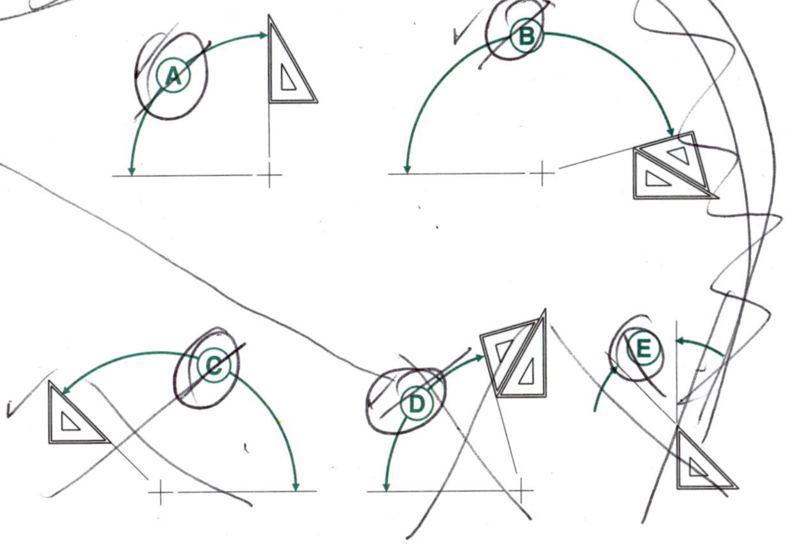

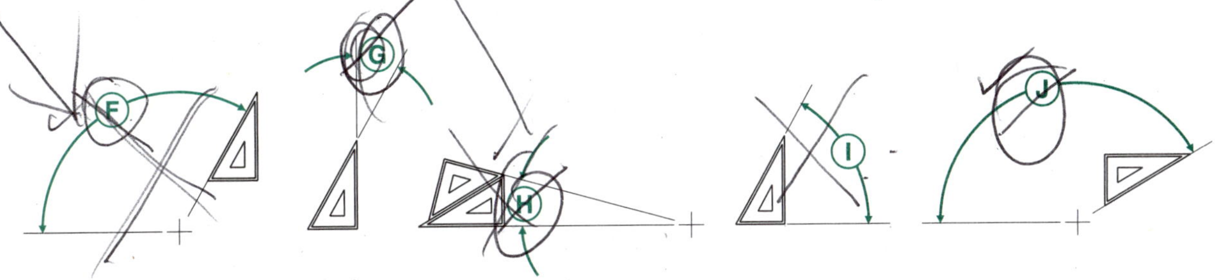

Number the following components of a drawing in the order in which they should be inked.

_______ 11. Vertical lines

_______ 12. Arrowheads

_______ 13. Horizontal lines

_______ 14. Curved lines

_______ 15. Inclined lines

EXERCISES

1. Use the provided endpoints to draw two horizontal lines by hand. Add intermediate points to ensure accuracy.

(A)

(B)

2. Using triangle(s), draw an inclined line at 75°.

3. Redraw the given curved lines by hand.

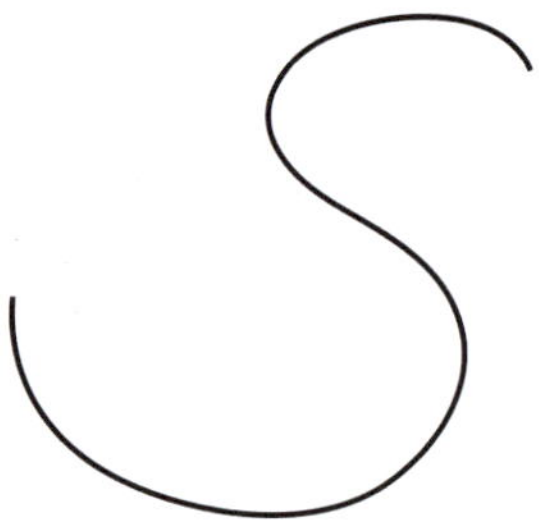

(A)

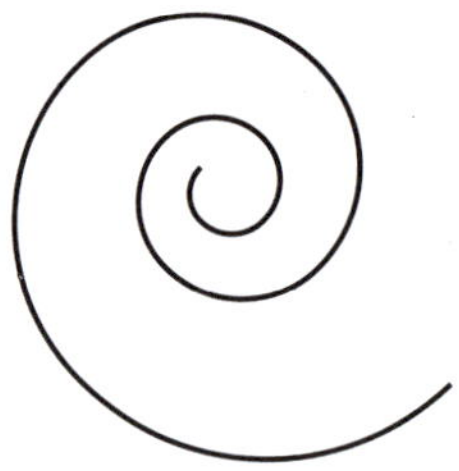

(B)

4. Using a straightedge and triangle(s), redraw the shape shown.

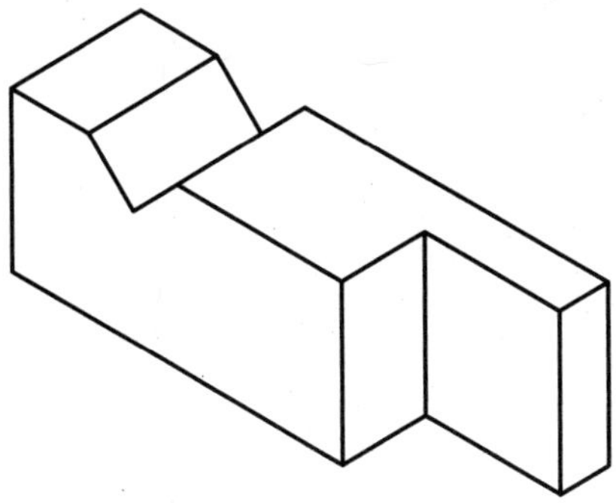

For each question, use a divider to transfer the distance of line A. Using the distance of line A, mark out the given number of equal line segments on line B.

5. Ten equal segments

A B

6. Five equal segments

A B

7. Four equal segments

A B

8. Seven equal segments

A B

Using a protractor, determine the measurements of the provided angles.

______________________ **9.**

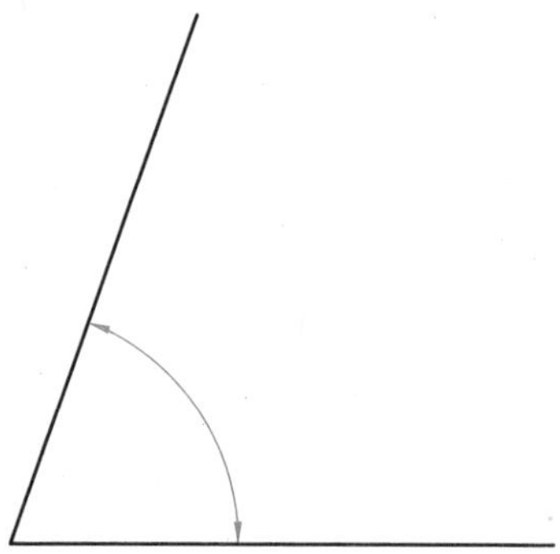

______________________ **10.**

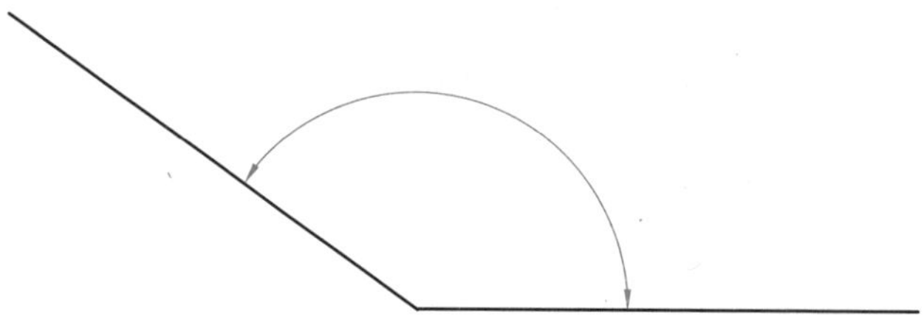

______________________ **11.**

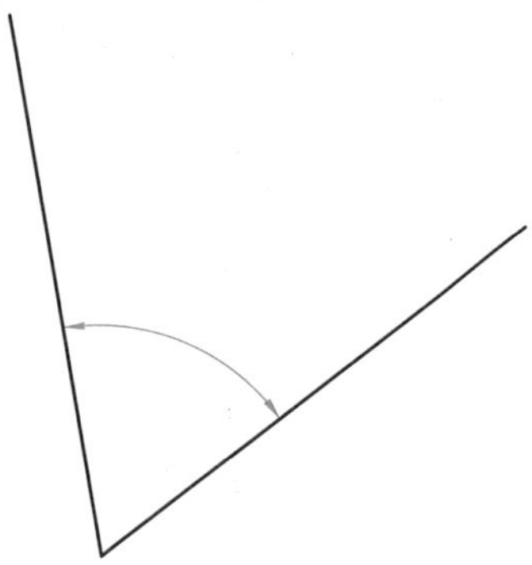

______________________ **12.**

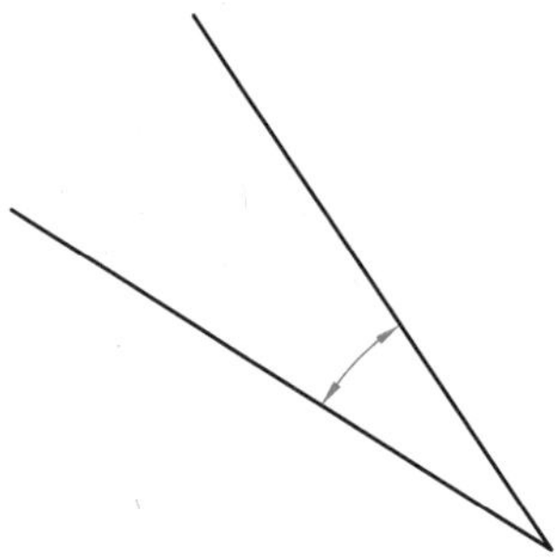

Using a protractor, draw each angle with the measurement given.

13. 55°

14. 150°

15. 137°

16. 12°

Sketch the following objects in the provided boxes. Shade the sketch using the method given. Draw in and use a light source for all drawings in the upper left-hand corner.

17. Apply block shading by thickening object lines.

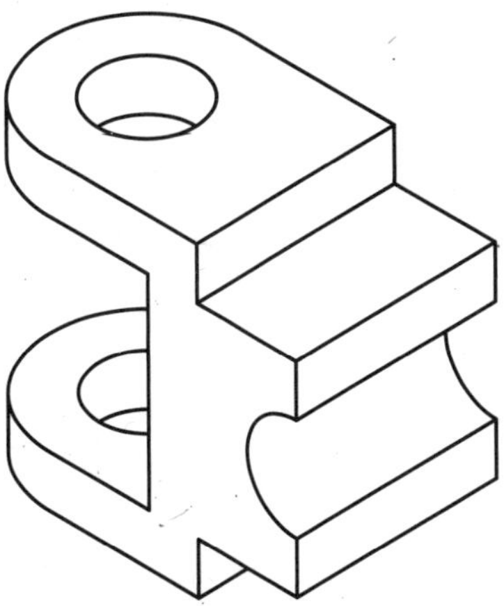

18. Apply block shading using areas of black and white.

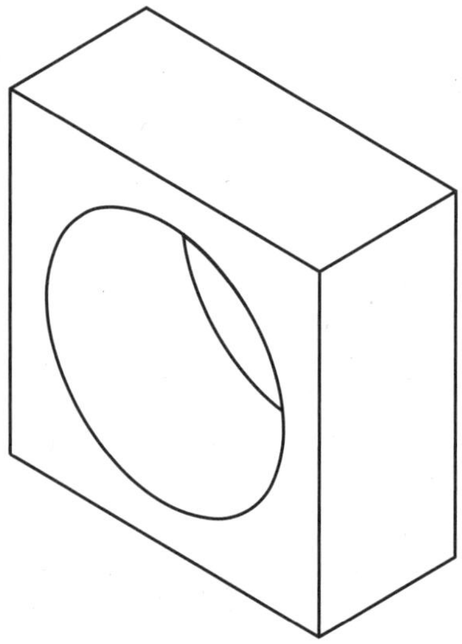

19. Apply straight-line shading using straight lines.

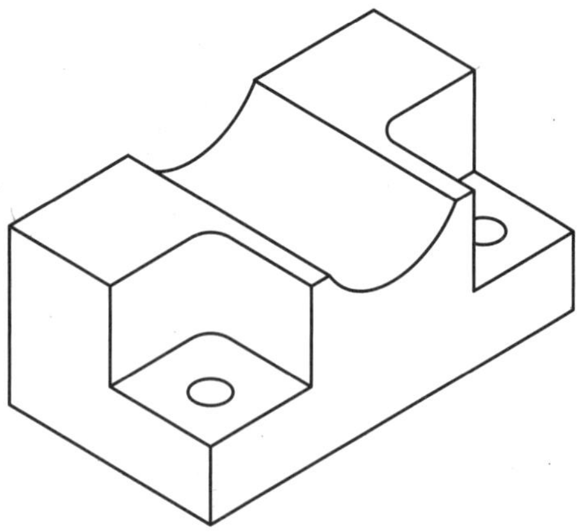

20. Apply straight-line shading using curved lines.

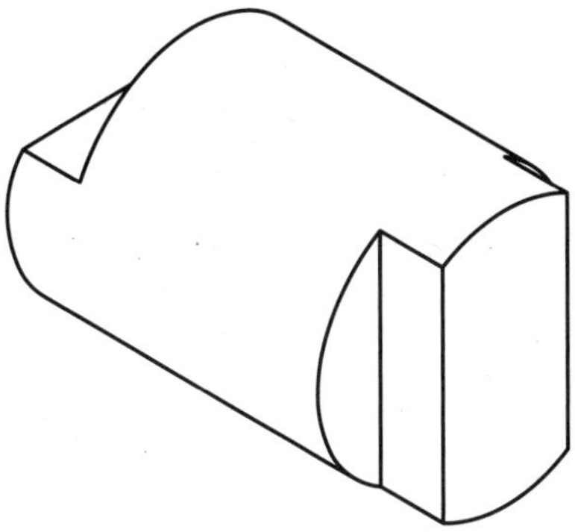

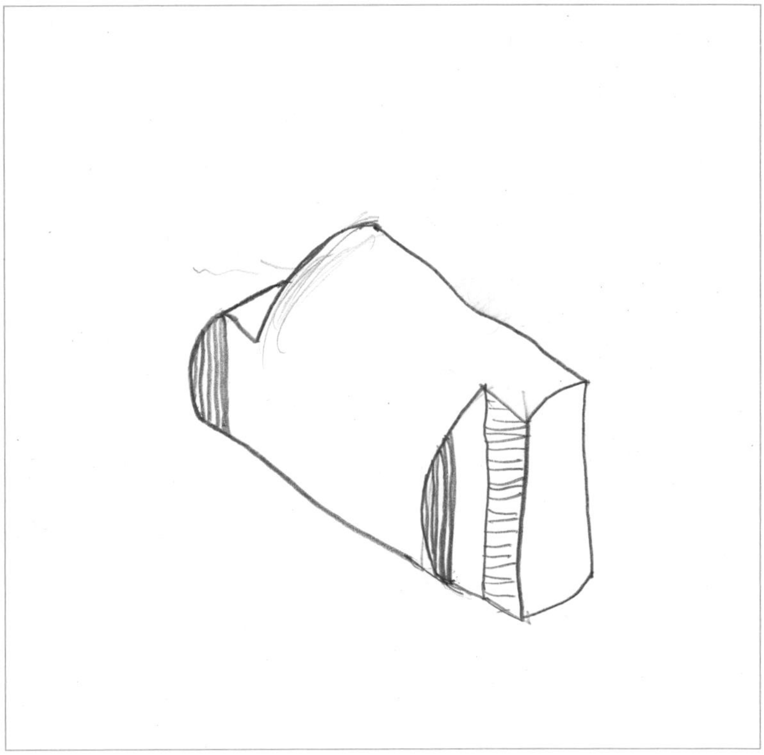

Using the proper lettering technique, write the phrase below in the given style.

"Lettering is added to a drawing to convey additional information."

21. Uppercase

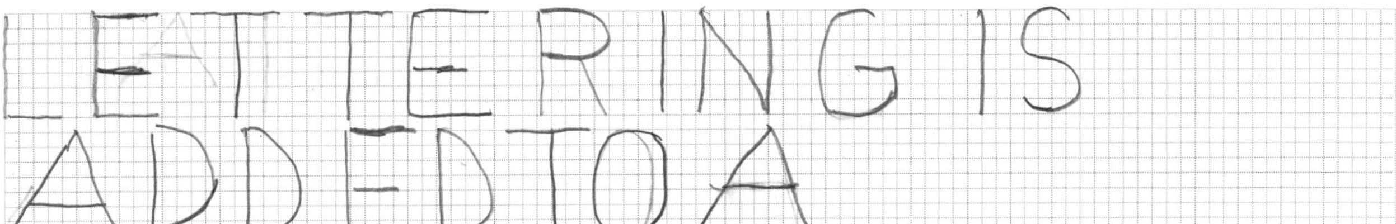

22. Italic uppercase

LETTERINGIS ADDED
TO A DRAWING TO
CONVEY ADDITON AL
INFORMATION

23. Lowercase

letteringis added
to a drawing to convey
additonal information

Technical Sketching and Drawing

3

Geometric Constructions

GEOMETRIC SHAPES AND TERMINOLOGY

Geometric shapes and constructions contain points, lines, and angles. A *point* is an exact location in space that has no dimension. A point is geometrically represented by the intersection of two lines. A point is shown on a drawing by two short, perpendicular lines that form a small cross. Points are not generally shown on a drawing using dots because they can be less accurate, but dots can be used for illustrative purposes or when technical accuracy is not necessary.

A line is a mark that has a continuous length and that connects one or more points. A line can be straight or curved. A straight line is the most direct connection between two points. A curved line is a line that bends but has no angles. An angle is formed by the intersection of two lines and is an element of many geometric shapes.

Points and lines exist on planes. A *plane* is a two-dimensional surface that has no depth. Planes define a shape when bound by a line or lines. Planes and lines have relationships to each other based on their placement in space. *Perpendicular* is when two lines or planes form a right angle. *Parallel* is when two lines or planes remain the same distance apart and never intersect. Technical drawings consist of points, lines, angles, and planes that combine in various ways to form two-dimensional shapes and three-dimensional solids.

Two-Dimensional Shapes

Two-dimensional shapes are plane figures that are formed with lines and include either angles or arcs. A *plane figure* is a two-dimensional shape bound by lines that exist on one plane. Plane figures can contain straight or curved lines. Types of plane figures include circles, polygons, triangles, and quadrilaterals. **See Figure 3-1.**

Circles. A *circle* is a plane figure that is a regular, closed curve with all points on the circle an equal distance from a center. A circle contains 360°. The circumference of a circle is the linear distance around the circle's outer edge. The diameter is a straight line through the center of the circle with each end terminating on the circumference. The radius is a straight line from the center of the circle to the circumference, or half the diameter. A *chord* is a straight line with two end points on the circumference of a circle. An elongated circle that is ovular in shape is referred to as an ellipse.

TECH TIP

Theoretically, a line has no width and can be infinitely long. A line can be drawn in any direction and can be a straight line, curved line, circle, or arc. A straight line is the shortest distance between two points. A straight line can also be considered an angle with no bend consisting of 180°, or one-half of a circle.

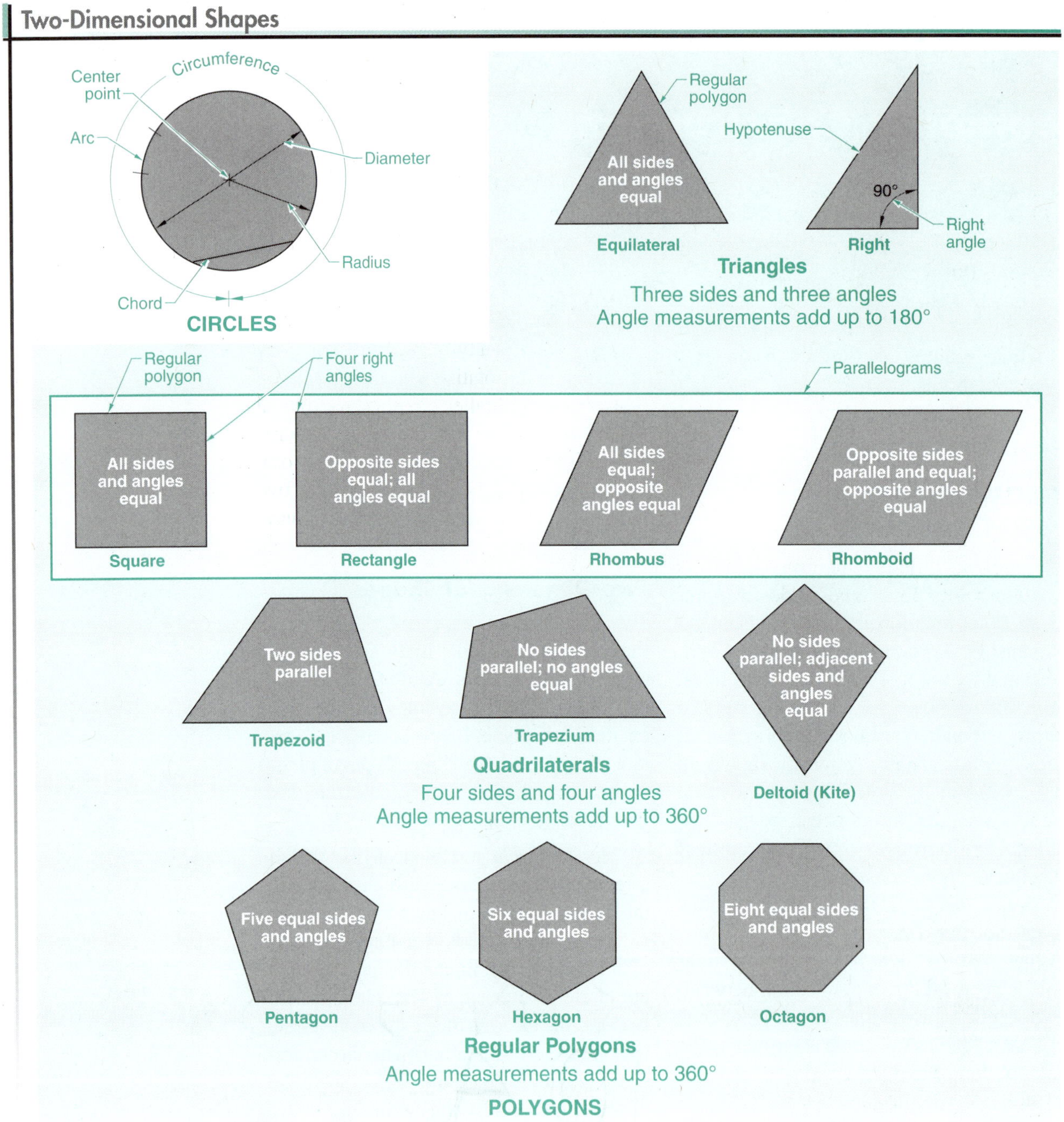

Figure 3-1. Plane figures are formed with lines and include circles, polygons, triangles, and quadrilaterals.

Polygons. A *polygon* is a plane figure that has at least three straight sides. A *regular polygon* is a polygon that has all angles of equal size and all sides of equal length. Polygons can include triangles and quadrilaterals, as well as other multisided figures including pentagons, hexagons, and octagons.

Triangles. A *triangle* is a polygonal plane figure with three straight sides that intersect to form three angles. All three angles have measurements that add up to 180°. An *equilateral triangle* is a triangle with three equal sides and three equal angles. A *right triangle* is a triangle with one right angle that measures 90°. The *hypotenuse* of a right triangle is the longest side that is opposite the right angle.

Quadrilaterals. A *quadrilateral* is a polygonal plane figure that has four straight sides. The four angle measurements in a quadrilateral add up to 360°. Quadrilaterals include squares, rectangles, rhombuses, rhomboids, trapezoids, trapeziums, and deltoids (kites). A *parallelogram* is a type of quadrilateral that has opposite sides parallel to each other. Parallelograms include squares, rectangles, rhombuses, and rhomboids.

Three-Dimensional Solids

Three-dimensional solids are plane figures that have depth and form. An axis is the center line in which a three-dimensional form rotates. Three-dimensional solids are often drawn on technical drawings by adding value to plane figures to give the illusion of depth. Three-dimensional solids can be spheres, ellipsoids, or polyhedrons. **See Figure 3-2.**

Spheres. A *sphere* is a three-dimensional form that is created when a circle is rotated on a single axis. All points of spheres are equidistant from the center point in all directions.

Ellipsoids. An *ellipsoid* is a three-dimensional form that is created by rotating an ellipse on a single axis. An ellipsoid is similar to a sphere but has an elliptical or egg-like shape and can be horizontal or vertical.

Polyhedrons. A *polyhedron* is a three-dimensional form that is constructed of plane figures. A *face* is the surface of a side in a polyhedron. A *base* is the face of a polyhedron that is situated at the top or bottom. A *regular polyhedron* is a polyhedron that has regular polygons for all of its faces. An example of a regular polyhedron is a cube, which is bound by six squares. A *right polyhedron* is a polyhedron with an axis perpendicular to the base(s). An *oblique polyhedron* is a polyhedron with an axis at an oblique angle to the base(s). Common polyhedrons include prisms, cylinders, cones, and pyramids.

A *prism* is a polyhedron with bases that are parallel equal polygons and three or more side faces that are parallelograms. Prisms are often named for the polygonal shape that forms their bases. For example, a triangular prism is a prism with two triangular bases, and a rectangular prism is a prism with two rectangular bases. A prism can be a right or oblique prism.

A *cylinder* is a type of prism created by a straight axis rotating one full revolution between two parallel circular or elliptical bases. A cylinder has only one face that wraps around the extent of the bases to form a closed three-dimensional shape. The two opposite points on a cylinder are always parallel to each other. A cylinder can be a right cylinder or an oblique cylinder.

A *cone* is polyhedron with a circular or elliptical base and one face that wraps around the form, meeting at a vertex. A *vertex* is the point opposite the base where the face(s) meet at a point. A cone can be a right cone or an oblique cone. The vertex of a right cone is a common distance from all points along the outer edge of the base. A cone can be a circular cone or an elliptical cone, depending on the shape of the base.

Objects, such as fastening devices, can be converted into two-dimensional shapes and three-dimensional solids to assist with drawing.

Three-Dimensional Solids

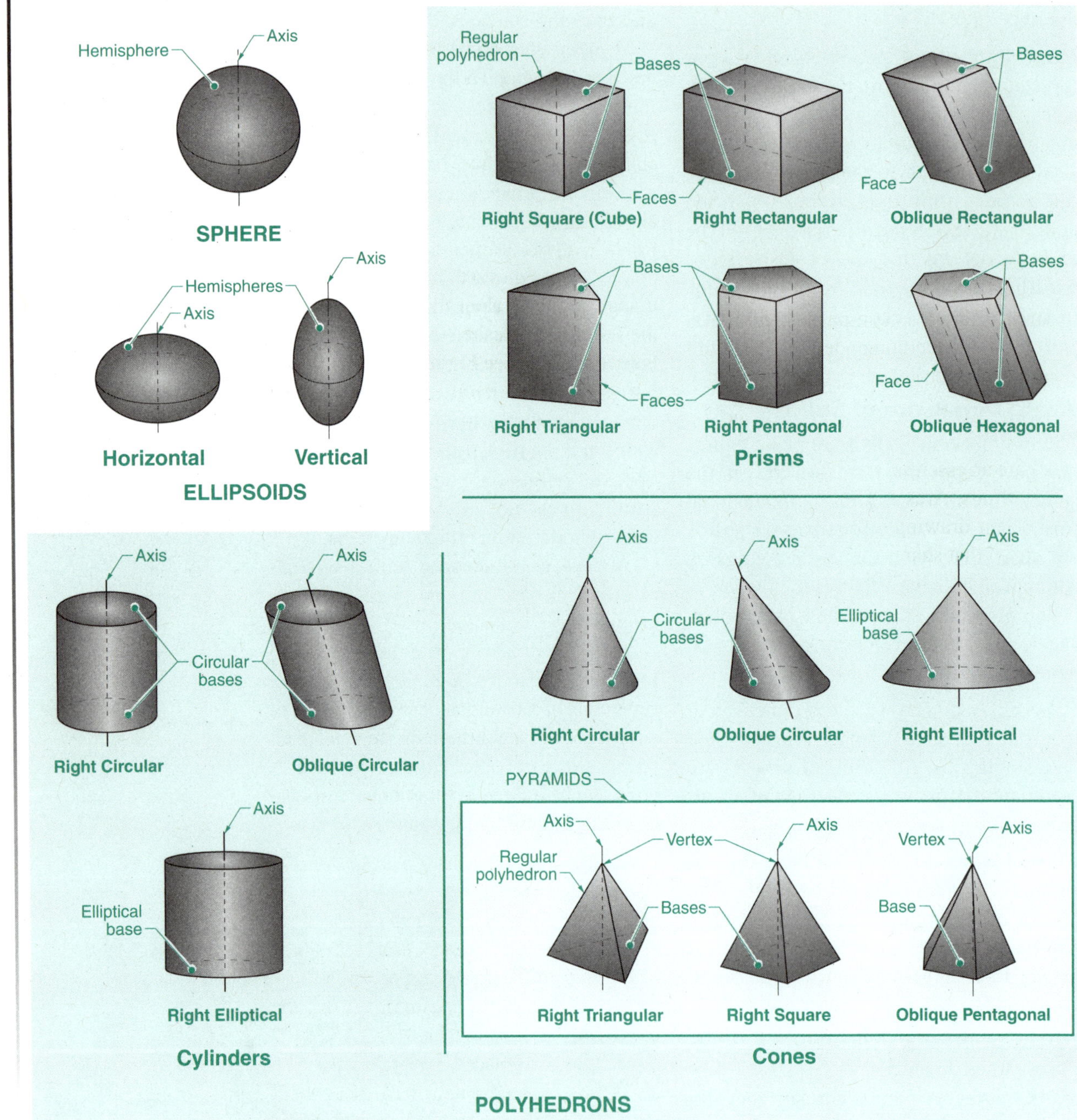

Figure 3-2. Three-dimensional solids are plane figures that have depth and form and include spheres or polyhedrons.

Pyramids are a type of cone with a polygonal base. A *pyramid* is a polyhedron with a polygon for a bottom base and three or more triangular faces that connect in common vertex. Pyramids are often named for the polygonal shape that forms their base. For example, a triangular pyramid is a pyramid with a triangular base and three faces, and a square pyramid is a pyramid with a square base and four faces. Pyramids can be right pyramids or oblique pyramids.

GEOMETRIC CONSTRUCTION PRINCIPLES

All geometric components are constructed following specific principles. Technical drawings may require an artist to draw lines, bisect geometric components, construct circular lines or forms, draw plane figures, enlarge or reduce shapes, or create complex line or shape constructions in a clear and precise manner. Step-by-step procedures allow geometric forms to be properly and consistently developed. Tools such as T-squares, triangles, compasses, dividers, and curves are recommended to assist with accurate geometric construction.

Drawing Lines

Lines are an essential component in technical drawings. Arcs and tangents are often used when drawing lines. A *tangent* is a line or arc that shares a point and direction with a given curve in at least one location. A *tangent point* is the point where a line and an arc or two arcs touch. **See Figure 3-3.** Drawing parallel and perpendicular lines and dividing lines into equal parts should be done by following step-by-step procedures to ensure accuracy.

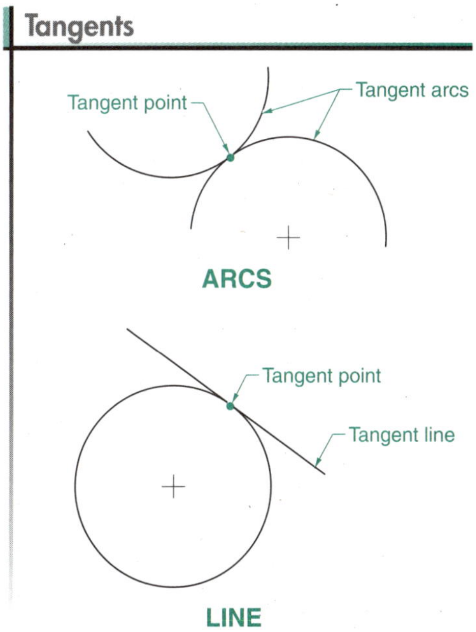

Figure 3-3. Arcs and tangents are often used when drawing lines.

Parallel Lines. The first step in drawing vertical or horizontal parallel lines using a triangle and T-square is to draw the first line. A triangle is then placed on the line with the base of the triangle resting against a T-square or other straightedge. The T-square is held in position and the triangle is slid to the desired distance from the first line. The second line is drawn along the triangle edge. To draw inclined parallel lines with a triangle or triangles and a T-square, the first line is drawn using a triangle or triangles at the desired incline. The triangle(s) are slid along the horizontal edge of the T-square the desired distance and to draw the second line. **See Figure 3-4.**

To draw two parallel lines at a given distance, apply the following procedure:

Given: Line AB at distance CD.

1. Set a compass with the radius equal to CD. Draw two arcs, EF and GH, from any point on the given line.
2. Draw line IJ tangent to arcs EF and GH.

Tangents
Media Clip

To draw two parallel curved lines at a given distance, apply the following procedure:

Given: Curved line AB at distance CD.

1. Set a compass with the radius equal to CD. Draw a series of arcs from multiple points on curved line AB.
2. Using a french curve or flexible curve, draw line EF tangent to these arcs.

Perpendicular Lines. Perpendicular lines are formed by drawing two lines that intersect at 90°. Horizontal and vertical lines that are perpendicular can be drawn using a T-square and a triangle. To create perpendicular lines that are at an angle, the first line is drawn at the desired incline using a straightedge. A 30°-60° or 45° triangle is placed with the longest side along the straightedge and one of the remaining sides along the first line. The triangle is slid up or down the straightedge and the second line is drawn along the opposite side of the triangle to create the perpendicular line.

TECH TIP

Parallel lines can be drawn more accurately when the points that form the lines are farther apart. When geometrically constructing parallel lines, if the arcs are spaced out evenly or drawn toward each end of the first line, there will be a smaller margin of error when the second line is drawn parallel.

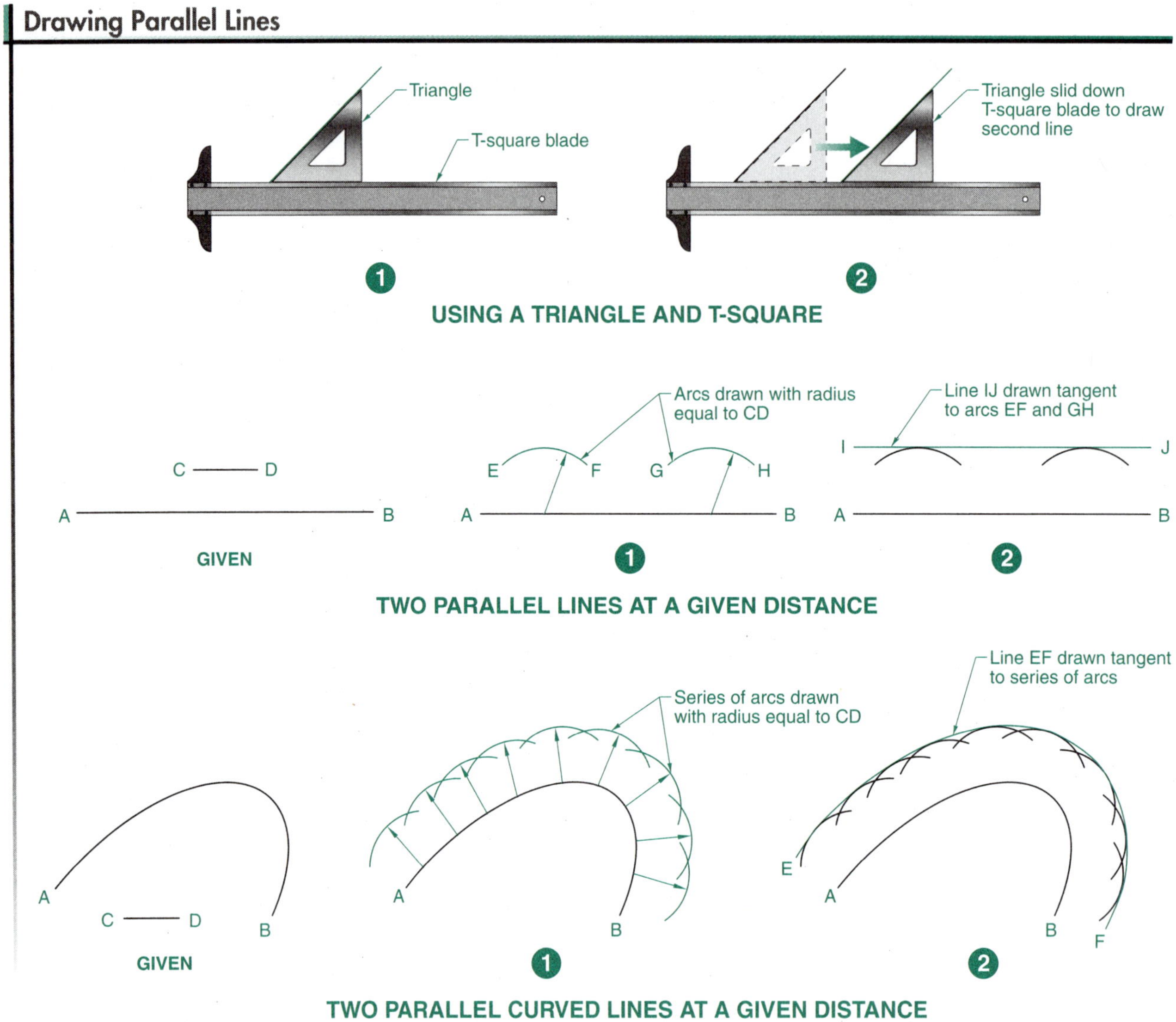

Figure 3-4. Parallel lines can be drawn using a triangle and T-square, or by drawing arcs at a given distance.

To draw a perpendicular line from a given point on the line, apply the following procedure:

Given: Line AB with point C.

1. With a center point of C, draw arc DE with any radius that intersects line AB at points D and E. **See Figure 3-5.**
2. Set the compass with the radius greater than the radius of arc DE. With a center point of D, draw arc FG. With a center point of E, draw arc GH. Arcs FG and GH intersect at point G.
3. Using a straightedge, draw line CG perpendicular to line AB.

To draw a perpendicular line from a given point not on the line, apply the following procedure:

Given: Line AB and point C.

1. With a center point of C, draw arc DE with any radius that intersects line AB at points D and E.
2. Set the compass with the radius greater than the radius of arc DE. With a center point of D, draw arc FG. With a center point of E, draw arc GH. Arcs FG and GH intersect at point G.
3. Using a straightedge, draw line CG perpendicular to line AB.

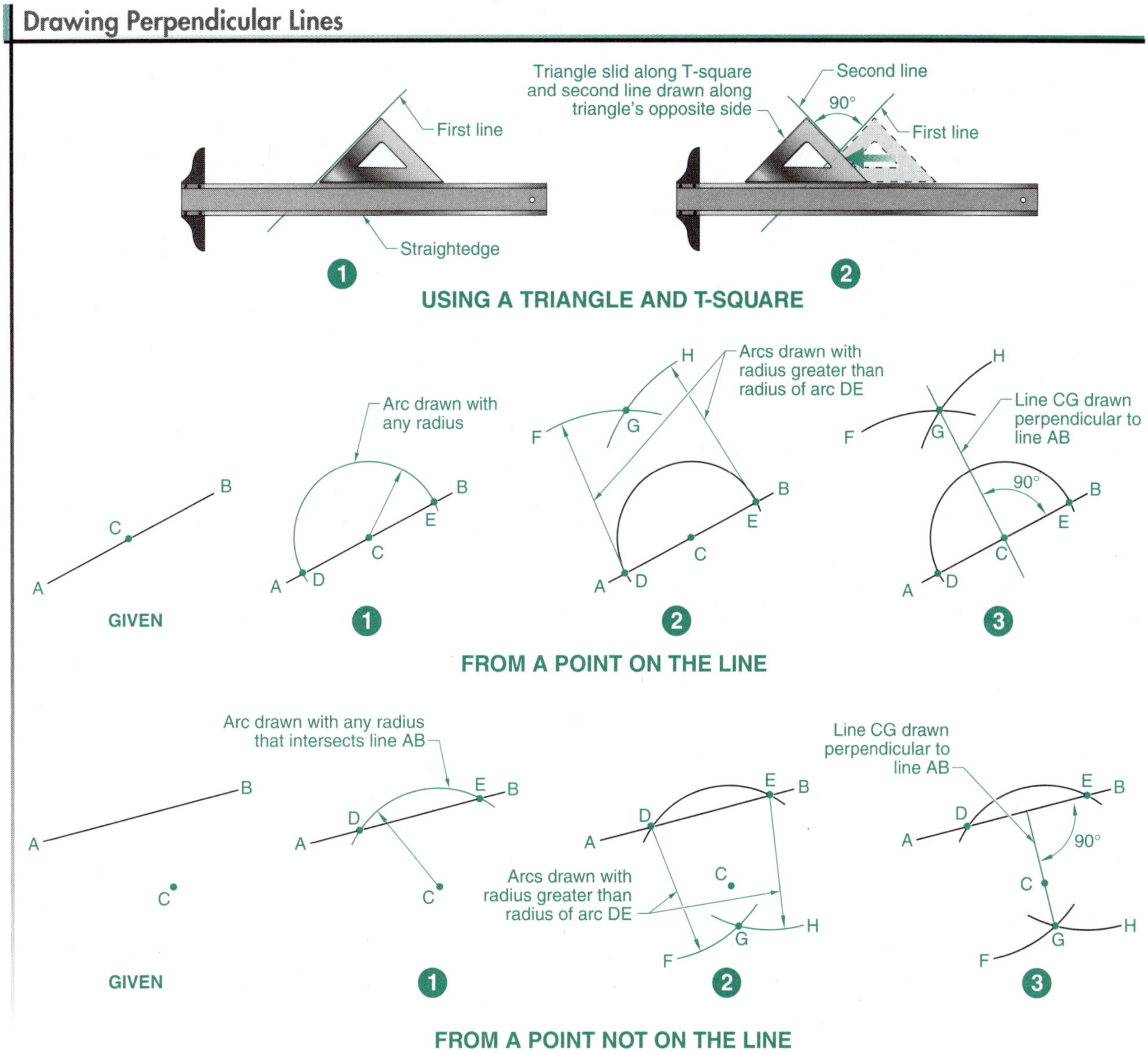

Figure 3-5. Perpendicular lines can be drawn using a triangle and T-square, or through a given point.

Dividing a Line into Equal Sections. Lines can be divided into equal parts using a ruler, scale, or divider. These tools are best used when the size of each section is known and measurable. Lines can also be divided geometrically. To divide a line into equal parts where the size of each section is unknown, apply the following procedure:

Given: Line AB.

1. Draw line AC at any acute angle from line AB. **See Figure 3-6.**
2. On line AC and originating from point A, mark out the desired equal sections using a selected unit on a ruler (inches, millimeters, centimeters, etc.) or a selected distance on a scale or divider. From the termination point D of the last equal section marked on AC, draw line BD.
3. Originating from the equal spaces marked on line AD, draw lines parallel to line BD and intersecting line AB. The intersections of the parallel lines and line AB are the equal division points on line AB.

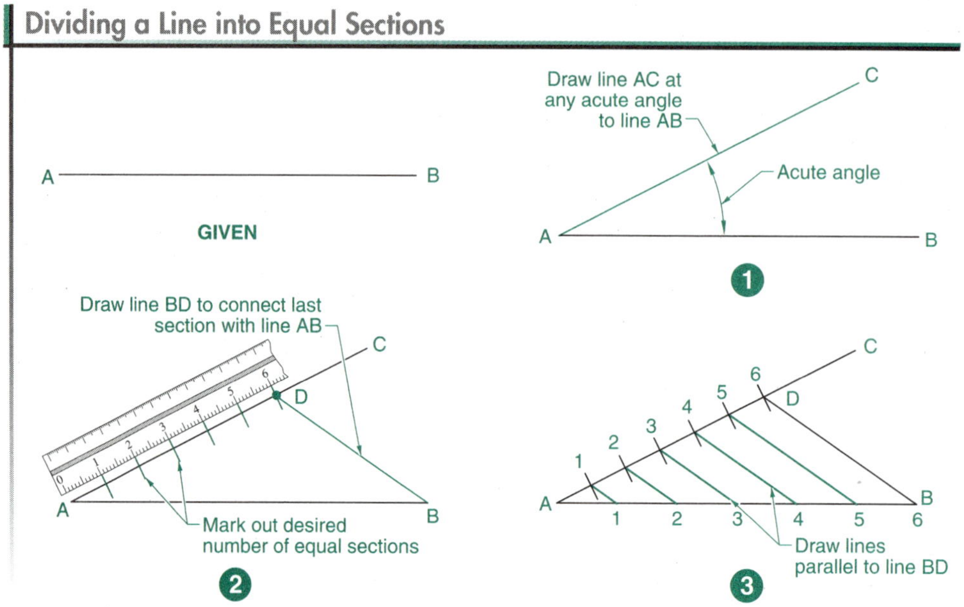

Figure 3-6. Lines can be divided into equal parts using a ruler, scale, or divider, or by drawing parallel lines.

Kolbe & Kolbe Millwork Co., Inc.

Bisecting lines and angles is helpful when drawing images with architectural elements such as peaked windows and railings.

Bisecting Geometric Components

Geometric components are bisected by locating the center point and dividing the line, angle, or arc in half. Lines, arcs, and angles are bisected by drawing arcs to find the center point. Lines can also be bisected by using a triangle.

Bisecting a Line or Arc. To bisect a line or arc geometrically, apply the following procedure:

Given: Line or arc AB.

1. Set a compass with the radius greater than half the length of line or arc AB. With a center point of A, draw arc CD. **See Figure 3-7.**
2. With a radius equal to arc CD and a center point of B, draw arc EF. Arc CD and EF intersect at points G and H.
3. Draw line GH. Lines AB and GH intersect at point I. I is the center point of the given line or arc.

To bisect a line using a 45° triangle, apply the following procedure:

Given: Line AB.

1. Line up one side of the triangle with line AB and the hypotenuse of the triangle at point A. Draw line AC. Line up one side of the triangle with line AB and the hypotenuse of the triangle at point B. Draw line BC.
2. Draw line CD perpendicular to line AB. Lines AB and CD intersect at point D. D is the center point of line AB.

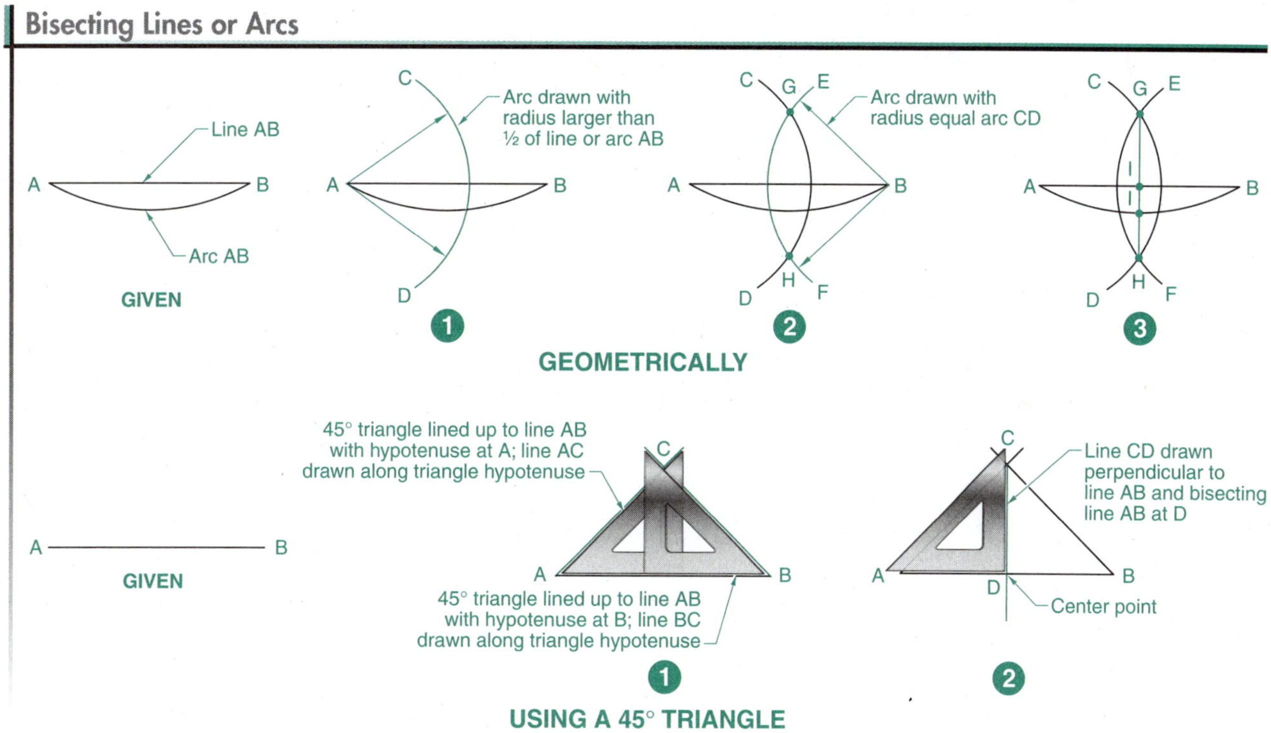

Figure 3-7. Lines and arcs can be bisected using a triangle or by drawing arcs to find the center point.

Bisecting an Angle. A bisected angle forms two equal angles with measurements that are one-half the total degree measurement of the original angle. To bisect an angle, apply the following procedure:

Given: Angle ABC.

1. Set a compass to any radius. With a center point of B, draw arc DE, which intersects angle ABC at points D and E. **See Figure 3-8.**
2. Set the compass with the radius greater than one-half the radius of arc DE. With a center point of D, draw arc FG. With a center point of E, draw arc HI. Arcs FG and HI intersect at point J. J is the center point of the angle.
3. Draw line BJ through points B and J to bisect angle ABC.

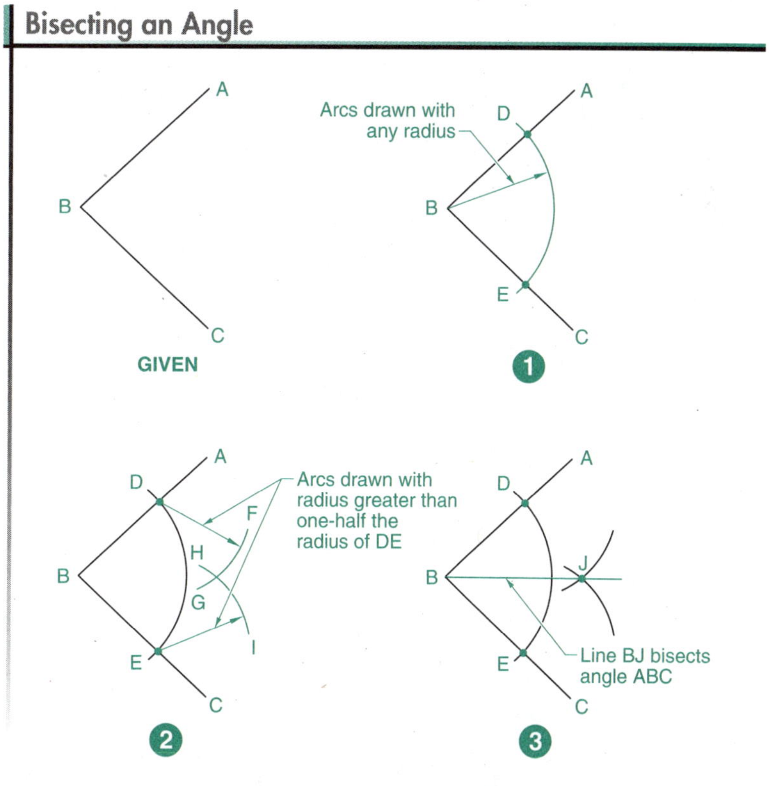

Figure 3-8. A bisected angle forms two equal angles that are one-half the total degree measurement of the original angle.

Circular Geometric Construction

Circular geometric construction often requires the use of lines, arcs, and tangents. Circles can be drawn and their center point found, and tangent lines, circles, and arcs can be drawn to circles or arcs using geometric means.

Circles. To sketch circles by hand, the center point is located and several intersecting diameter lines are drawn. The radius on these lines is marked off, and the radiuses are connected with a series of arcs.

To draw an arc or circle through three given points, apply the following procedure:

Given: Points A, B, and C.

1. Draw line AB. Draw line BC. **See Figure 3-9.**
2. Bisect line AB with line DE. Bisect line BC with line EF. Line DE and EF intersect at point E. E is the center of the desired circle.
3. Set a compass with the radius equal to the distance between given points A and E, B and E, or C and E. With a center point of E, draw a circle through given points A, B, and C.

To find the center of a circle, apply the following procedure:

Given: Circle A.

1. Within the circle, draw any chord (named AB). **See Figure 3-10.**
2. Originating at point A, draw line AD so that it is perpendicular to line AB. Originating at point B, draw line BC so that it is perpendicular to line AB.
3. Draw a straight line connecting points A and C. Draw another straight line connecting points B and D. Lines AC and BD intersect at point E. E is the center of the circle. *Note:* Lines AC and BD are equal to the diameter of the circle.

Drawing Circles

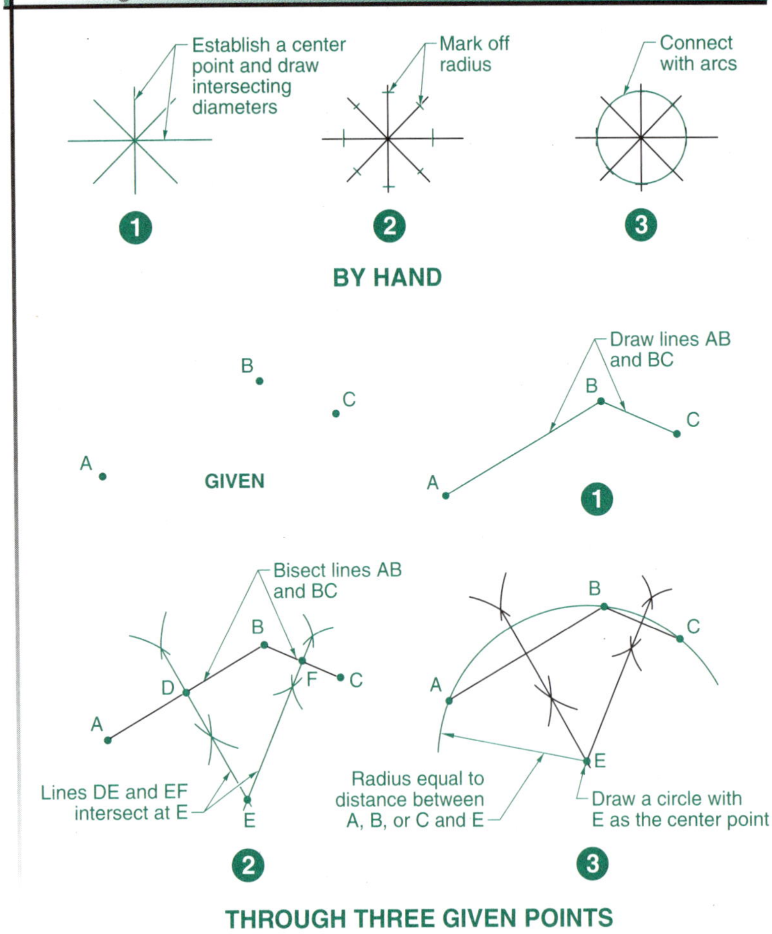

Figure 3-9. Circles can be sketched by hand or through three given points.

Finding the Center Point of a Circle

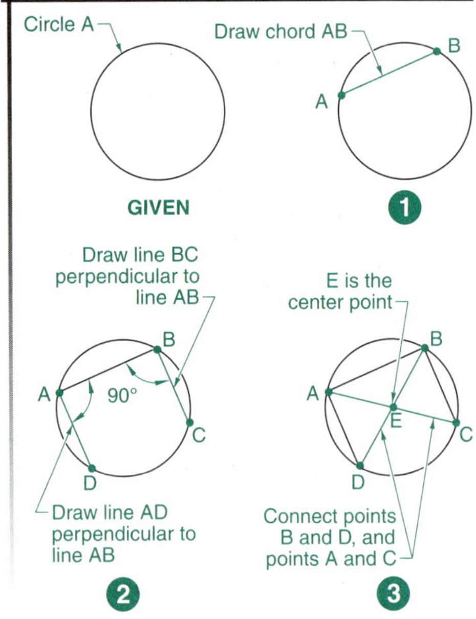

Figure 3-10. The center point of a circle is found by drawing chords.

Tangents. A tangent point is the point where a line and an arc or two arcs touch. In other words, when a line and an arc or two arcs share a tangent point, they are considered tangent. When a line is tangent to an arc or circle, the tangent point is perpendicular to the center point of the arc or circle. For two arcs or circles, the tangent point is the point that intersects both arcs or both circles when the center points are connected with a line.

To draw a circle tangent to a line at a given point, apply the following procedure:

Given: Line AB and tangent point C.

1. Originating at point C, draw line CD perpendicular to line AB. **See Figure 3-11.**
2. Set a compass to the desired radius. Originating from point C, mark the radius on line CD at point E.
3. With a center point of E, draw a circle through point C.

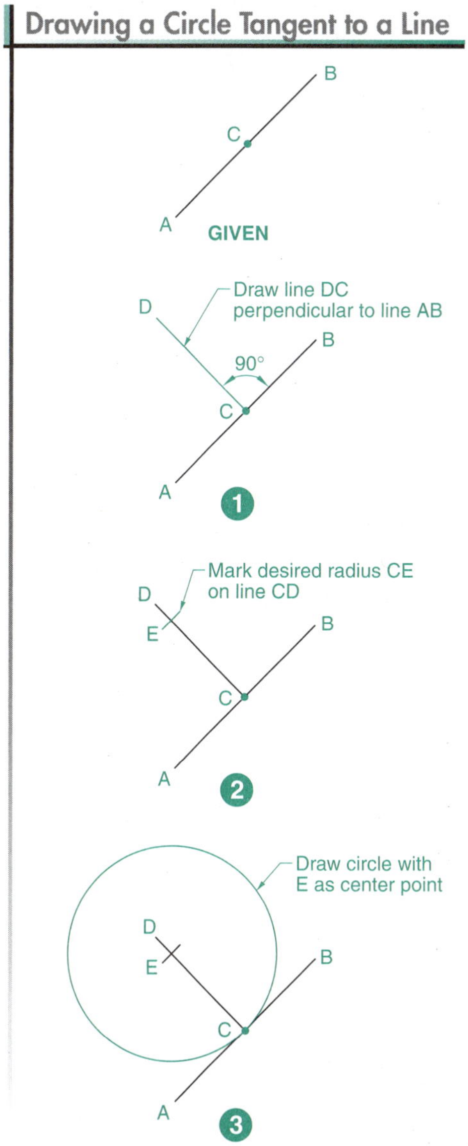

Figure 3-11. A circle drawn tangent to a line has a center point perpendicular to the line at the tangent point.

To draw a line tangent to a circle through a point on the circle, apply the following procedure:

Given: Circle with center point A and given point B.

1. Set the compass with the radius equal to AB. With B as a center point, draw arc AC, which intersects the circle at D. **See Figure 3-12.**
2. With a radius equal to arc AC and a center point of D, draw arc EF, which intersects arc AC at G.
3. Set the compass with a radius greater than AB and a center point of D. Draw arc HI. With a radius equal to arc HI and G as the center point, draw arc IJ. Arcs HI and IJ intersect at point I.
4. Draw a tangent line through points B and I.

To draw a line tangent to a circle through a point not on the circle, apply the following procedure:

Given: Circle with center point A and given point B.

1. Draw line AB. Bisect line AB at point C.
2. Set a compass with the radius equal to AC. With a center point of C, draw arc AD. D is the point of tangency.
3. Draw line BD tangent to the given circle.

To draw a line tangent to a circle through a point using a 45° triangle and a T-square, apply the following procedure:

Given: Circle with center point A and given point B.

1. With the hypotenuse of the triangle placed against the T-square, line up one side of the triangle with points A and B. Draw line AB.
2. Slide the triangle along the T-square until the other side is lined up with point B. Draw the required tangent line.

Note: If the point is outside the circle, begin by lining up one side of the triangle with point B and what appears to be the tangent point on the edge of the circle. Slide the triangle along the edge of the T-square to line up the other side to point A. The point where the triangle edge crosses the circle's outer edge is the point of tangency, or point C. Draw line AC. Return the triangle to its starting position and draw the tangent line BC.

> **TECH TIP**
>
> *A tangent point is where two lines appear to touch at the place where one line stops and another begins. The point where a curved line stops and a straight line begins is a tangent point.*

Drawing a Line Tangent to a Circle

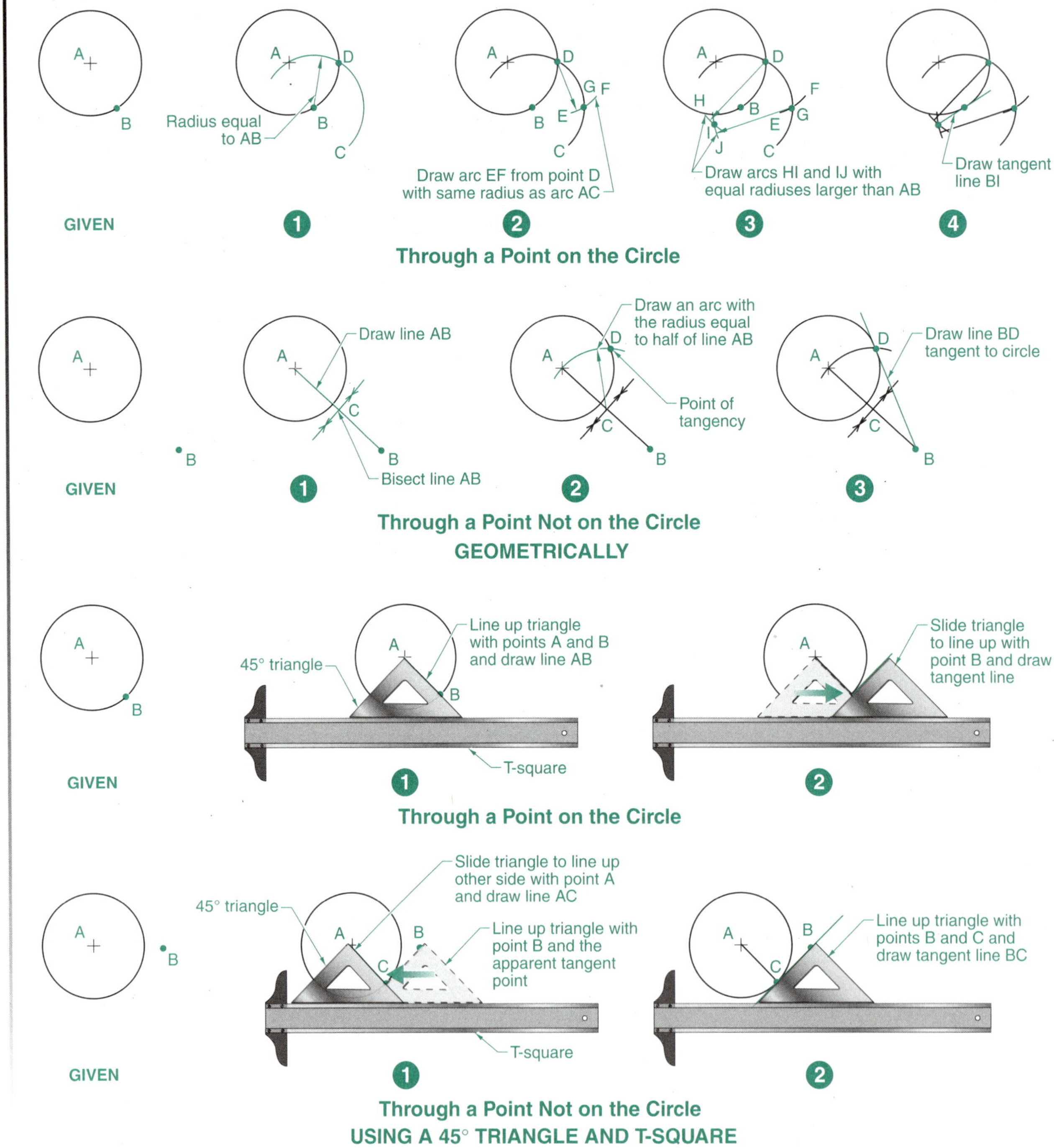

Figure 3-12. A line can be drawn tangent to a circle through a given point.

To draw an arc tangent to two lines at right (90°) angles, apply the following procedure:

Given: Lines AB and BC, which form a right angle.

1. Set a compass to the desired radius of the arc. With a center point of B, draw an arc, which intersects line AB at D and line BC at E. **See Figure 3-13.**
2. With the desired radius and a center point of D, draw an arc. With the desired radius and a center point of E, draw another arc. Where these arcs intersect (at point F) is the center point of the desired arc.
3. With the desired radius and a center point of F, draw an arc tangent to lines AB and BC.

To draw an arc tangent to two lines at an angle greater than or less than 90°, apply the following procedure:

Given: Lines AB and BC that form an angle greater than or less than 90°.

1. Draw line DE parallel to line AB on the inside of the angle. Draw line EF parallel to line BC on the inside of the angle. DE and EF intersect at point E. *Note:* The radius of the arc will be equal to the distance of parallel lines DE and EF from lines AB and BC.

Drawing an Arc Tangent to Two Lines

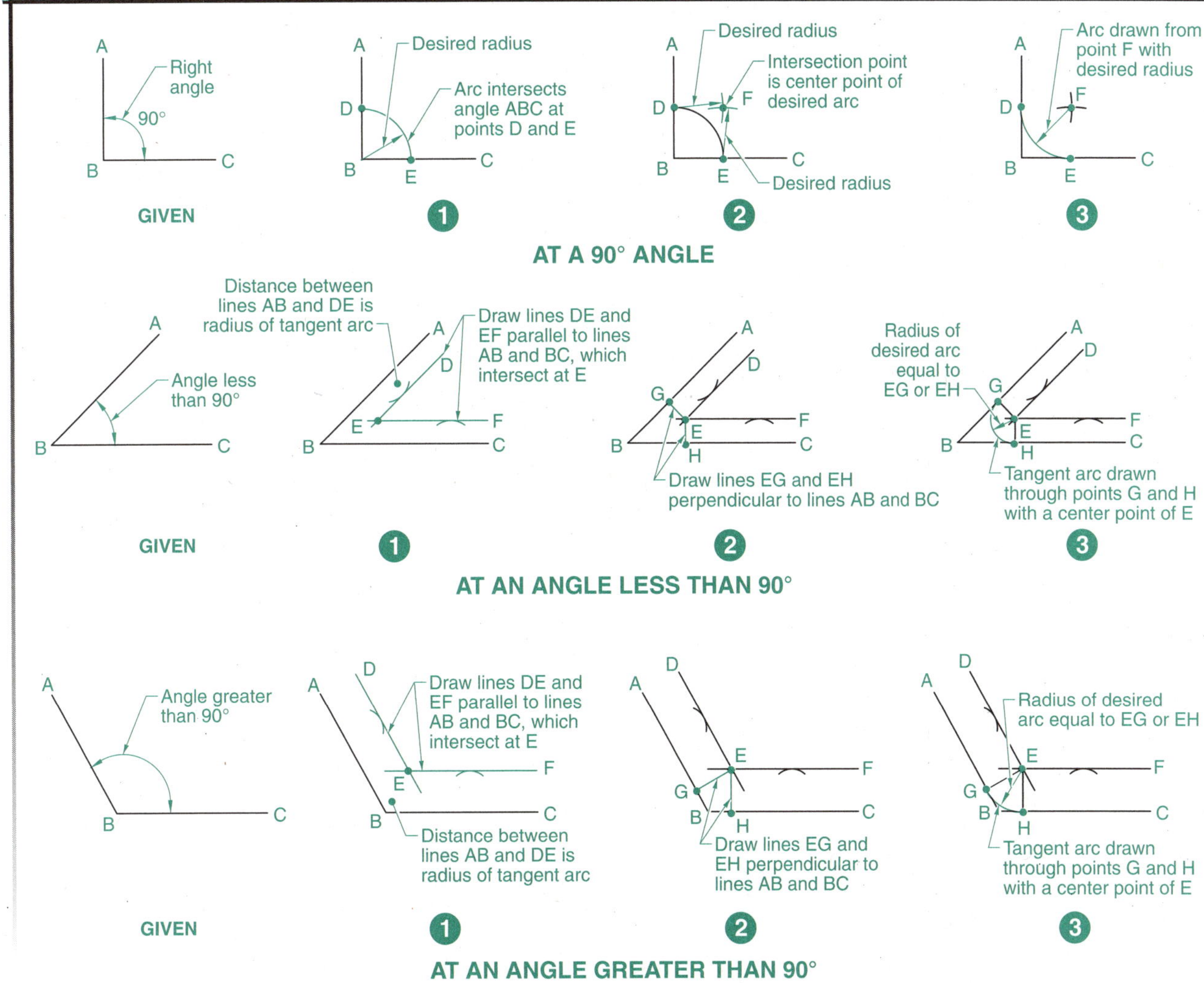

Figure 3-13. An arc can be drawn tangent to a right angle, an angle less than 90°, or an angle greater than 90°.

2. Originating at point E, draw line EG perpendicular to line AB. Originating at point E, draw line EH perpendicular to line BC. Points G and H are the tangent points and lines EG and EH are equal to the radius of the desired arc.
3. Set a compass to the radius of the desired arc. With a center point of E, draw an arc through points G and H.

To draw an arc tangent to a straight line and an arc, apply the following procedure:

Given: Line AB and arc CD with a known center point E.

1. Draw line FG parallel to line AB. *Note:* The radius of the arc will be equal to the distance of parallel line FG from line AB.**See Figure 3-14.**
2. Draw arc HI parallel to arc CD, which intersects with line FG at point J.
3. Originating at point J, draw line JK perpendicular to line AB. Originating at point J, draw line EJ, which intersects arc CD at point L. Points K and L are the tangent points and lines JK and JL are equal to the radius of the desired arc.
4. With J as a center point, draw arc KL.

Note: This method can be applied to create an arc tangent to two arcs by creating two parallel arcs to the given arcs, then connecting the intersection point of the two parallel arcs with the center points of the given arcs to create the points of tangency and center point of the required arc.

TECH TIP

The radius of an arc tangent to two lines or a line and an arc is related to the parallel lines drawn in the construction. The distance of the parallel lines from the original given lines or arcs is the same as the radius of the tangent arc.

To draw a straight line tangent to two arcs, apply the following procedure:

Given: Arc AB with center point C and a known radius; arc DE with center point F with a known radius.

1. Draw a straight line (CF) connecting center points C and F. Bisect line CF to find center point G. **See Figure 3-15.**
2. Subtract the radius of the smaller arc from the radius of the larger arc to find the difference in radius. Set the compass to the difference. With a center point of C, draw arc HI. Set a compass with the radius equal to the distance between points C and G. With a center point of G, draw arc CF. Arc CF intersects arc HI at point J.

Drawing an Arc Tangent to a Straight Line and an Arc

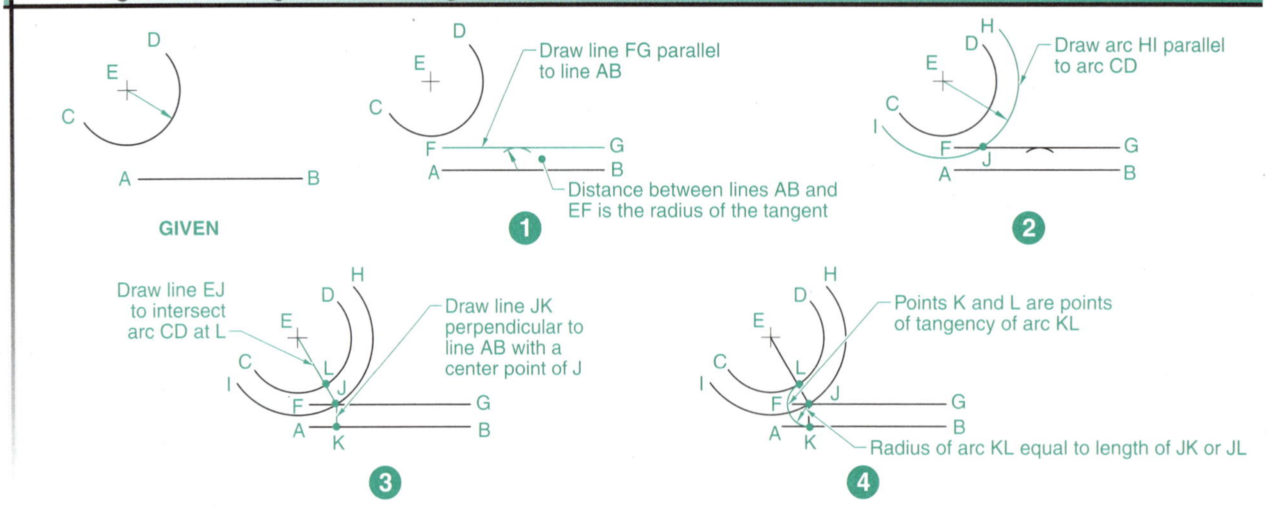

Figure 3-14. An arc is drawn tangent to a straight line and another arc by drawing parallel and perpendicular lines to find the center point of the tangent arc.

Drawing a Straight Line Tangent to Two Arcs

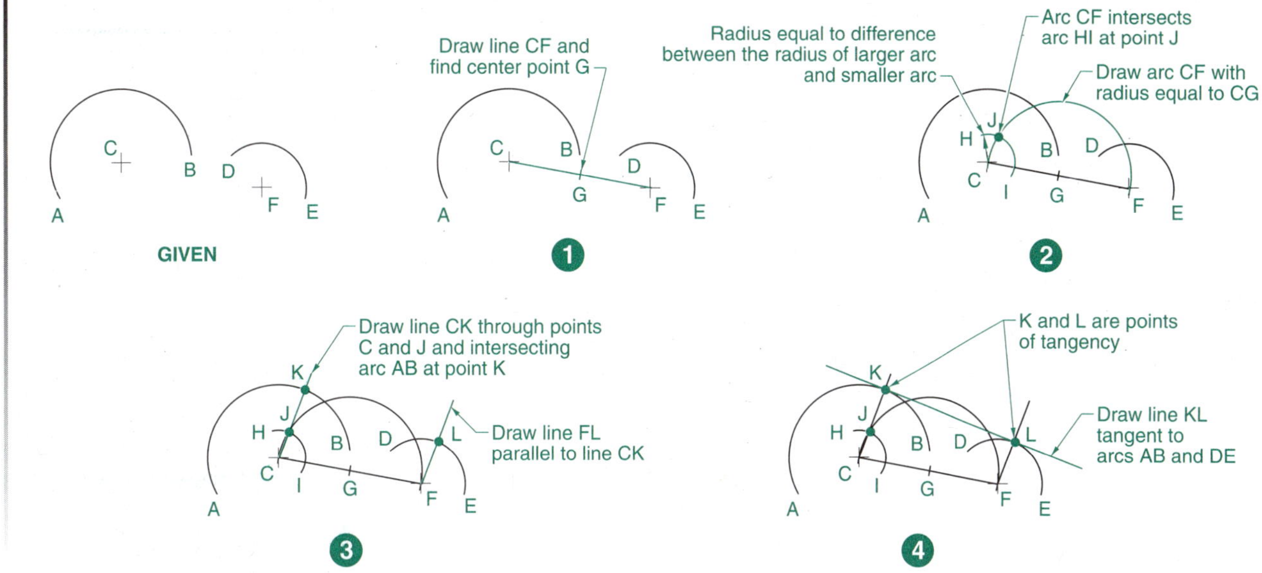

Figure 3-15. A straight line can be drawn tangent to two arcs by using the center points of the circle.

3. Originating at point C, draw a straight line through point J, which intersects arc AB at point K. K is the point of tangency on arc AB. Originating from point F, draw line FL parallel to line CK, which intersects arc DE at point L. L is the point of tangency on arc DE.
4. Draw line KL tangent to arcs AB and DE.

Plane Figures

In technical drawings, plane figures can be drawn by hand using basic principles. However, it is often necessary to draw plane figures using geometric construction principles to ensure accuracy of angles and sides. Triangles, quadrilaterals, and polygons can all be drawn quickly by hand or using more complex geometric methods.

Triangles. To draw triangles by hand, the artist draws the base, determines and marks the angles of the sides, and draws straight lines to complete the triangles. Triangles can be geometrically drawn using several methods and can be completed with varying given information. **See Figure 3-16.**

To draw a triangle with known lengths of sides, apply the following procedure:

Given: Sides AB, BC, and AC.

1. Draw side AB in the desired position.
2. Set a compass with the radius equal to side AC. With a center point of A, draw an arc. Set a compass with the radius equal to side BC. With a center point of B, draw an arc.
3. Where both arcs intersect at point C is the intersection of sides BC and AC. Draw side BC with a straight line connecting points B and C. Draw side AC with a straight line connecting points A and C.

To draw a right triangle with the hypotenuse and one side given, apply the following procedure:

Given: Hypotenuse AB and side AC

1. Draw a semicircle with the diameter as the length of hypotenuse AB.
2. Set a compass with the radius equal to side AC. With a center point of A, draw an arc. Where the semicircle and arc intersect (point C) is the point where sides AC and BC meet.
3. Draw side AC with a straight line connecting points A and C. Draw side BC with a straight line connecting points B and C.

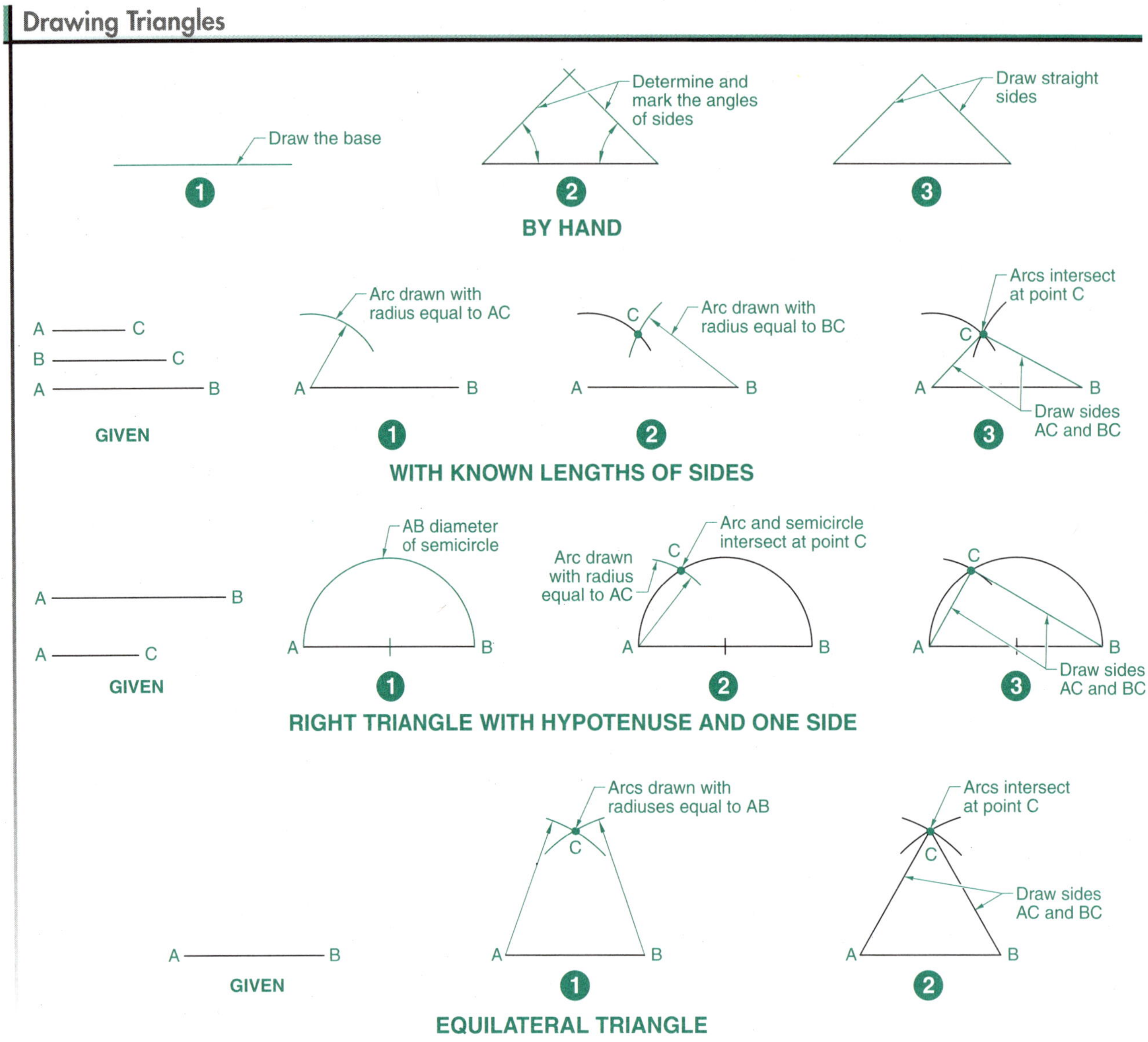

Figure 3-16. Triangles can be drawn geometrically using several methods, and can be completed with varying given information.

To draw an equilateral triangle, apply the following procedure:

Given: Side AB.

1. Set a compass with the radius equal to side AB. With a center point of A, draw an arc. With a center point of B, draw a second arc. Where these two arcs intersect (at point C) is the meeting point for sides BC and AC.
2. Draw side AC with a straight line connecting points A and C. Draw side BC with a straight line connecting points B and C.

Note: An equilateral triangle can also be drawn using a 30°-60° triangle because each angle in an equilateral triangle is 60°.

Quadrilaterals. To draw quadrilaterals by hand, the baseline is drawn and the angles of the sides are marked. The corner points are determined and they are connected with straight lines to complete the quadrilateral. Dimensions of at least two sides and two angles, depending on the type of quadrilateral, are required to fully describe the proper shape. Quadrilaterals can also be quickly drawn using the right angle on a triangle. **See Figure 3-17.**

To draw a square with the sides given, apply the following procedure:

Given: Side AB.

1. Originating at point A, draw a line perpendicular to side AB.
2. Set a compass with the radius equal to side AB. With a center point of A, draw arc BC.
3. With a center point of B, draw another arc. With a center point of C, draw a third arc. These arcs intersect at point D.
4. Point D is the intersection of sides BD and CD. Draw sides BD and CD using a straightedge.

To draw a square inscribed within a circle, apply the following procedure:

Given: Circle A.

1. Draw two diameters in circle A that are perpendicular to each other.
2. The intersections of these diameters with the outer edge of the circle are the corners of the square. Connect the intersections with straight lines using a straightedge.

Note: A polygon can be inscribed (with the sides "across corners") within the outer line of the circle, or circumscribed (with the sides "across flats") with the sides of the polygon drawn tangent to the outer line of the circle.

Drawing Quadrilaterals

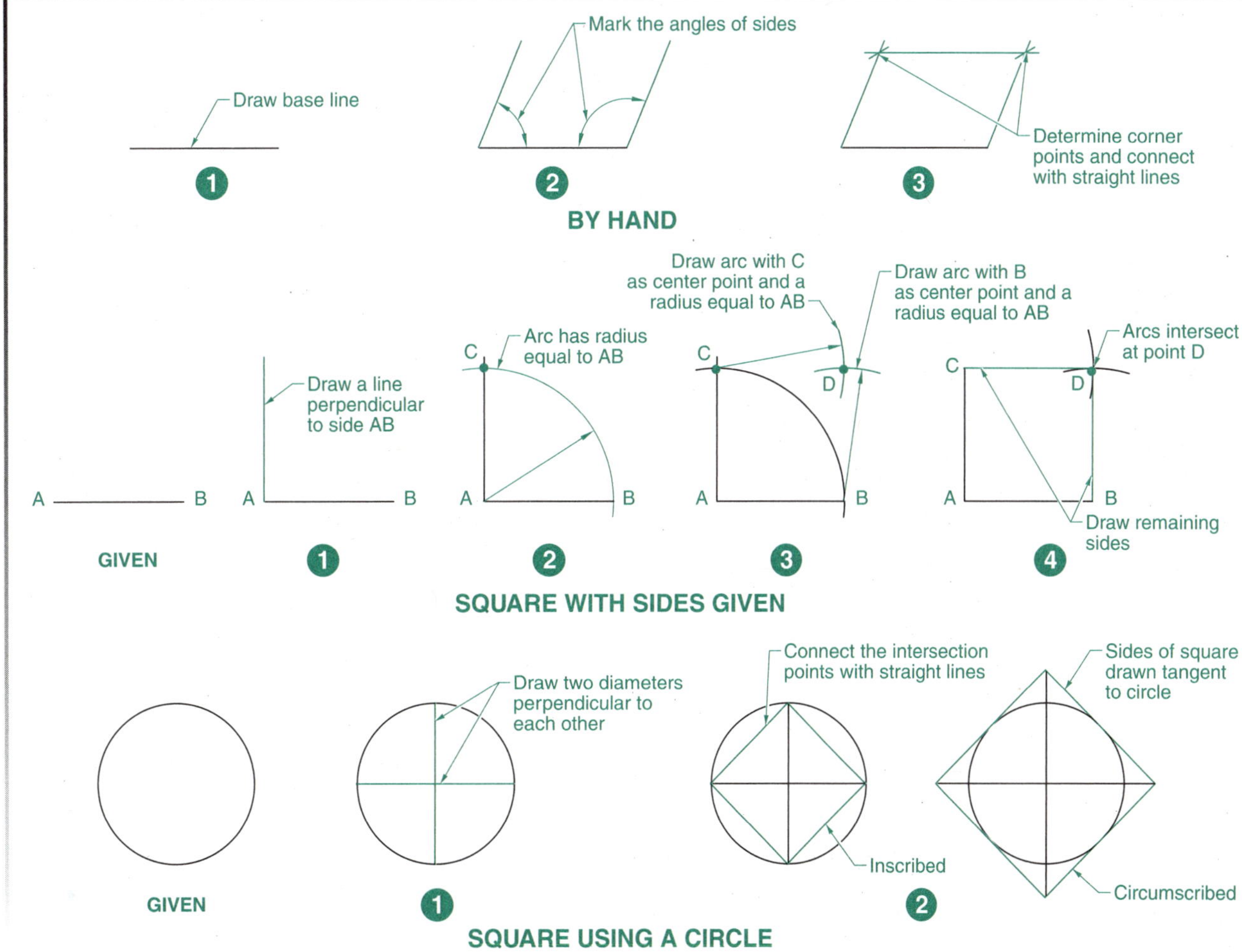

Figure 3-17. Quadrilaterals can be drawn by hand or by using a triangle. Squares can also be geometrically drawn if one side or a circle is provided.

Polygons. Common regular polygons are the pentagon, the hexagon, and the octagon. A pentagon has five equal sides with no sides parallel and five equal angles. A hexagon has six equal sides with opposite sides parallel and six equal angles. An octagon has eight equal sides with opposite sides parallel and eight equal angles. **See Figure 3-18.**

To draw polygons with more than four sides by hand, the center point is located, and a circle is drawn the approximate size of the desired polygon. Dividers can be used to create the required number of equal segments and mark off the length of each side on the circumference of the circle, or the angle can be found.

The angle needed to determine the length of each side is found by dividing 360° by the number of desired sides. From the center point of the circle, the angles can be drawn to intersect with the circumference, giving the length and location for each side.

To draw a regular pentagon using the geometrical method, apply the following procedure:

Given: Circle A.

1. Draw perpendicular center lines AB and CD. AB and CD intersect at point E. **See Figure 3-19.**
2. Bisect radius BE to find point F.
3. Set the compass with the radius equal to CF. With a center point of C, create arc GH. Arc GH intersects the circle at point I.
4. Connect points C and I with a straight line. Use dividers to determine the length of the side. Mark the remaining sides with straight lines.

Note: If dividers are not available, the remaining sides can be found by repeating steps 3 and 4 and working counterclockwise with the center point for each arc as the intersection point created with the previously drawn arc.

Common Regular Polygons

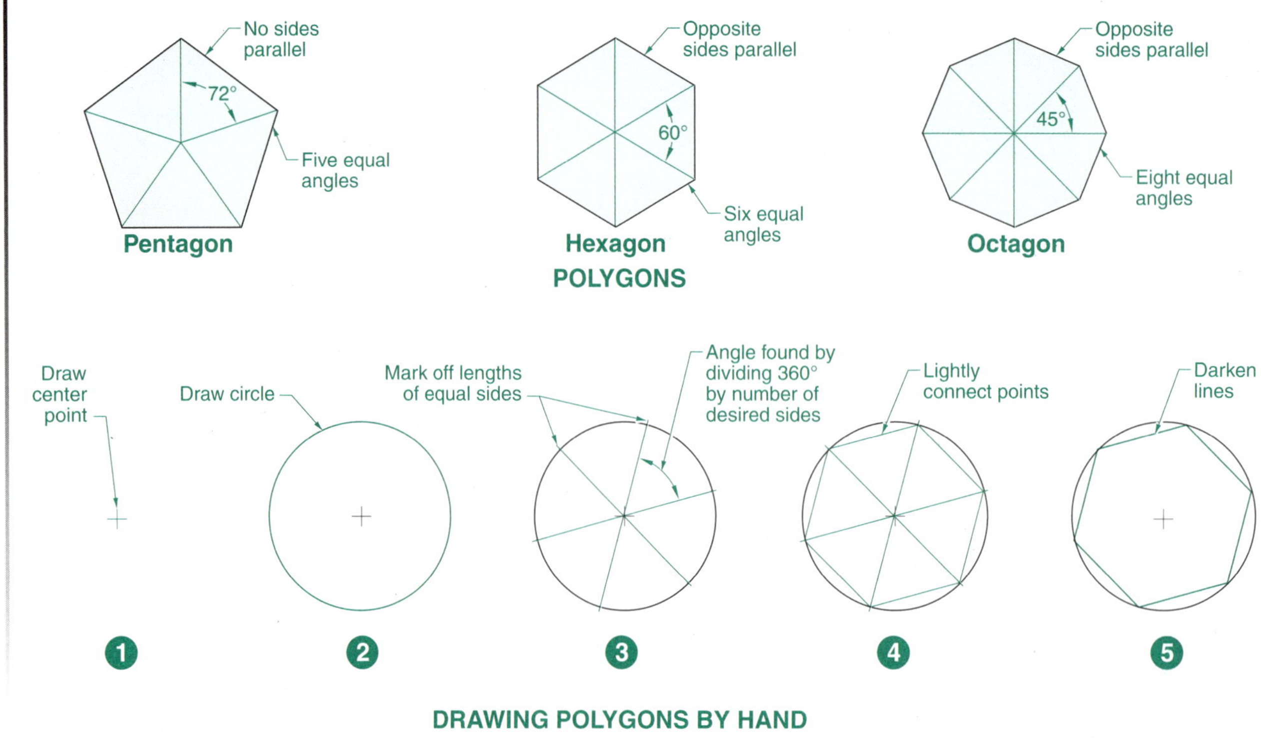

Figure 3-18. Common regular polygons are the pentagon, the hexagon, and the octagon.

Drawing a Pentagon

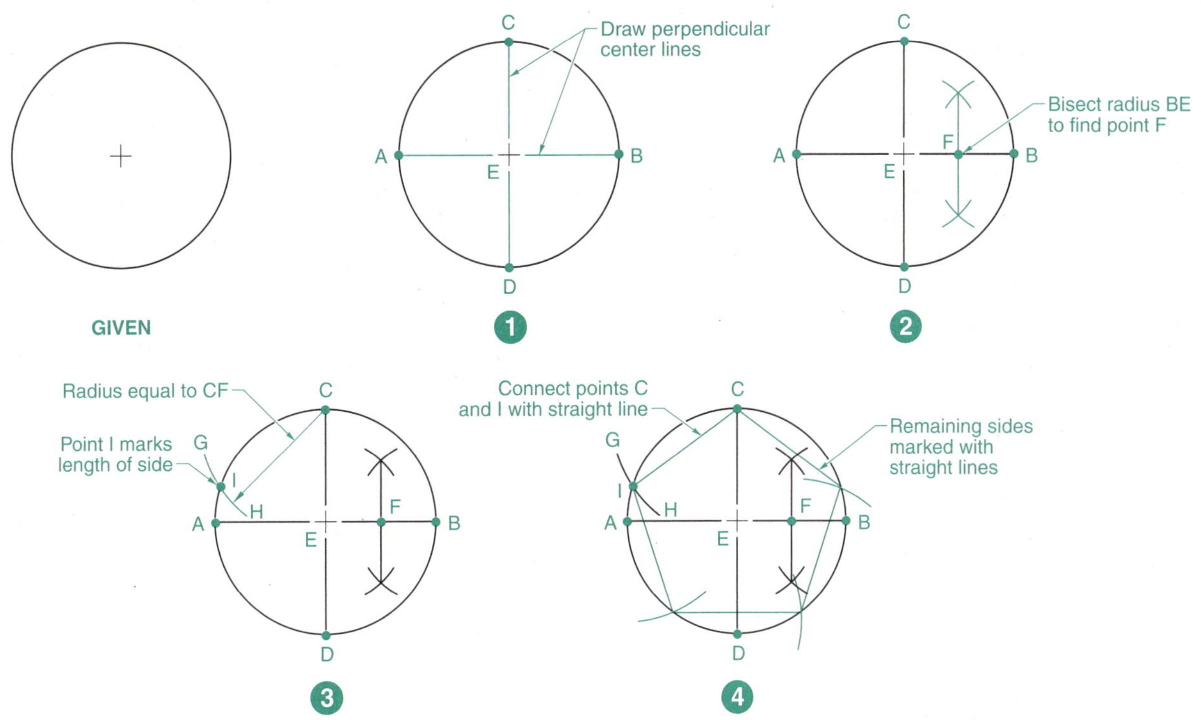

Figure 3-19. A regular pentagon is drawn using a circle.

To draw a regular hexagon using geometric construction principles, apply the following procedure:

Given: Circle with center point A and known radius.

1. Set a compass with the radius equal to the known radius of the circle. With any point B on the circumference of the circle as a center point, draw an arc intersecting with the circumference of the circle at point C. With a center point of point C, draw an arc intersecting the circumference of the circle at point D. **See Figure 3-20.**
2. Repeat step 1 with each successive intersection point as a center point for the remaining arcs.
3. Draw a straight line to connect all intersection points.

Note: A circumscribed hexagon can be drawn using a 30°-60° triangle. Start with a circle with perpendicular center lines and draw two sides perpendicular to one center point. Use the triangle to draw the six sides tangent to the circle. If the hexagon is situated so that two parallel sides are vertical, the remaining sides are at an angle of 30°. If the hexagon is situated so that two parallel sides are horizontal, the remaining sides are at an angle of 60°.

TECH TIP

When making technical drawings, it is sometimes necessary to draw objects such as nuts and bolts. Because the shaft of the bolt is a circle, drawing a circumscribed hexagon around the circle of the shaft is an effective way to create the shape needed for the bolt head.

To draw a regular octagon using geometric construction principles, apply the following procedure:

Given: Square ABCD.

1. Draw a diagonal line through points A and C. Draw another diagonal line through points B and D. These two lines intersect at point E. **See Figure 3-21.**

2. Set a compass with the radius equal to AE. With a center point of A, draw an arc that intersects the outer edges of the square. With a center point of B, draw a second arc. With a center point of C, draw a third arc. With a center point of D, draw a fourth arc.
3. The intersection points of all four arcs with the outer edge of the square mark the lengths of the eight sides. Connect the points with straight lines.

Note: An inscribed octagon can be drawn using a 45° triangle. Start with a circle and draw four lines spanning the diameter of the circle, 45° apart. The intersection points of these lines with the circle are the corners of the sides. Connect the points with straight lines.

Radial Lines Media Clip

TECH TIP

Polygons can be drawn by hand by calculating the angle measurements and using triangles to draw the sides. However, it is important to learn geometric construction principles in order to create more exact shapes and when measurement tools are not available.

Enlarging or Reducing Shapes

Shapes are enlarged or reduced using radial lines. *Radial lines* are a series of lines that radiate from the same point. Radial lines are drawn from a common corner point and extend through a surface characteristic, such as a corner or feature, on a shape. These points can then be transferred to a larger or smaller size by lining up the same surface characteristic along the radial lines at the new size. **See Figure 3-22.**

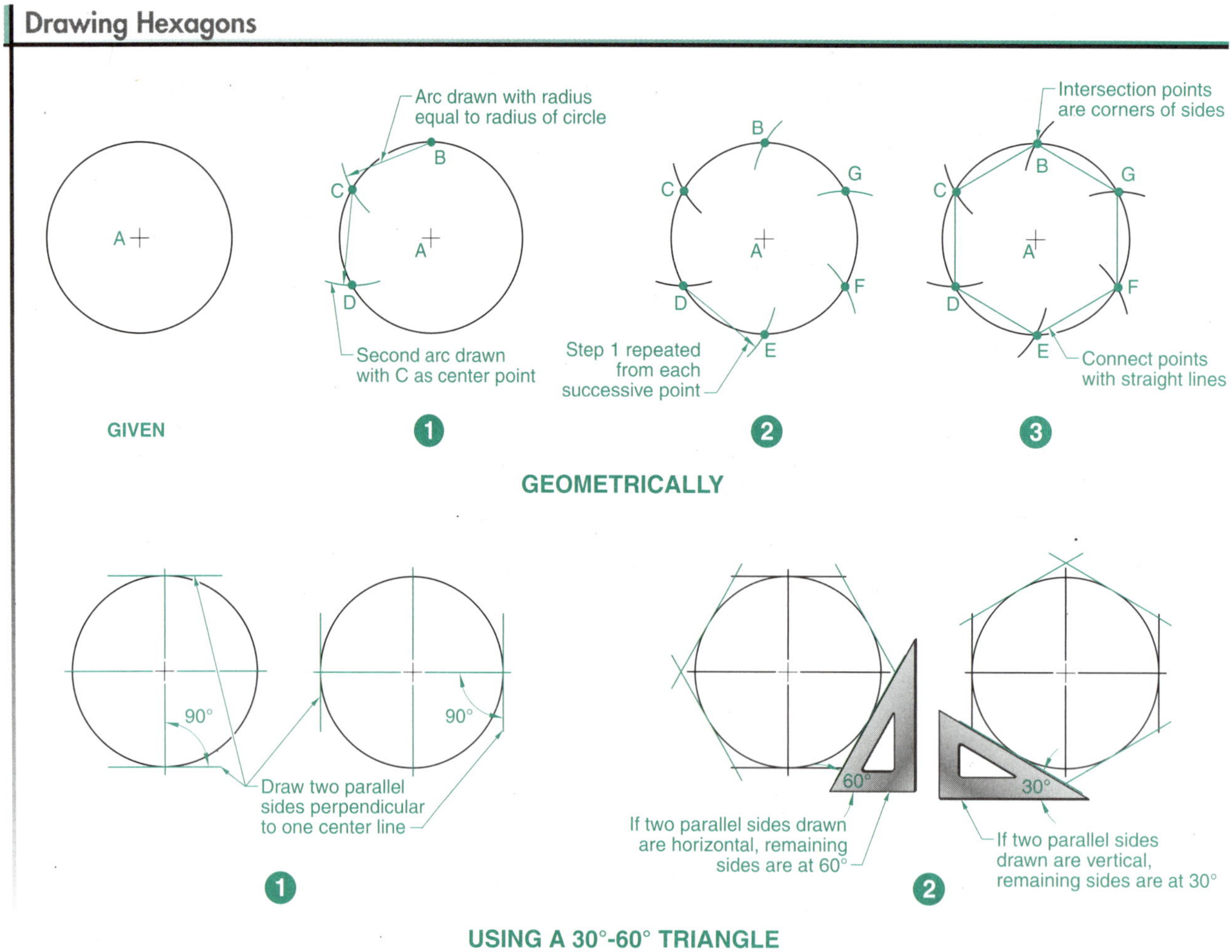

Figure 3-20. A regular hexagon can be drawn inscribed or circumscribed.

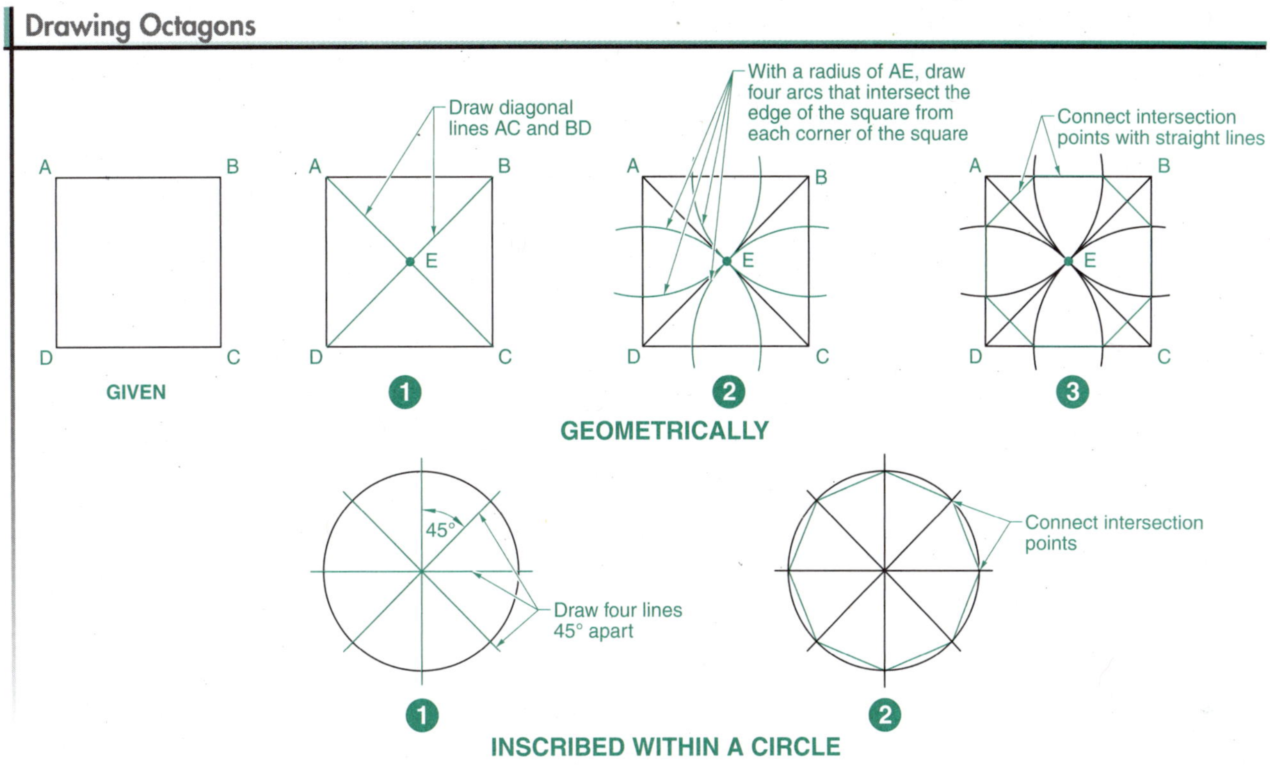

Figure 3-21. Regular octagons can be drawn using a square or circle.

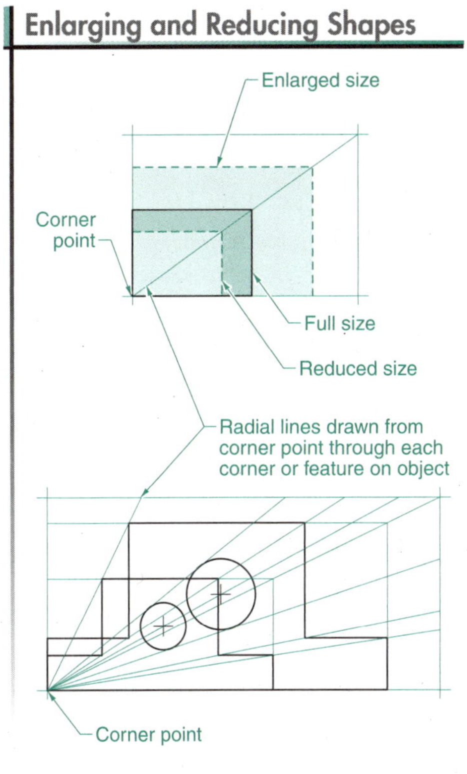

Figure 3-22. Shapes are enlarged or reduced using radial lines.

Complex Geometric Construction

Some geometric shapes, such as conic sections, utilize more complex construction procedures. A *conic section* is a curve produced by an intersecting plane on a three-dimensional cone form. The shape formed when these planes intersect can be a circle, but can also be more complex shapes such as an ellipse, parabola, or hyperbola. **See Figure 3-23.** The shape of the cone and the angle of the intersecting plane determine the conic section formed.

Ellipses. An *ellipse* is an elongated circle that is formed when the sum of the distances from a point on the curve to two fixed focus points remains the same. Ellipses have two axes of different sizes. The *major axis* is the axis in an ellipse that is longer. The *minor axis* is the axis in an ellipse that is shorter. A *focus* is a point on an ellipse that is found by drawing arcs with a radius equal to half the major axis and with a center at one end of the minor axis. Each ellipse has two foci that can be used to geometrically construct the shape.

Conic Sections

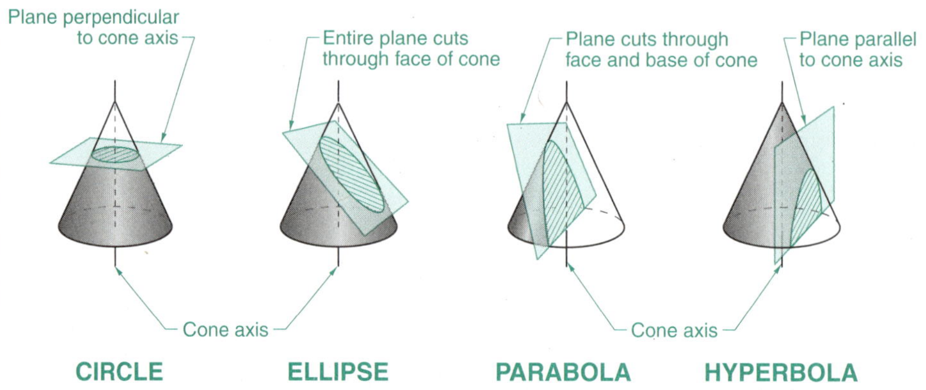

Figure 3-23. Conic sections include a circle, ellipse, parabola, and hyperbola.

Ellipse templates can be useful tools to help create quick and accurate ellipses of varying sizes. However, sometimes it is necessary to draw an ellipse using the geometric construction methods of foci or parallelograms. **See Figure 3-24.**

To draw an ellipse using foci, apply the following procedure:

Given: Ellipse with major axis AB, minor axis CD, and center point E.

1. Set a compass with the radius equal to AE. With a center point of C, draw an arc that intersects AB at points F and G. Points F and G are the foci of the ellipse.
2. Between points E and F, mark out any number of points.

   Note: The spacing between points need not be equal, but the points closer to the outside of the desired ellipse should be closer together than those toward the middle. The larger the size of the desired ellipse, the more points should be used.
3. Choose one point. Determine the distance from this point to point A, and set the compass radius to this distance. With a center point of F, draw one arc above and one arc below the major axis. With a center point of G, draw one arc above and one arc below the major axis.
4. Determine the distance from point B to the first selected point and set the compass radius to this distance. With G as a center point, draw one arc above and one arc below the major axis, which intersect the first two arcs drawn. These intersection points will be points on the ellipse edge.
5. Repeat steps 3 and 4 for all points marked on length EF.
6. Lightly connect the arc intersection points by hand. Use a french curve to finish and darken the lines.

To draw a parallelogram ellipse, apply the following procedure:

Given: Major axis line AB, minor axis line CD, and center point E.

1. Draw a parallelogram with sides FG and HI parallel to line AB, and sides FI and GH parallel to line CD.
2. Divide AF into a number of equal parts. Divide AE into the same number of equal parts.
3. Originating from point D, draw light lines through each division point on line AE. Originating from point C, draw light lines through each division point on line AF. The intersection points of these lines form points on the ellipse.
4. Repeat step 3 for the remaining three quadrants of the ellipse.
5. Once a sufficient number of points are found and marked, lightly connect the points by hand. Use a french curve to finish and darken the lines.

Note: This method can be used to easily draw slanted ellipses by beginning with a rhombus.

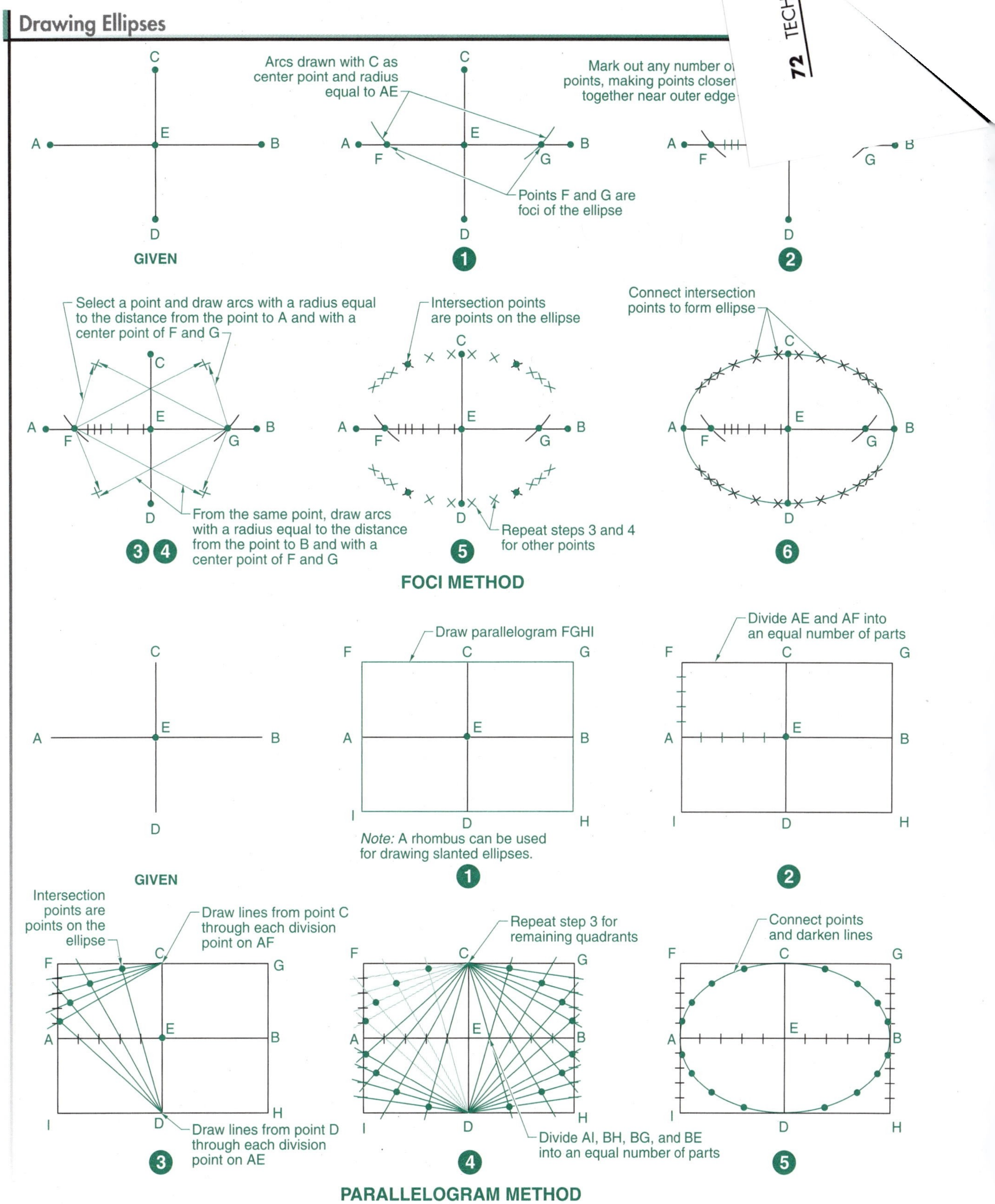

Figure 3-24. An ellipse can be drawn geometrically using the foci or parallelogram methods.

Parabolas. A *parabola* is a curve that is formed by all points being equidistant from a given point, focus, and line. Parabolas are formed when a cone is truncated at a diagonal.

To draw a parabola, apply the following procedure:

Given: Line AB and line CD with focus point E.

1. Draw a line parallel to line AB at any distance past point E, which intersects line CD at point F. **See Figure 3-25.**
2. Set a compass with the radius equal to the distance between point C and point F. With a center point of E, draw two arcs that intersect the parallel line above and below line CD. These intersection points are points on the parabola.
3. Repeat steps 1 and 2 by drawing any number of parallel lines between line AB and the first parallel line. Draw arcs with the radius equal to the distance between the point where the parallel line intersects line CD and point C. Using each parallel line, mark additional points for the parabola. *Note:* The number of parallel lines drawn equals the number of points for the top and bottom of the parabola that will be created.
4. Once a sufficient number of points are found and marked, lightly connect the points by hand. Use a french curve to finish and darken the lines.

Hyperbolas. A *hyperbola* is a curve that is formed when the difference of the distances from the foci remains the same. Hyperbolas are formed when a cone is truncated parallel to the axis of the cone. The size of the hyperbola is determined by how close to the vertex the cone is truncated.

To draw a hyperbola, apply the following procedure:

Given: Points A and B with center point C. Point C is the vertex of the cone. Point B is the vertex of the hyperbola. *Note:* The distance from points A and B to point C is determined by how close the plane truncating the cone is to the vertex of the cone.

1. Draw line AB through point C and extending past points A and B. **See Figure 3-26.**
2. Draw line DE perpendicular to line AB through point A. Draw line FG perpendicular to line AB through point B.
3. Draw line EF at 45° from line AB and through point C. Draw line DG at 45° from line AB and through point C. DG should be perpendicular to EF.

Drawing a Parabola

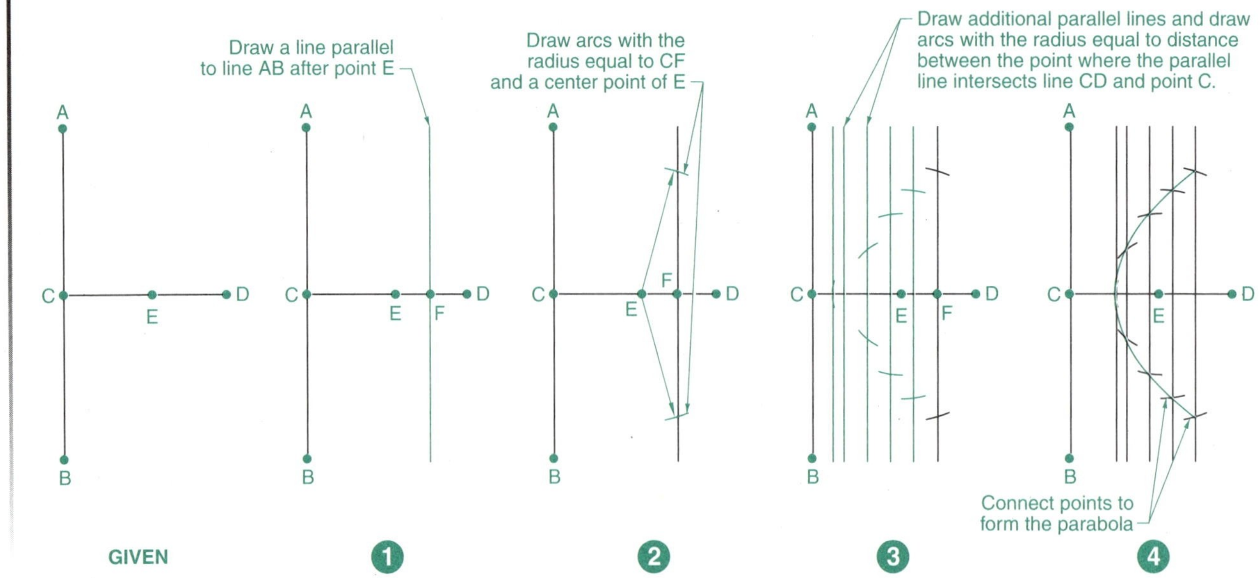

Figure 3-25. Points on a parabola are found using parallel lines and arcs.

4. Draw lines DF and EG perpendicular to lines DE and FG to form square DFGE.
5. Set the compass with a radius equal to CD. With a center point of C, draw an arc and mark points H and I on line AB. H and I are the foci of the hyperbola.
6. Mark off any number of equally spaced points on line AB after point I.
7. Set the compass with the radius equal to the distance between point A and the first marked point. With a center point of H, draw two arcs. Set the compass with the radius equal to the distance between point B and the first marked point. With a center point of I, draw two arcs that intersect the other two arcs. These intersection points are points on the hyperbola.
8. Repeat step 7 using each of the marked points to determine the radiuses and using the found radiuses to draw two arcs from each focus.
9. Once all marked points have been used, lightly connect the arc intersection points by hand. Use a french curve to finish and darken the lines.

Helixes. A *helix* is the curve traced on a cylinder or cone by a point rotating at a right angle to the axis. The *lead* is the distance along the axis for one revolution around the cylinder or cone. The direction of rotation for a helix can be right-handed or left-handed. A right-handed helix will slope up to the right, or counterclockwise, when viewed from the side. A left-handed helix will slope up to the left, or clockwise, when viewed from the side.

Drawing a Hyperbola

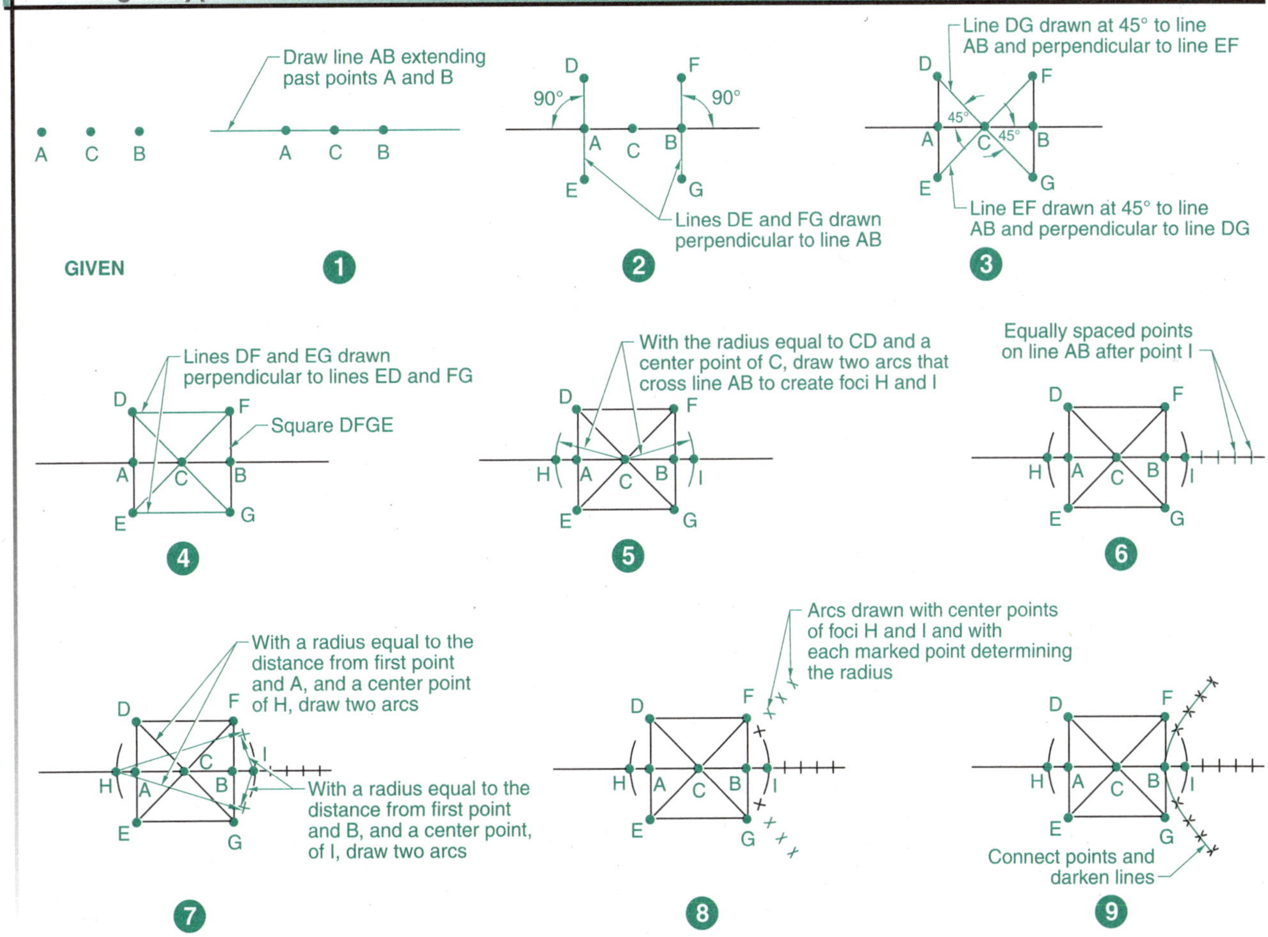

Figure 3-26. Points on a hyperbola are found using arcs drawn from points on a straight line.

To draw a right-handed helix, apply the following procedure:

Given: Cylinder A that bounds the desired helix with a known diameter and lead.

1. Draw the rectangular side view and the circular top view of the cylinder in a column. **See Figure 3-27.**
2. Divide the circular top view into any number of equal parts by drawing straight diameter lines. Each division point will equal one point on the helix. Number the division points of the circle counterclockwise starting with zero in the nine o'clock position. Divide the height of the lead into the same number of equal parts. Number the divisions starting with zero farthest from the circular top view. *Note:* For a left-handed helix, the circle is numbered clockwise with zero in the three o'clock position.
3. Extend a vertical parallel line from the first numbered division point on the circle into the rectangle. Extend a horizontal parallel line from the first numbered division point in the rectangle. Where these lines intersect is a point on the helix. Repeat for all numbered division points.
4. Connect the intersection points with a curved line. Use a french curve to finish and darken the lines.

Chapter 3 Quick Quiz®

Drawing a Helix

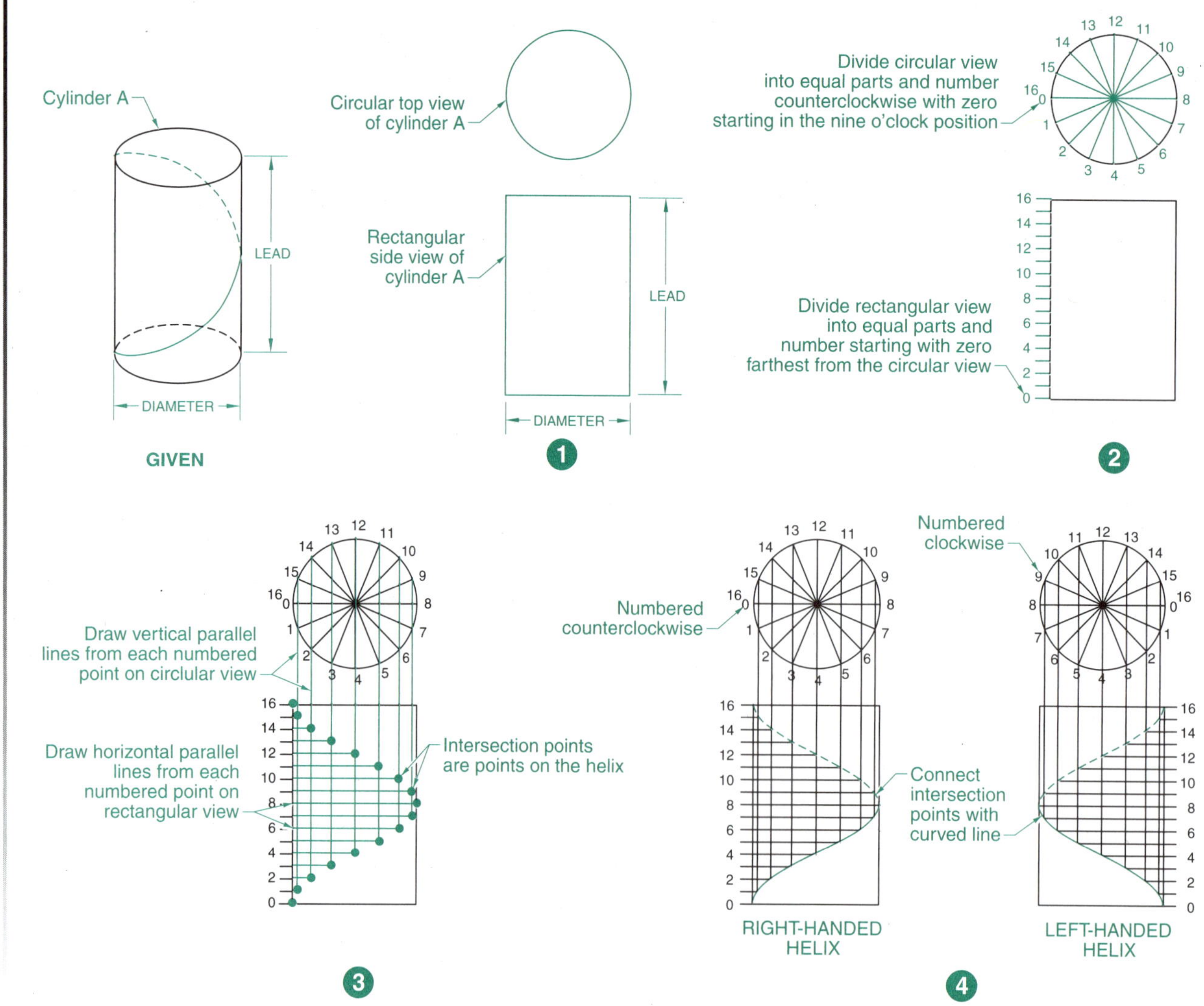

Figure 3-27. A helix is drawn using the diameter and lead of the given cylinder or cone.

Geometric Constructions

Review 3

Name ____________________ Date ____________

REVIEW QUESTIONS

________ 1. A point is geometrically represented by the intersection of two ___.

________ 2. A(n) ___ is a two-dimensional surface that has no depth.

T F 3. Parallel is when two lines or planes form a right angle.

T F 4. Plane figures include spheres, pyramids, and cones.

________ 5. A circle contains ___°.

________ 6. A(n) ___ is a straight line with two end points on the circumference of a circle.

T F 7. A regular polygon has all angles of equal size and all sides of equal length.

________ 8. A(n) ___ is a polygonal plane figure with three straight sides that intersect to form three angles.

T F 9. The hypotenuse is the shortest side of a right triangle.

________ 10. A(n) ___ is a quadrilateral that has opposite sides parallel to each other.

________ 11. A(n) ___ is a three-dimensional form that is created when a circle is rotated on a single axis.

________ 12. A(n) ___ is the surface of a side in a polyhedron.

T F 13. A prism is a polyhedron with bases that are parallel equal polygons and three or more side faces that are parallelograms.

________ 14. The two opposite points on a ___ are always parallel to each other.

________ 15. A(n) ___ is the point on a cone opposite the base where the face(s) meet at a point.

T F 16. Pyramids are a type of cylinder with a polygonal base.

________ 17. A(n) ___ is a line or arc that shares a point with a given curve in at least one location but does not intersect it.

________ 18. Shapes are enlarged or reduced using ___ lines.

T F 19. The minor axis is the axis in an ellipse that is longer.

________ 20. A(n) ___ is the curve traced on a cylinder or cone by a point rotating at a right angle to the axis.

IDENTIFICATION – Parts of a Circle

______________ 1. Circumference

______________ 2. Center point

______________ 3. Arc

______________ 4. Chord

______________ 5. Radius

______________ 6. Diameter

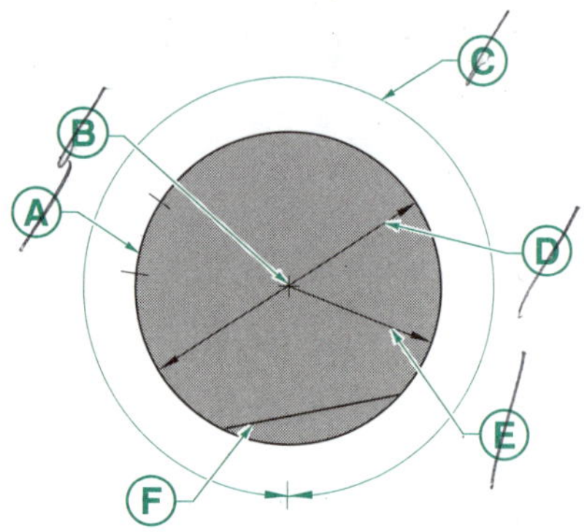

IDENTIFICATION – Two-Dimensional Shapes

______________ 1. Trapezoid

______________ 2. Right triangle

______________ 3. Octagon

______________ 4. Hexagon

______________ 5. Rectangle

______________ 6. Rhombus

______________ 7. Equilateral triangle

______________ 8. Pentagon

______________ 9. Square

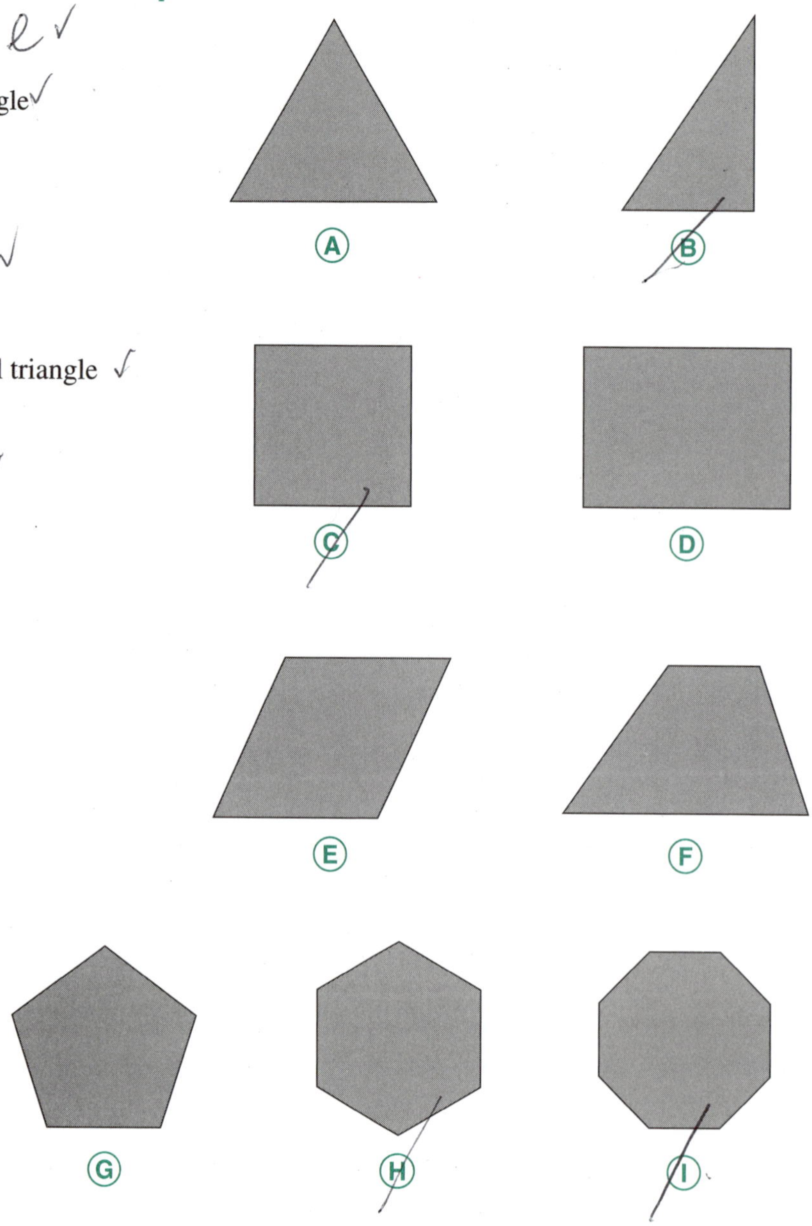

EXERCISES

Using geometric construction principles, complete the following exercises with the given information.

1. Draw a line parallel to the given line at the given distance.

DISTANCE: ________

2. Draw a line perpendicular to the given line through the given point.

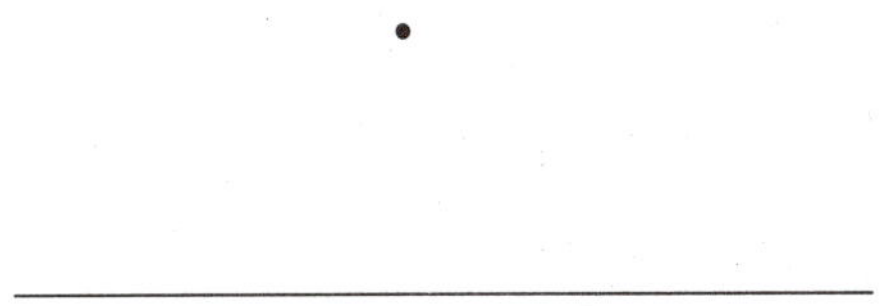

3. Divide the given line into five equal sections.

4. Bisect the given arc.

5. Find the center point of the given circle.

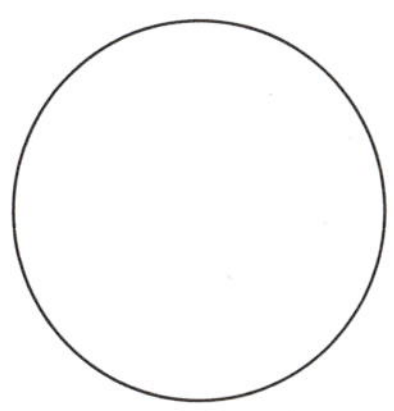

6. Draw a line tangent to the given circle through the given point.

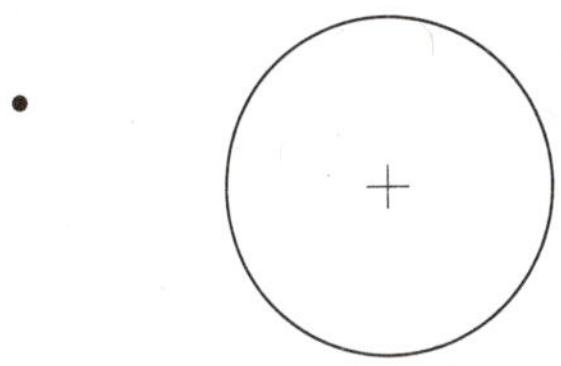

7. Draw an arc tangent to the two given lines that form an acute angle.

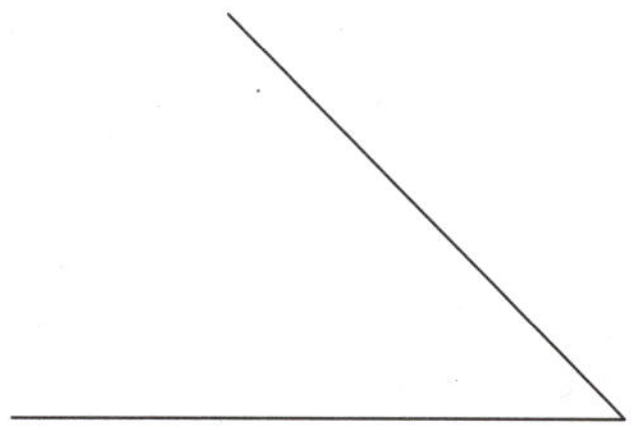

8. Draw a right triangle using the two given sides.

GIVEN SIDES:

9. Draw a square using the given side.

GIVEN SIDE:

10. Draw a hexagon using the given circle.

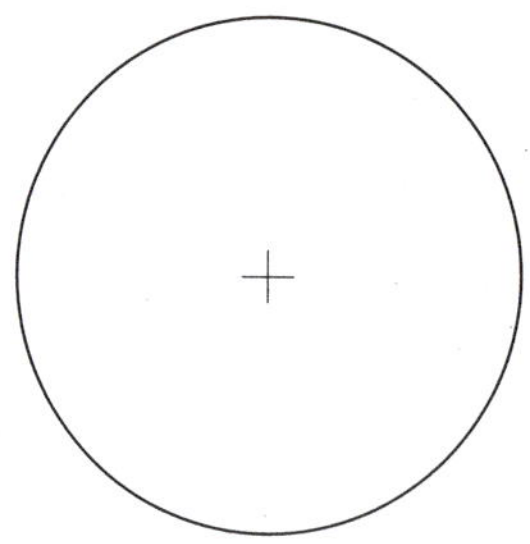

11. Enlarge the given shape using radial lines.

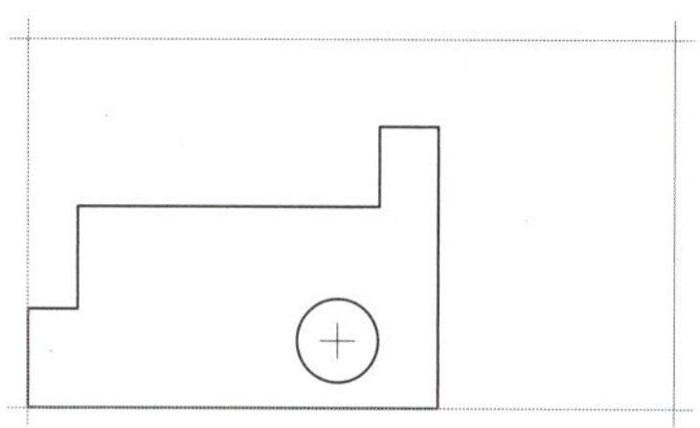

12. Draw an ellipse using the parallelogram method and using the given major and minor axes.

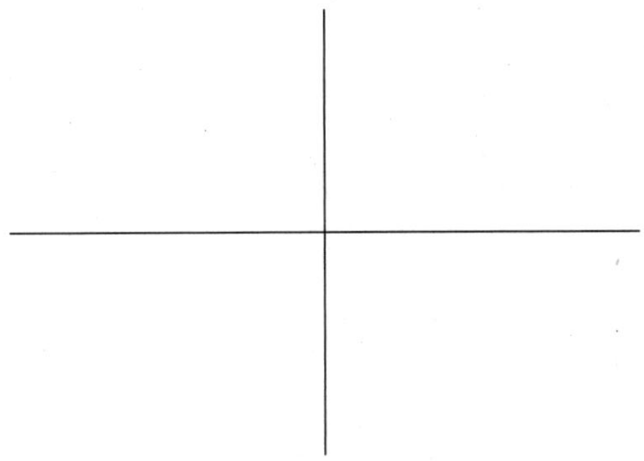

13. Draw a parabola using the given lines and focus.

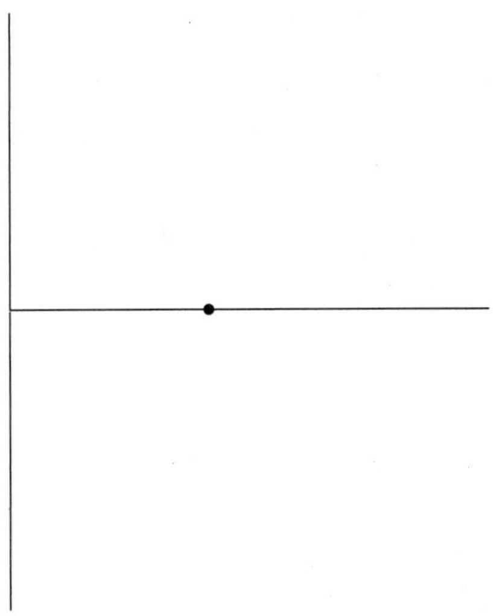

14. Draw a hyperbola using the given points.

15. Draw a left-handed helix using the given cylinder, diameter, and lead.

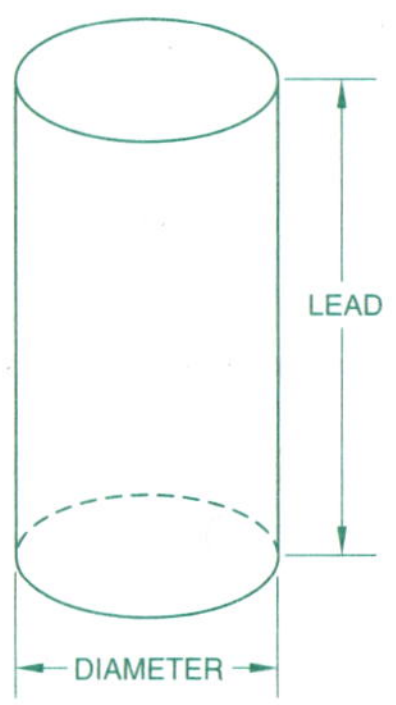

Technical Sketching and Drawing

Pictorial Drawings

4

PICTORIAL DRAWINGS

The ability to quickly make a pictorial drawing is useful for communicating technical information. A *pictorial drawing* is a three-dimensional representation of an object. Pictorial drawings are easy to understand because they convey a sense of realism by combining height, width, and depth in one drawing.

Three basic types of pictorial drawings are axonometric, oblique, and perspective. **See Figure 4-1.** Each type of pictorial drawing represents a different relationship between the viewer and the object and depicts the object in a different way.

Pictorial drawings are often made from multiview drawings. A *multiview drawing* is a two-dimensional representation of a three-dimensional object shown with no distortion of lines or angles. **See Figure 4-2.** Multiview drawings show each unique view separately and in true view. A *true view* is a view where the face of the object being shown is parallel to the picture plane and has no distortion. A true view shows all angles and features as their actual shape and size.

Pictorial drawings are drawn using visible lines. Hidden lines are not often included in pictorial drawings. Pictorial drawings are based on the conventions described in ANSI/ASME Standard Y14.4M, *Pictorial Drawing*.

Pictorial Drawings

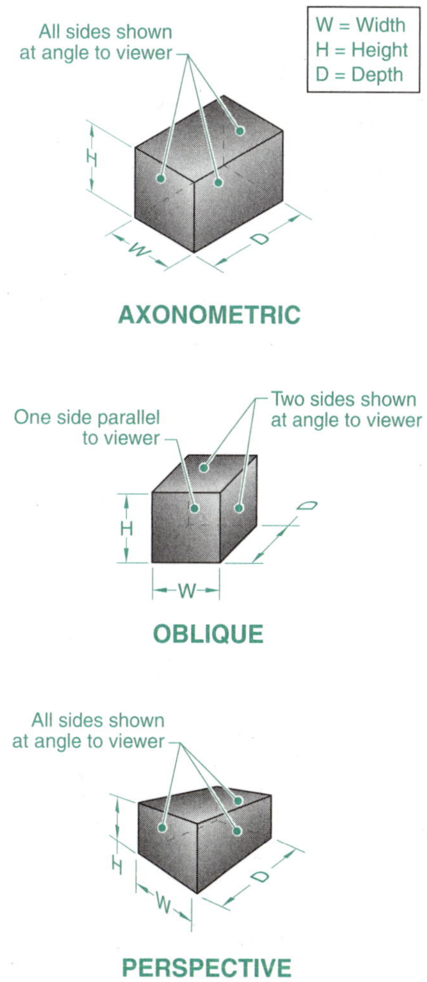

Figure 4-1. Axonometric, oblique, and perspective are three types of pictorial drawings.

Pictorial Drawings Media Clip

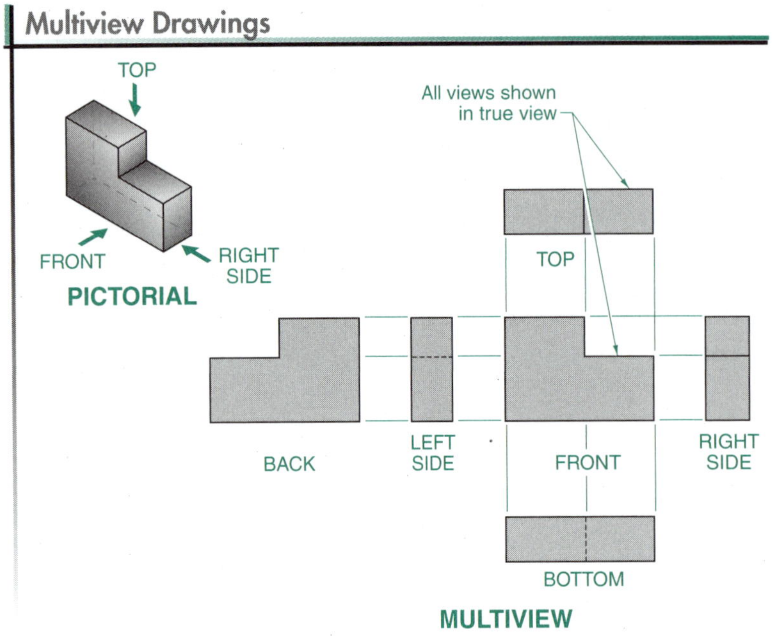

Figure 4-2. A multiview drawing is a two-dimensional representation of an object with all sides drawn in true view.

Axonometric Drawings

An *axonometric drawing* is a pictorial drawing that shows three sides of an object and contains no true view of any side. In an axonometric drawing, planes are created by the three axes of height, width, and depth. A *frontal plane* is a plane that is parallel to the height and width axes. A *horizontal plane* is a plane that is parallel to the width and depth axes. A *profile plane* is a plane that is parallel to the height and depth axes. An *angled plane* is a plane that is not parallel to any of the axonometric axes. **See Figure 4-3.**

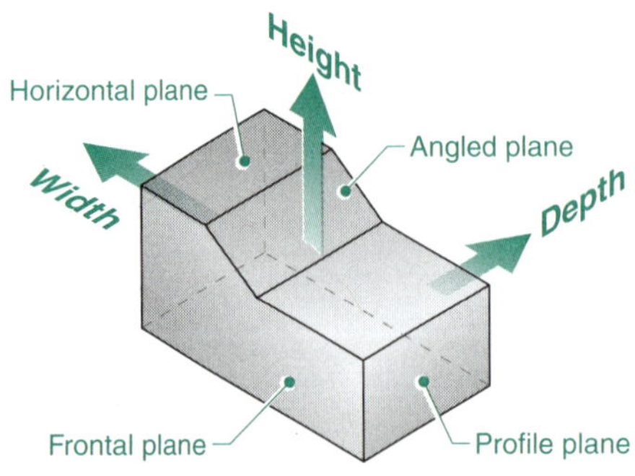

Figure 4-3. Horizontal, frontal, profile, and angled planes are determined based on the axonometric axes they are parallel to.

Foreshortening is a technique used to create an illusion that a surface or distance on an object not seen in true view is shorter than actual size. Foreshortening the surfaces and lengths of an object that recede in space makes the object appear more realistic in a drawing. All lengths that are not parallel to the viewing plane in an axonometric drawing are foreshortened to appear more realistic. Drawn lines that signify the object's width and depth are usually drawn at an angle and are foreshortened, while drawn lines that signify the object's height remain at full size. Parallel lines remain parallel whether or not they are foreshortened.

Three types of axonometric drawings include isometric, dimetric, and trimetric. The type of axonometric drawing selected for use is the one that most clearly portrays the object. Isometric is the most commonly used type.

Isometric Drawings. An *isometric drawing* is an axonometric drawing with the axes equally spaced 120° apart. Isometric drawings clearly show three surfaces of an object with no surfaces in true view. **See Figure 4-4.** Lines that signify the object's width and depth are drawn at 30° from the horizon, while lines that signify the object's height are drawn vertically.

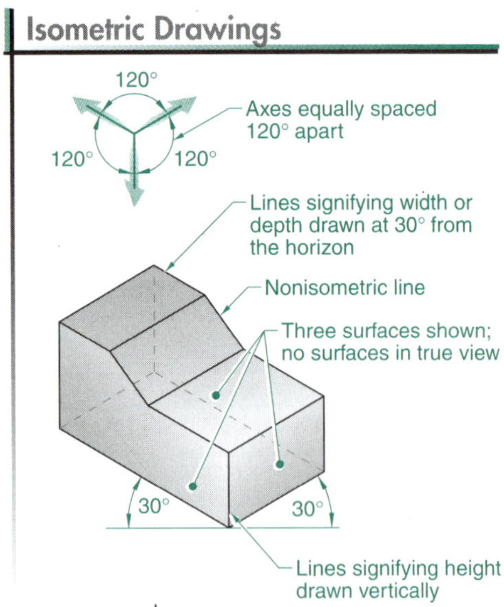

Figure 4-4. An isometric drawing is an axonometric drawing with the axes drawn equally spaced at 120° apart.

Any feature that is not parallel to the height, width, or depth of the object, and therefore not parallel to one of the three isometric axes, is considered *nonisometric*. Nonisometric features are drawn by measuring and marking their end points on isometric lines. Irregularly shaped objects can be drawn in or around a rectangular box to assist with shape and placement. Grids or numbered lines can also be used to measure and transfer irregularly shaped features from a true view to an isometric view. **See Figure 4-5.**

Transferring Irregularly Shaped Features

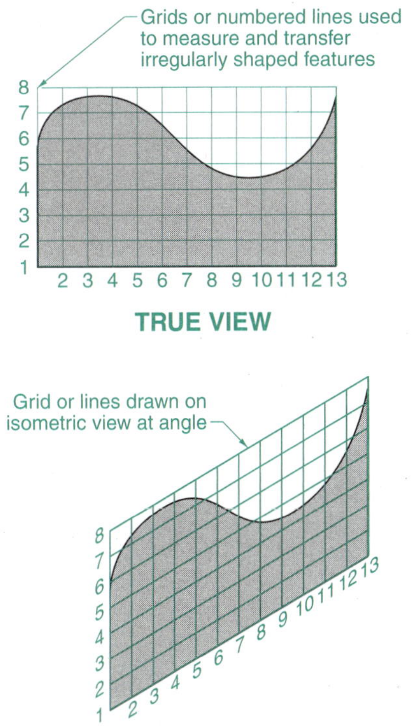

Figure 4-5. Grids or numbered lines can be used to measure and transfer curves or other irregular shaped features from a true view to an isometric view.

The shapes of angles and circles are distorted on receding planes. Angles appear skewed. Circles appear as ellipses. The minor axis of an ellipse is equal to the isometric axis that is perpendicular to the isometric axis of the plane on which the circle is located. For example, an ellipse on the horizontal plane always has a minor axis parallel to the height axis, an ellipse on the frontal plane always has a minor axis parallel to the depth axis, and an ellipse on the profile plane always has a minor axis parallel to the width axis. **See Figure 4-6.**

Isometric in Print and Drawing Library

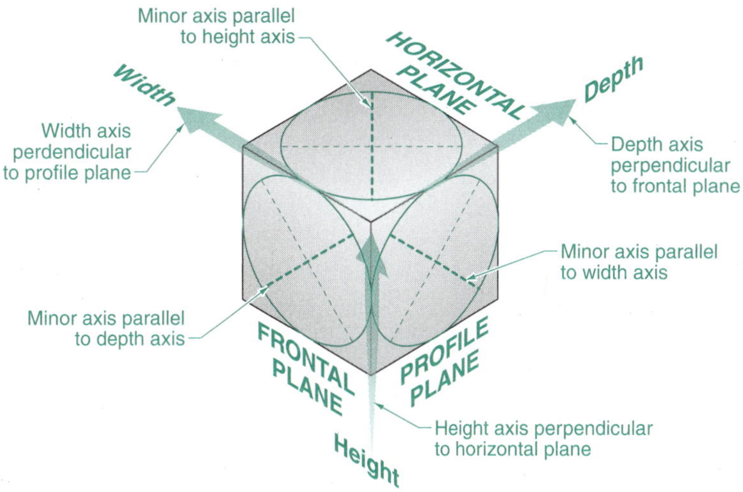

Figure 4-6. Circles or other round features appear as ellipses on receding planes and are drawn with the minor axis matching the isometric axis the plane is perpendicular to.

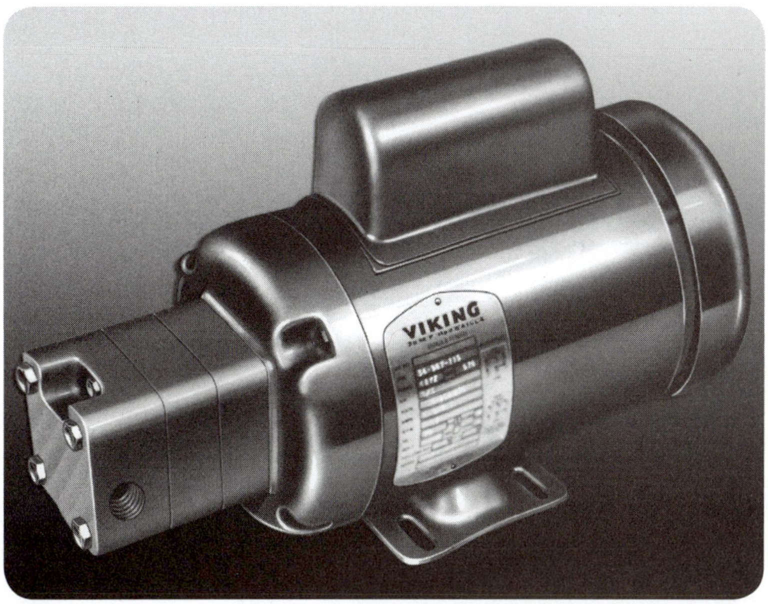

Viking Pump, Inc., A Unit of IDEX Corporation

Irregular shapes on an object can be drawn using grids aligned to the axes in both the multiview and the isometric view.

In dimetric and trimetric drawings, the angle of the axes can be changed to show more of the top view or side view of the object.

Dimetric Drawings. A *dimetric drawing* is an axonometric drawing with two axes drawn on equal angles to the third, but with an unequal angle between them. **See Figure 4-7.** The main axis is generally placed perpendicular to the horizon with the other two axes varying. Adjusting the angles between the axes will allow more or less of the top view to be visible. Lines that run parallel to one of the two dimetric axes with equal angles are foreshortened at the same scale, whereas lines that run parallel to the third dimetric axis are foreshortened at a different scale.

Trimetric Drawings. A *trimetric drawing* is an axonometric drawing with all axes drawn at different angles. The axes can be positioned to allow any side of an object to be more or less visible. In a trimetric drawing, each axis has a different scale of foreshortening. The use of three scales makes this method more time-consuming and, therefore, it is seldom used.

Drawing Axonometrics. Isometric, dimetric, and trimetric drawings are created in a similar way. When creating an axonometric drawing, each horizontal and vertical surface on the object is drawn parallel to one of the chosen axonometric axes, with angled or irregular surfaces drawn by determining their shape based on other lines.

To create an isometric drawing, complete the following procedure:

1. Draw the isometric axes and mark the height and width measurements of the frontal plane on the axes. Draw the outline of the frontal plane surface(s). **See Figure 4-8.**
2. Draw the receding lines for the horizontal and profile planes parallel to the isometric axes and mark the depth of the object. Draw lines parallel to the isometric axes to complete the horizontal and profile plane surfaces. Nonisometric lines are drawn by marking and connecting their end points on isometric lines.
3. Locate the center points of any circles and arcs. Draw the circles and arcs as ellipses.
4. Darken visible lines to complete the isometric drawing.

Dimetric and Trimetric Drawings

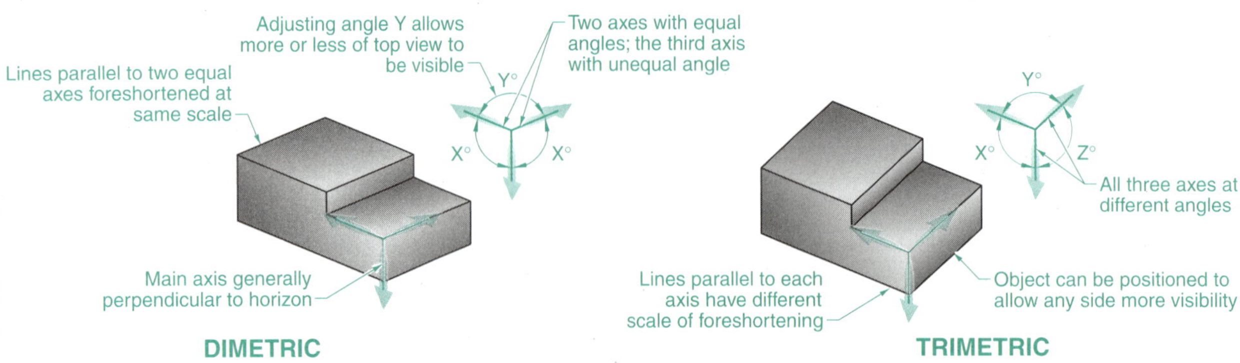

Figure 4-7. Dimetric and trimetric drawings are axonometric drawings with axes at varying angles.

Drawing Isometrics

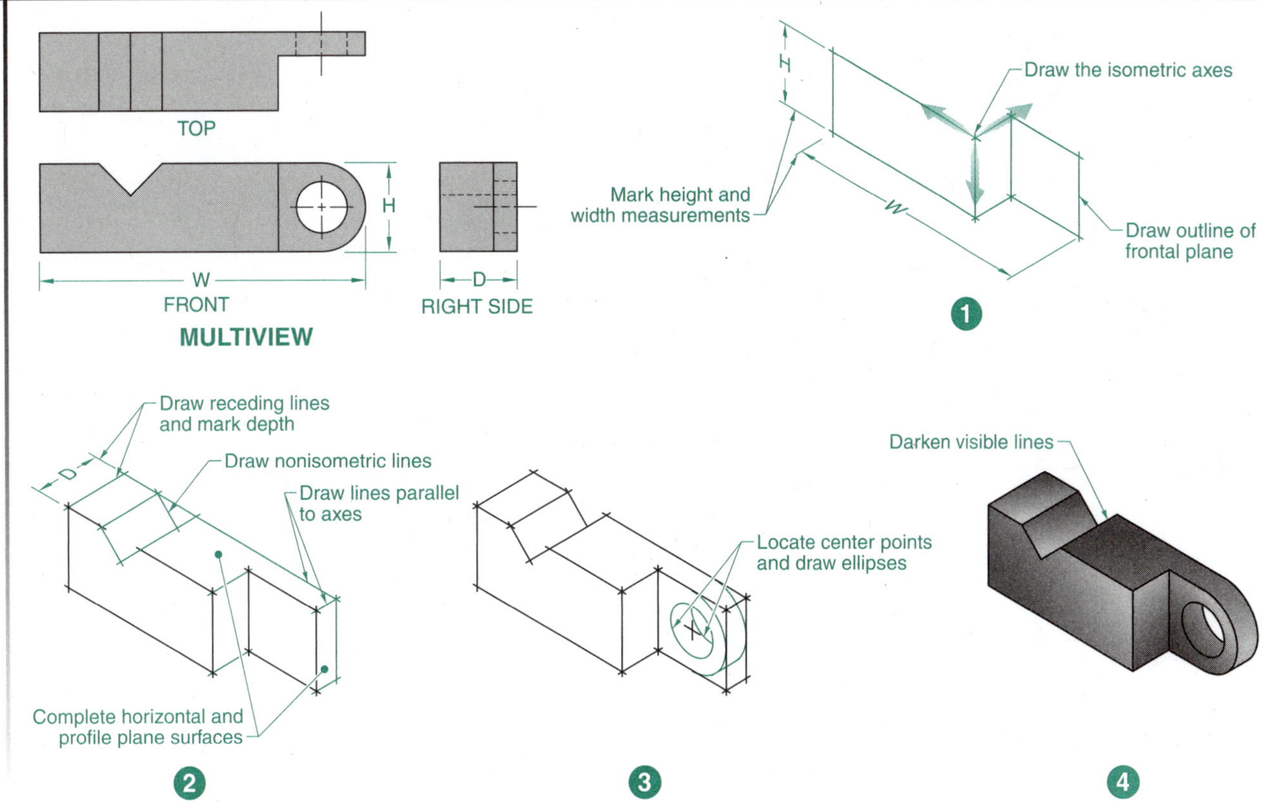

Figure 4-8. In an isometric drawing, each surface of the object is drawn with lines parallel to the axes.

Drawing Circles in Axonometrics. In axonometric drawings, circles are drawn as ellipses and arcs are drawn as portions of ellipses. For isometric drawings, a 35° ellipse template can be used, or ellipses can be constructed using the parallelogram method. To draw a circle in an isometric view using the parallelogram method, complete the following procedure:

Given: Circle with known diameter.

1. In the desired location of the ellipse, draw parallelogram ABCD with the sides equal to the diameter of the circle and at the angle of the isometric axes. **See Figure 4-9.**
2. Bisect sides AB and CD with line EF. Bisect sides AD and BC with line GH. Lines AB, CD, and GH should be parallel. Lines AD, BC, and EF should be parallel.
3. Draw line DE perpendicular to side AB. Draw line BF perpendicular to side CD. Draw line BG perpendicular to side AD. Draw line DH perpendicular to side BC. Lines DE and BG intersect at point I. Lines BF and DH intersect at point J.
4. Set a compass with the radius equal to line segment EI. With a center point of I, draw arc EF tangent to parallelogram ABCD. With a center point of J, draw arc GH tangent to parallelogram ABCD.
5. Set a compass with the radius equal to line segment DE. With a center point of D, draw arc EG tangent to parallelogram ABCD. With a center point of B, draw arc FH tangent to parallelogram ABCD.
6. Darken visible lines to complete the ellipse.

Drawing Circles in Isometric View

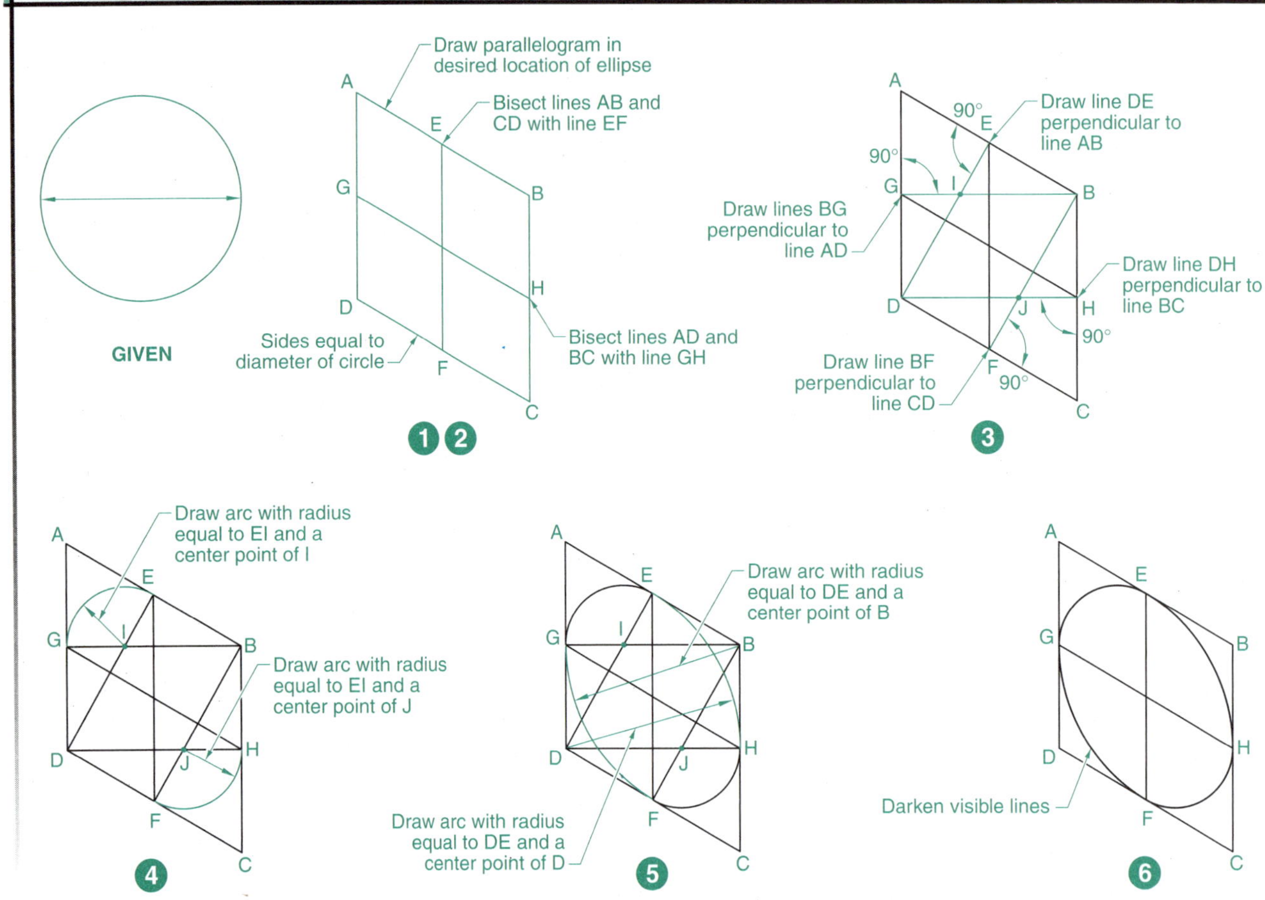

Figure 4-9. Circles can be constructed in isometric drawings using the parallelogram method.

Oblique Drawings

An *oblique drawing* is a pictorial drawing that shows one side of an object in true view with the remaining sides receding back in space. **See Figure 4-10.** The true view face is situated parallel to the picture plane with two axes at 90° and portrays the height and the width of the object. The third axis portraying depth can be at any oblique angle to the first two axes, which in turn distorts receding faces. Choosing a larger angle for the third axis gives a more prominent top view, where choosing a smaller angle gives a more prominent side view. Circles appear in true shape on the true view face and as ellipses on receding faces. Oblique drawings can be cabinet, general, or cavalier drawings.

Oblique Drawings

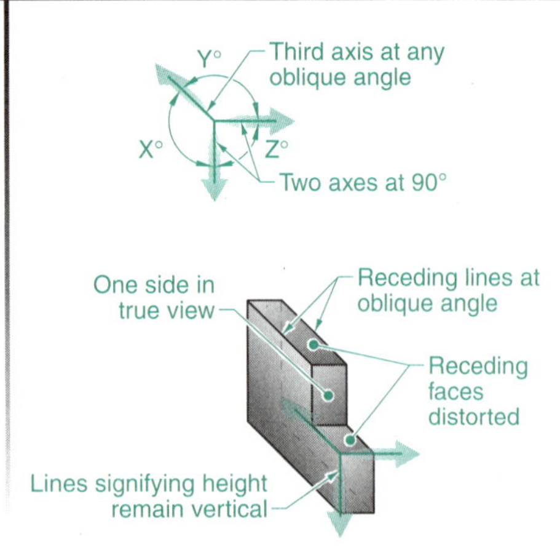

Figure 4-10. An oblique drawing shows one side of an object in true view and the remaining sides receding back in space.

A *cabinet drawing* is an oblique drawing with receding lines foreshortened to one-half the scale of lines in the true view. Cabinet drawings are the most common type of oblique drawing and are often made with the receding lines at a 30° or 45° angle. **See Figure 4-11.**

A *general oblique drawing* is a drawing with receding lines foreshortened at any scale between half and full. General oblique drawings are often considered cabinet drawings because a scale of foreshortening is used.

A *cavalier drawing* is an oblique drawing with receding lines drawn to the same scale as the lines in the true view. Cavalier drawings use no foreshortening, which results in an unnatural and distorted appearance.

Drawing Obliques. Oblique drawings are created by first choosing the angles of the axes and the scale of foreshortening that best conveys the object. The front surface is drawn in true view, with receding lines drawn at the desired scale. Nonoblique lines and angles are drawn by marking their end points on oblique lines and connecting the marks. Circles and arcs on the front view appear in true shape. Circles and arcs on receding views are drawn as ellipses by using an ellipse template or by constructing them within a parallelogram or other simple shape.

To create an oblique cabinet drawing, complete the following procedure:

1. Draw the oblique axes and mark the height and width measurements on the axes. Draw the front view. **See Figure 4-12.**
2. Draw receding lines following the oblique axes and mark the depth of the object. Draw remaining lines parallel to the axes to complete the top and side surfaces. Nonoblique lines are drawn by marking and connecting their end points on oblique lines. *Note:* The depth is foreshortened to one-half the depth dimension.
3. Locate the center points of any circles and arcs. Draw the circles and arcs.
4. Darken all visible lines to complete the oblique cabinet drawing.

General oblique drawings are drawn in a similar manner but with a foreshortening scale at any amount between one-half and full. Cavalier drawings are also drawn using the above procedure, but with all measurements drawn at full-scale.

Oblique in Print and Drawing Library

TECH TIP

Oblique drawings are ideal for complex objects containing circular or curved shapes that are more difficult to draw on an axonometric axis. Oblique drawings allow the most complex face of an object to be selected as the front view and drawn in true view without any distortion of the features.

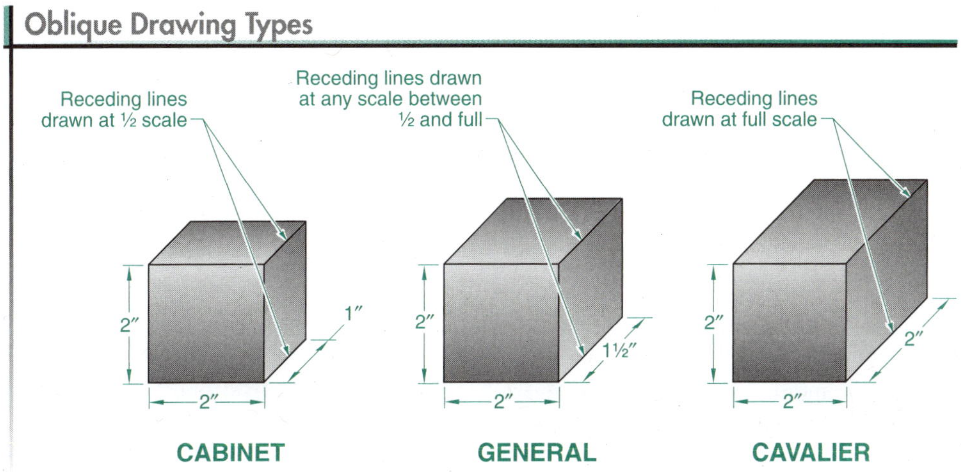

Figure 4-11. Oblique drawings include cabinet, general, or cavalier drawings.

Drawing Cabinet Obliques

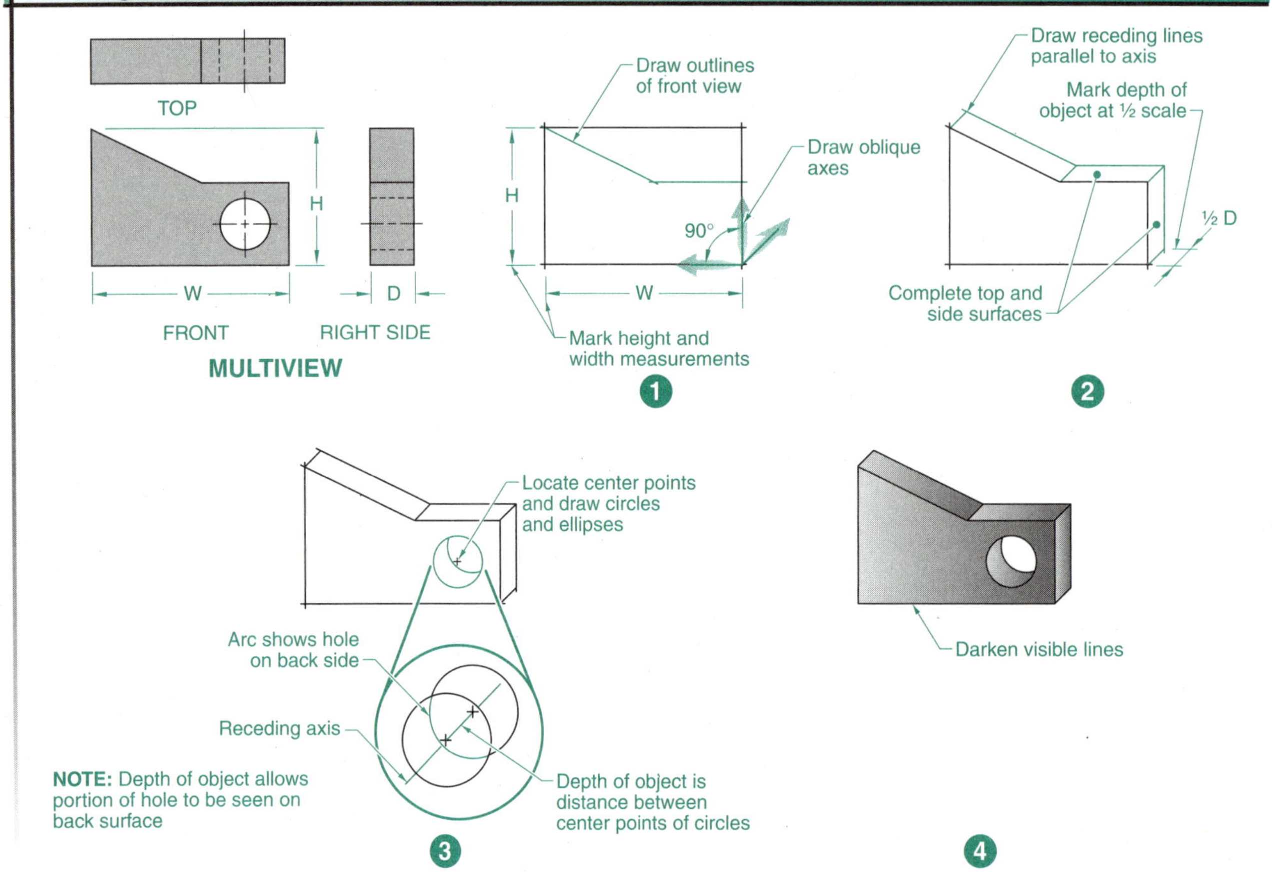

Figure 4-12. In an oblique cabinet drawing, the true shape of the front surface is drawn to scale with receding lines drawn at one-half scale.

Vanishing Point Media Clip

1-point Perspective in Print and Drawing Library

Perspective Drawings

Perspective is based on the phenomenon of objects appearing smaller when farther away from the viewer. A *perspective drawing* is a pictorial drawing where all receding lines converge at one or more vanishing points. A *vanishing point* is a point on the horizon where parallel lines appear to meet. The *horizon* is the eye level of the viewer. Vanishing points can be to the left, right, or center of a drawing, depending on the point of view. The object being drawn may be located above (worm's-eye view), on (eye-level view), or below (bird's-eye view) the vanishing point and horizon. Eye-level view is most commonly used. **See Figure 4-13.**

Perspective drawings are named based on the number of vanishing points and are considered one-point, two-point, or three-point. Lines that run horizontally on an object and recede in space appear to merge at vanishing point(s) on the horizon. Lines that run vertically on an object maintain their appearance in drawings with one or two vanishing points, but recede when there are three vanishing points. **See Figure 4-14.**

One-Point Perspective. A *one-point perspective* is a drawing with one vanishing point on the horizon where parallel lines appear to recede. Parallel lines that are perpendicular to the picture plane are foreshortened and recede to the vanishing point. Parallel lines that are parallel to the picture plane remain parallel but are foreshortened in height and spacing as they recede. In a one-point perspective, one face of a cube will be parallel to the picture plane and drawn in true view, while the remaining faces are foreshortened and recede toward the vanishing point.

Perspective Drawings

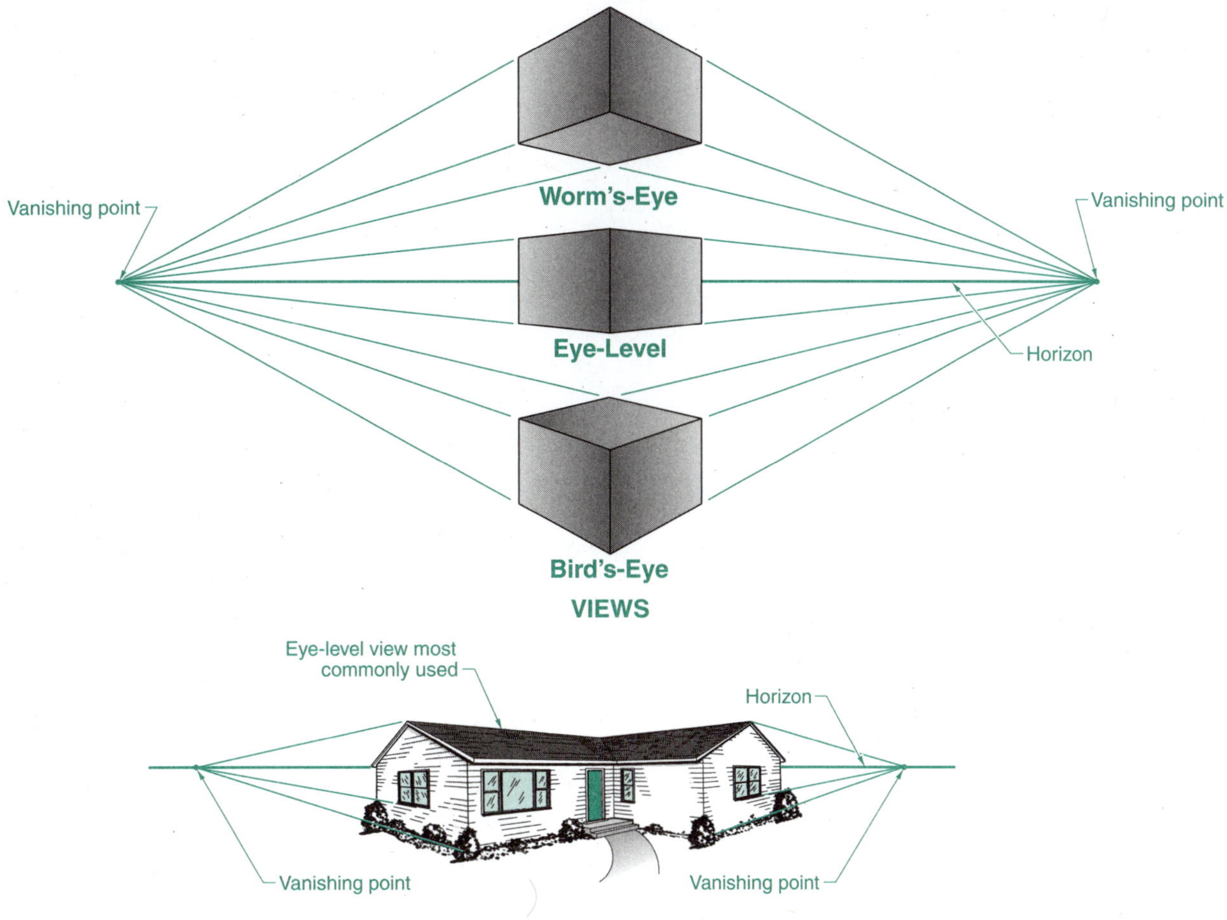

Figure 4-13. Perspective drawings can have a worm's eye, eye-level, or bird's-eye point of view.

Types of Perspective Drawings

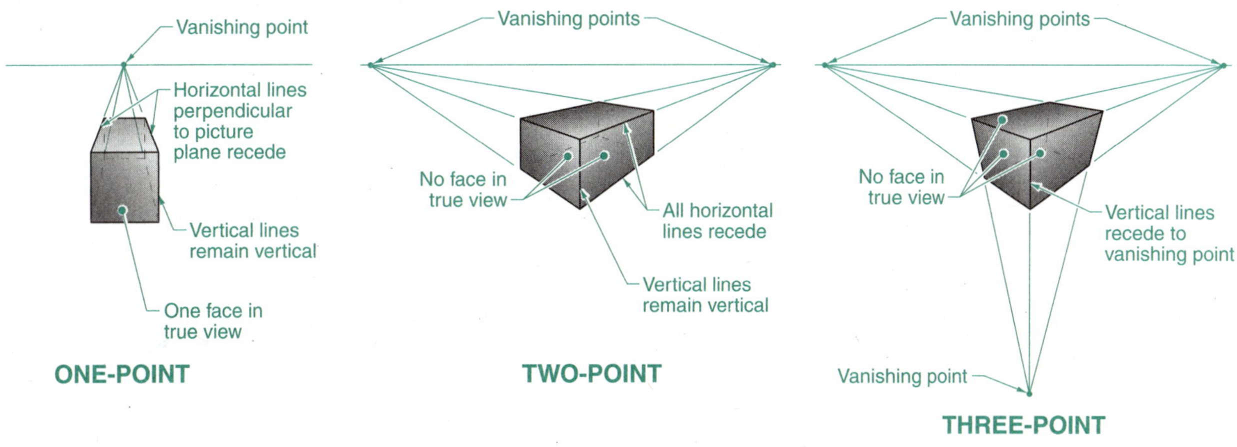

Figure 4-14. Perspective drawings are considered one-point, two-point, or three-point.

2-point Perspective in Print and Drawing Library

Two-Point Perspective. A *two-point perspective* is a drawing with two vanishing points on the horizon where parallel lines appear to recede. A two-point perspective depicts objects at an angle with no side in true view. Each vanishing point represents one set of parallel lines moving back in space. In a two-point perspective, no face of a cube is parallel to the picture plane. Lines that run horizontally on an object are foreshortened as they recede toward two vanishing points while lines that run vertically remain vertical.

3-point Perspective in Print and Drawing Library

Three-Point Perspective. A *three-point perspective* is a drawing with two vanishing points on the horizon and a third vanishing point above or below the horizon where parallel lines appear to recede. A three-point perspective is similar to a two-point perspective but has lines that run vertically on an object recede to the third point. A three-point perspective shows no side in true view and is generally used for objects viewed from above or below. In a three-point perspective, no face of a cube is parallel to the picture plane. All lines in an object are foreshortened as they recede toward three vanishing points.

Drawing Perspectives. When beginning a perspective drawing, the type of view, location of horizon, and number and location of the vanishing point(s) should be determined. All receding lines converge at the vanishing point(s) marked, while vertical lines on an object remain vertical in one- and two-point perspectives, and converge at a vanishing point in three-point perspectives.

In a one-point perspective, lines indicating depth recede to a vanishing point. To create a one-point perspective drawing, complete the following procedure:

1. Locate and mark the horizon and vanishing point. **See Figure 4-15.**
2. Draw a front view of the object in true view.
3. Draw light construction lines from the corners of the front view that converge at the vanishing point. Mark the depth of the object on the construction lines. *Note:* The depth will be foreshortened.
4. Draw surfaces that are parallel to the front surface by lining up the surface with the construction lines. Draw the side and top surfaces. Receding lines should recede to the vanishing point.
5. Locate the center points of any circles or arcs. Draw the circles and arcs. *Note:* Circles and arcs on receding planes distort and become smaller as they recede.
6. Erase the construction lines and darken all visible lines.

In a two-point perspective, the lines indicating width and depth recede to vanishing points. To create a two-point perspective drawing, complete the following procedure:

1. Locate and mark the horizon and two vanishing points. **See Figure 4-16.**
2. Draw the front corner of the object and establish the height of the object.
3. From the front corner, draw light construction lines that converge at the vanishing points. Mark the width and depth of the object on the construction lines. *Note:* The width and depth will be foreshortened.
4. Draw the surfaces of the sides along the construction lines. Draw the top surface. *Note:* All lines parallel to each other recede to the same vanishing point.
5. Locate the center points of any circles or arcs. Draw the circles and arcs as ellipses. *Note:* Circles and arcs on receding planes distort and become smaller as they recede.
6. Erase construction lines and darken all visible lines.

In a three-point perspective, the lines indicating height, width, and depth recede to vanishing points. To create a three-point perspective drawing, complete the following procedure:

1. Locate and mark the horizon and three vanishing points. **See Figure 4-17.**

2. Draw the line for the front corner to establish the height of the object, which recedes to the third vanishing point. *Note:* The height will be foreshortened.
3. From the front corner line, draw light construction lines that converge at each of the two vanishing points. Mark the width and depth of the object on the construction lines. *Note:* The width and depth will be foreshortened.
4. Draw the side surfaces along the construction lines from all three vanishing points. Draw the top surface. *Note:* All lines that are parallel to each other recede at the same vanishing point.
5. Locate the center points of any circles or arcs. Draw the circles and arcs. *Note:* Circles and arcs on receding planes distort and become smaller as they recede.
6. Erase the construction lines and darken all visible lines.

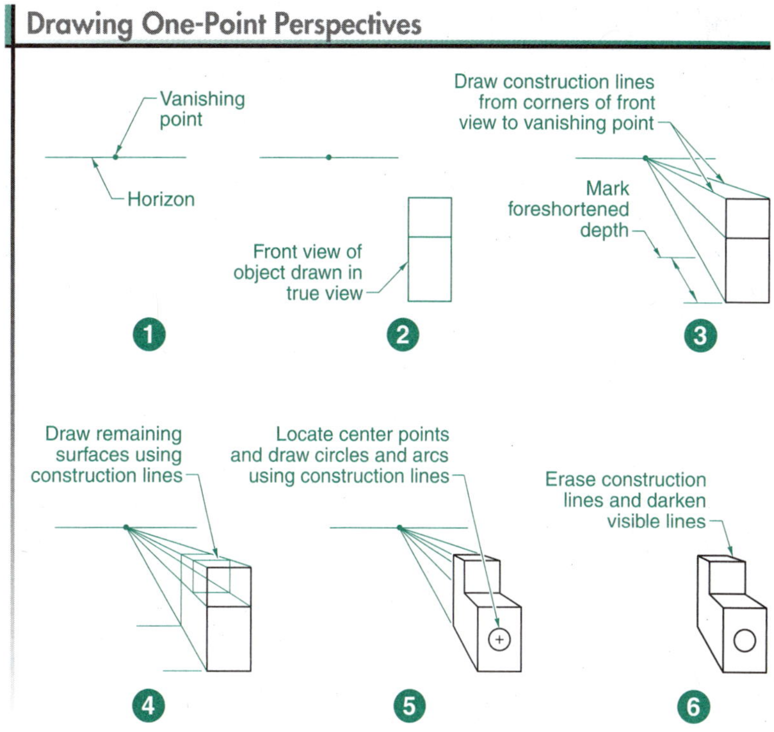

Figure 4-15. One-point perspectives are drawn with receding lines converging at one vanishing point.

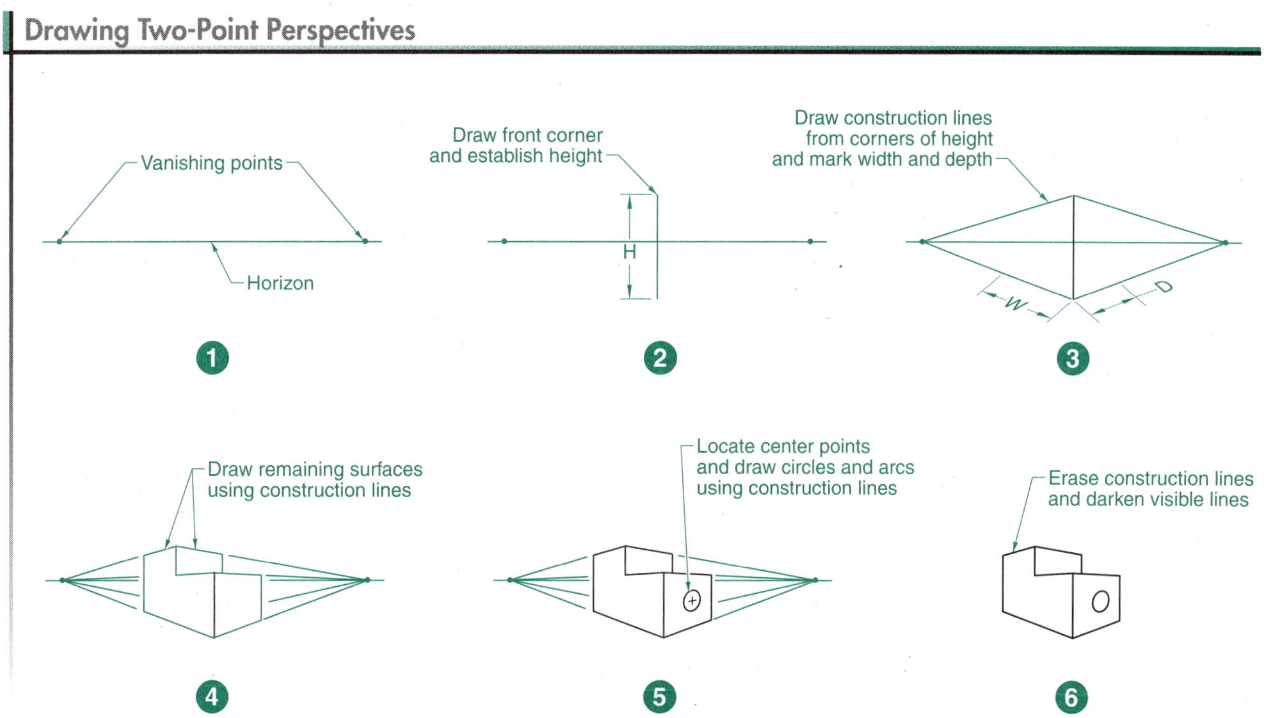

Figure 4-16. Two-point perspectives are drawn with receding lines converging at two vanishing points.

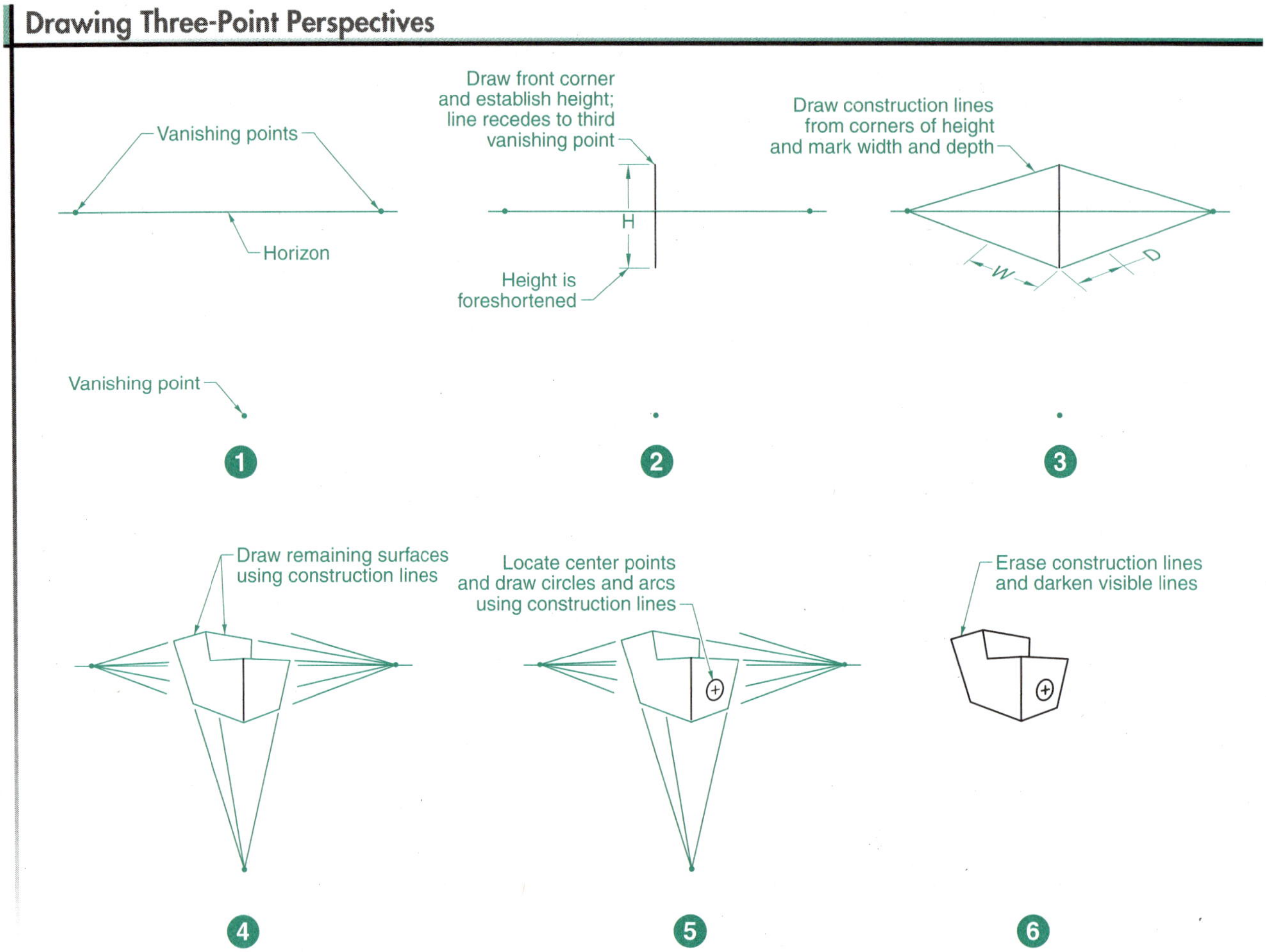

Figure 4-17. Three-point perspectives are drawn with the receding lines for width and depth converging at two vanishing points and the receding lines for height converging at a third vanishing point.

COMPLEX PICTORIAL DRAWINGS

Assemblies that contain multiple parts often require more complex pictorial drawings for a complete understanding of the object as a whole. Interior arrangement, cutaway, and exploded view drawings are often used in industry to show interior and exterior components of assemblies and constructions.

Interior Arrangement Drawings

An *interior arrangement drawing (phantom drawing)* is a pictorial drawing where the exterior of an object or assembly is drawn transparent or lightly shaded in order to reveal interior parts. Interior arrangement drawings show the relative location and scale of parts in an assembly exactly as they are assembled. The exterior of the object is drawn as an outline or with a light, flat tone, where the interior parts are fully described. **See Figure 4-18.**

Drawing Interior Arrangements. When creating an interior arrangement drawing, select a pictorial drawing type that will clearly convey the interior and exterior of the object. Draw the exterior outline following the appropriate pictorial axes or vanishing points. The part or parts closest to the viewing plane should be drawn first, working back in space to draw additional parts. Parts that are in front will overlap the parts in back. Interior assembly drawings are usually drawn in one view, but several views may to needed to clearly show all parts in the assembly.

Cutaway Drawings

A *cutaway drawing* is a pictorial drawing of an object or assembly that is cut into sections to show interior parts. The exterior shell and all interior parts are fully described. Cutaway drawings show parts in an assembly in their proper location with areas removed in order to show other parts that are behind or within. **See Figure 4-19.**

Drawing Cutaways. When creating a cutaway drawing, select a pictorial drawing type that will clearly convey the object. Mark the appropriate pictorial axes or vanishing points and begin with objects that are closest to the viewing plane. Only objects obscuring internal details that must be shown should be cut away. Cutaway section locations should be chosen to fully convey each aspect of the assembly while maintaining the assembly composition as a whole.

Exploded View Drawings

An *exploded view drawing* is a pictorial drawing that shows an assembly pulled apart and its parts aligned to show their location within the whole assembly. In an exploded view drawing, the exterior and all parts of the assembly are fully described. To show all parts, interior components are removed and laid out along an alignment line. Parts that fit together are aligned on the same line. Parts are generally positioned on the drawing in their proper order and placement in the assembly. **See Figure 4-20.**

Drawing Exploded Views. Exploded view drawings are best conveyed using an axonometric drawing technique. The main exterior portion of the assembly is drawn first using the axonometric axes chosen. The axes of the alignment lines are generally drawn at the same angle as the axonometric axes chosen for the drawing. Any number of alignment lines can be used on a drawing depending on the number of interior parts and their placement on the assembly.

Parts are generally drawn in the order in which they are assembled, working outward from the main exterior portion of the assembly and along the axis or axes that align the parts to their place on the assembly. Parts on the interior of the assembly are generally drawn closer to the main exterior portion of the drawing, while the exterior parts are drawn the farthest out along the alignment line. To avoid overlapping other lines or parts, alignment lines can be offset along the appropriate axis for improved readability.

Cutaway in Print and Drawing Library

Exploded View in Print and Drawing Library

Interior Arrangement Drawings

Figure 4-18. An interior arrangement drawing shows the exterior of an object or assembly as transparent or lightly shaded in order to reveal interior parts.

Cutaway Drawings

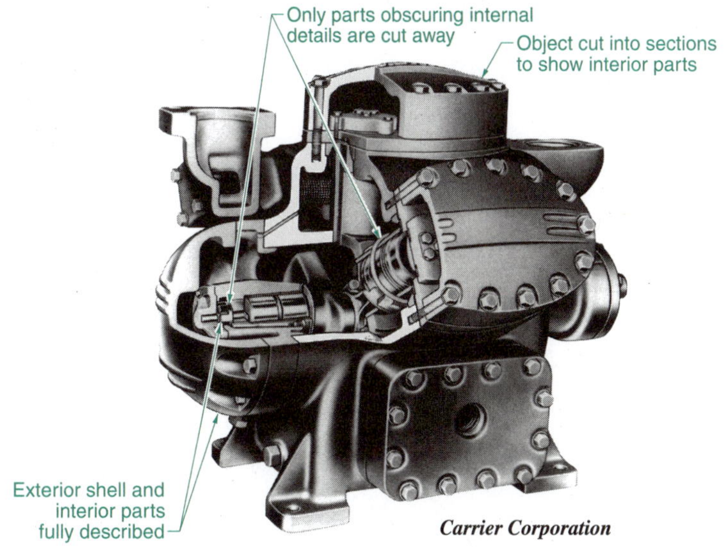

Figure 4-19. A cutaway drawing is cut into sections to show interior parts of an object or assembly.

Exploded View Drawings

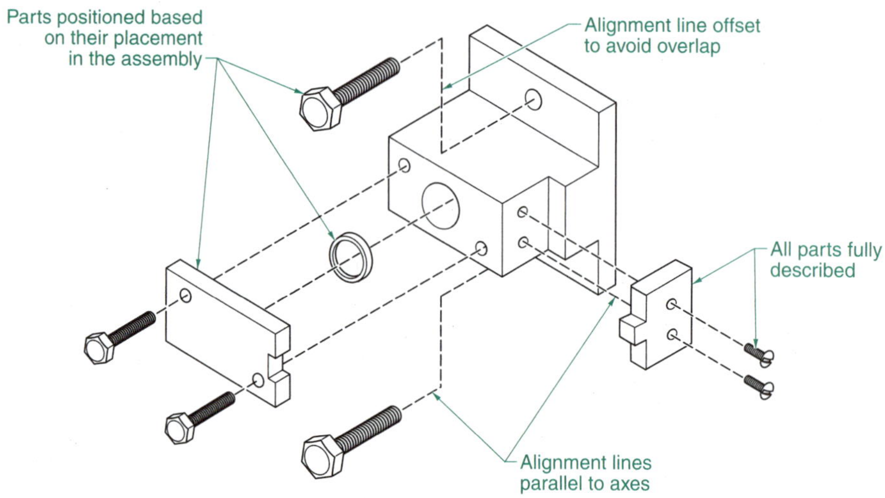

Chapter 4 Quick Quiz®

Figure 4-20. An exploded view drawing shows an assembly pulled apart and its parts aligned to show their location within the whole assembly.

Technical Sketching and Drawing

Pictorial Drawings

Review 4

Name ______________________________ Date ____________________

REVIEW QUESTIONS

________________ **1.** Three basic types of pictorial drawings are axonometric, ___, and perspective.

________________ **2.** A(n) ___ view shows the face of the object parallel to the picture plane and with no distortion.

________________ **3.** A(n) ___ drawing is a pictorial drawing that shows three sides of an object and contains no true view of any side.

T F **4.** An angled plane is a plane that is parallel to one axonometric axis.

________________ **5.** ___ is a technique used to create an illusion that a surface on an object that is not parallel to the picture plane is shorter than actual size.

T F **6.** An isometric drawing has three axes equally spaced 120° apart.

T F **7.** In isometric drawings, the shapes of angles and circles are shown in true shape on receding planes.

________________ **8.** A(n) ___ drawing is an axonometric drawing with two axes drawn on equal angles to the third, but with an unequal angle between them.

________________ **9.** A(n) ___ drawing is a pictorial drawing that shows one side of an object in true view with the remaining sides receding back in space.

T F **10.** Cavalier drawings are the most common type of oblique drawing.

________________ **11.** A(n) ___ oblique drawing has receding lines foreshortened at any scale between half and full.

________________ **12.** A(n) ___ drawing is a pictorial drawing where all receding lines converge at one or more vanishing points.

T F **13.** The horizon is the eye level of the viewer.

________________ **14.** In perspective drawings, lines that run horizontally on an object and recede in space appear to merge at the ___ on the horizon.

________________ **15.** ___-point perspective drawings show one face of a cube in true view, while the remaining faces are foreshortened and recede toward the vanishing point.

T F **16.** In a three-point perspective, no face of a cube is parallel to the picture plane.

________________ **17.** In a ___-point perspective, the lines indicating height, width, and depth recede to vanishing points.

________________ **18.** A(n) ___ drawing shows the exterior of an object or assembly transparent or lightly shaded in order to reveal interior parts.

T F **19.** In a cutaway drawing, all objects should be sectioned and cut away, whether or not they have internal details.

________________ **20.** In exploded view drawings, interior components are removed and laid out along a(n) ___ line.

IDENTIFICATION—Drawing Types

________________ **1.** One-point perspective

________________ **2.** Cabinet oblique

________________ **3.** Isometric

________________ **4.** Two-point perspective

________________ **5.** Cavalier oblique

________________ **6.** Dimetric

________________ **7.** Trimetric

________________ **8.** Three-point perspective

________________ **9.** General oblique

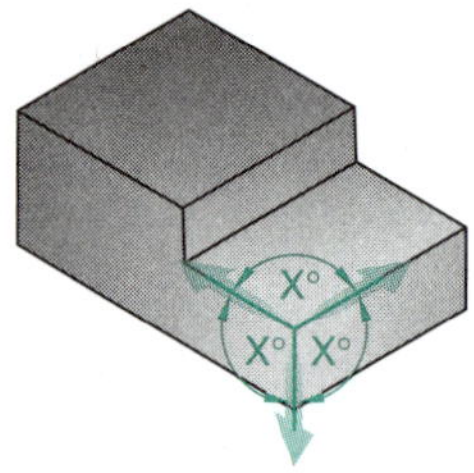

A

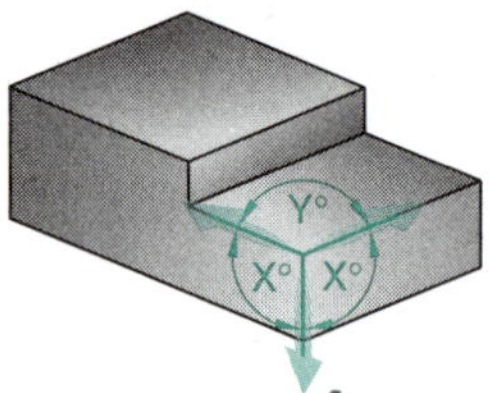

B

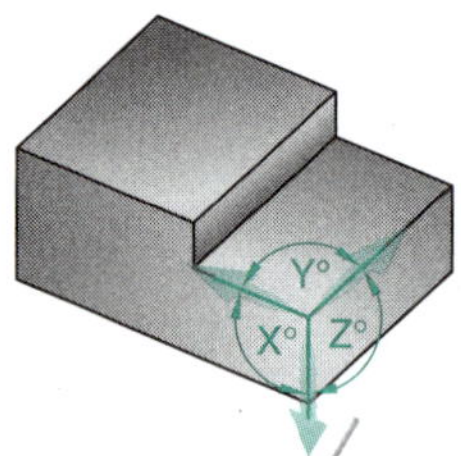

C

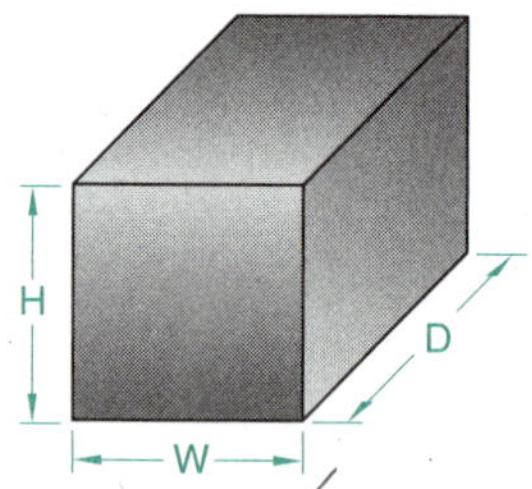

D

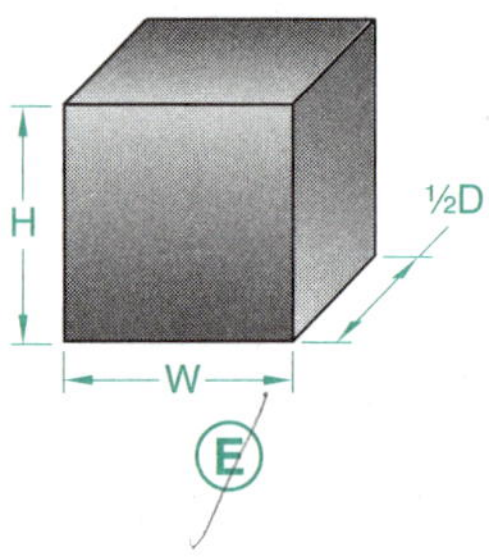

E

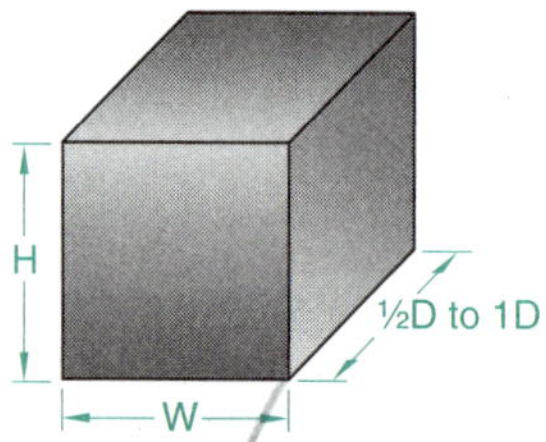

F

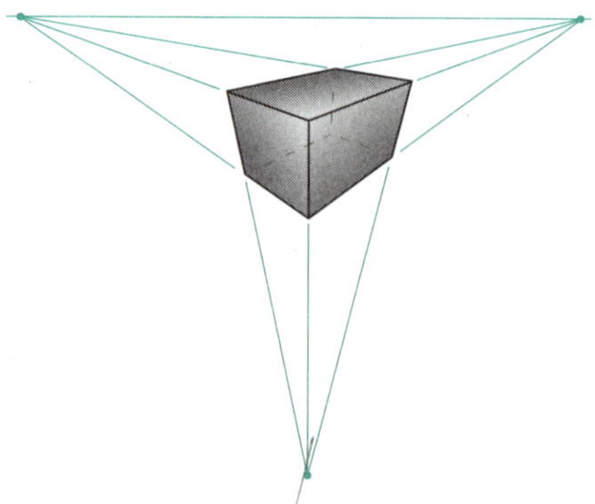

I

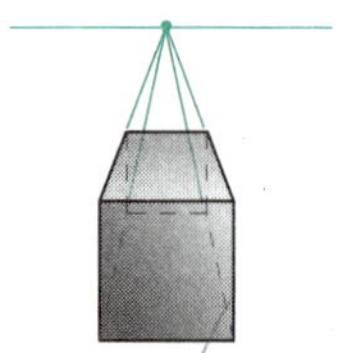

G

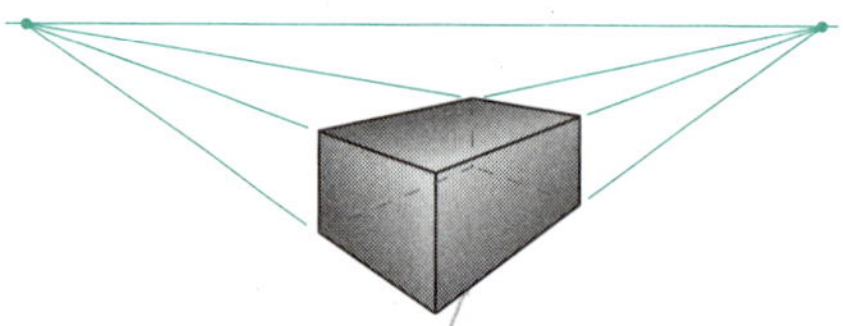

H

IDENTIFICATION—Complex Drawing Types

c ________ **1.** Cutaway

b ________ **2.** Exploded view

a ________ **3.** Interior arrangement

Ⓐ

Ⓑ

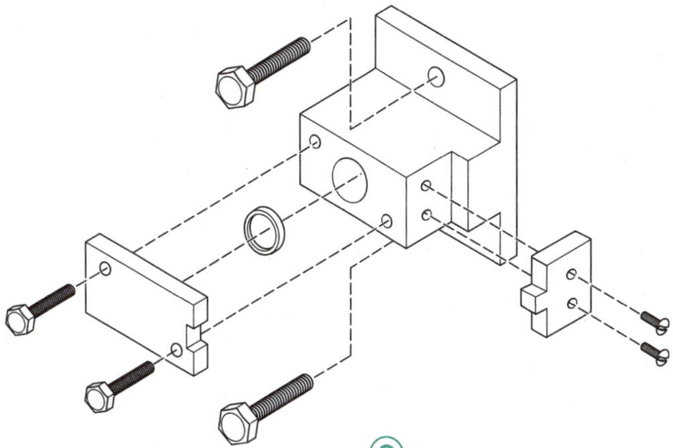

Ⓒ

EXERCISES

Using the multiview drawing given below, create the pictorial drawing type indicated.

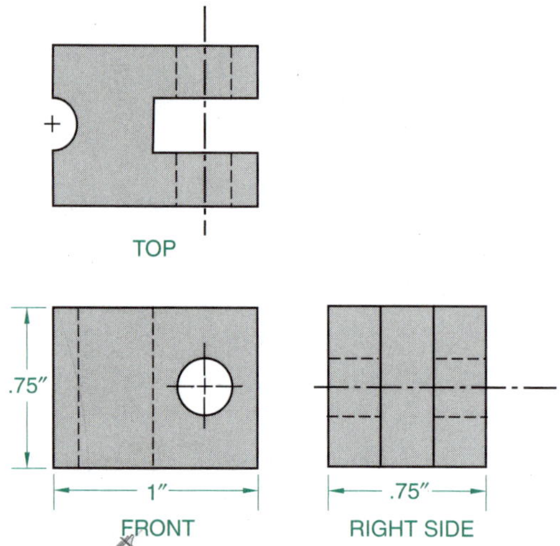

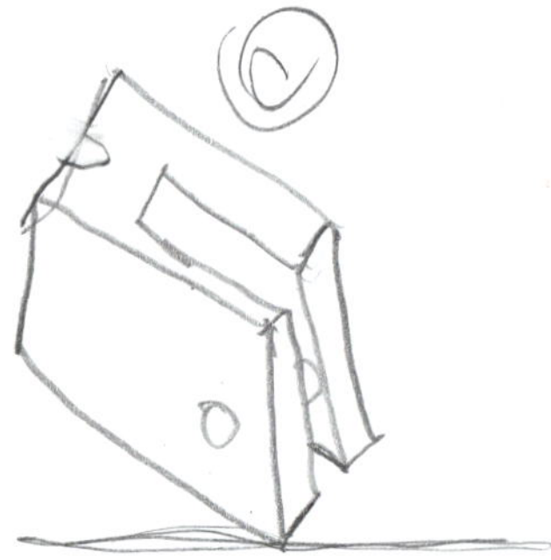

1. Isometric

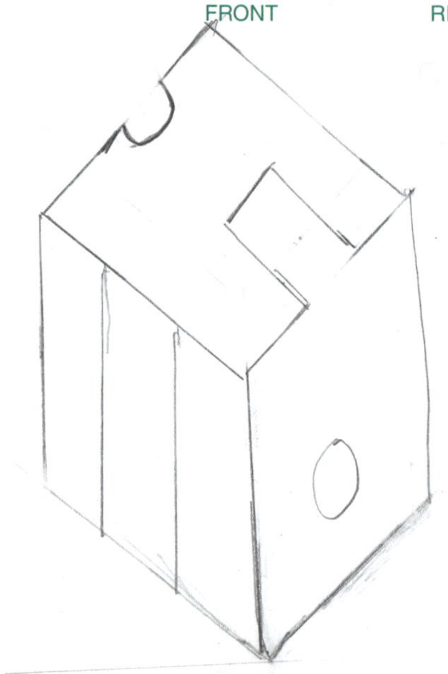

2. Dimetric with axes at the angles indicated below

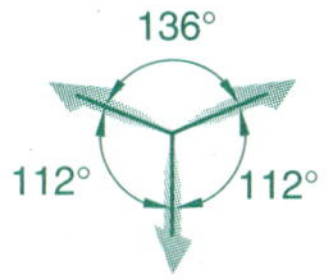

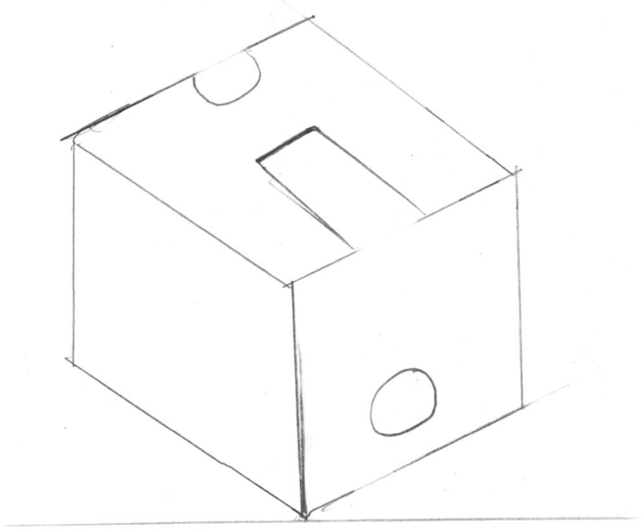

3. Trimetric with axes at the angles indicated below

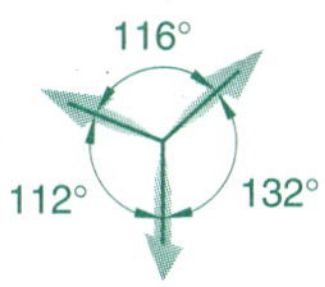

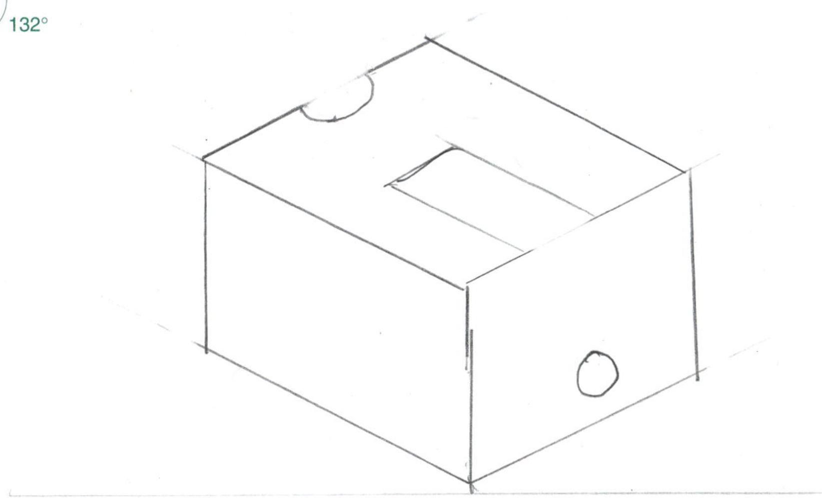

4. Cavalier oblique with receding lines at a 30° angle

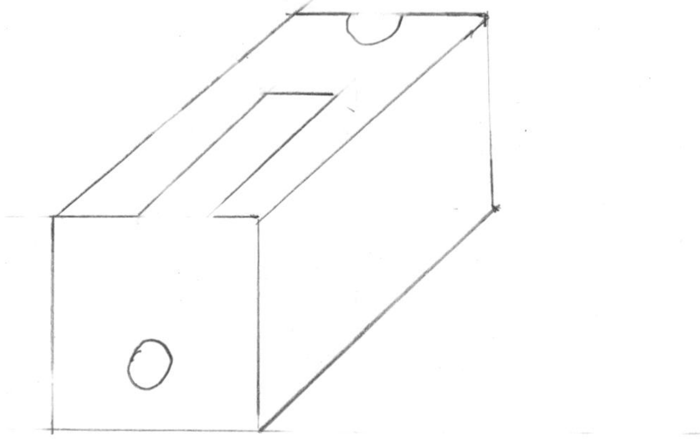

5. Cabinet oblique with receding lines at a 45° angle

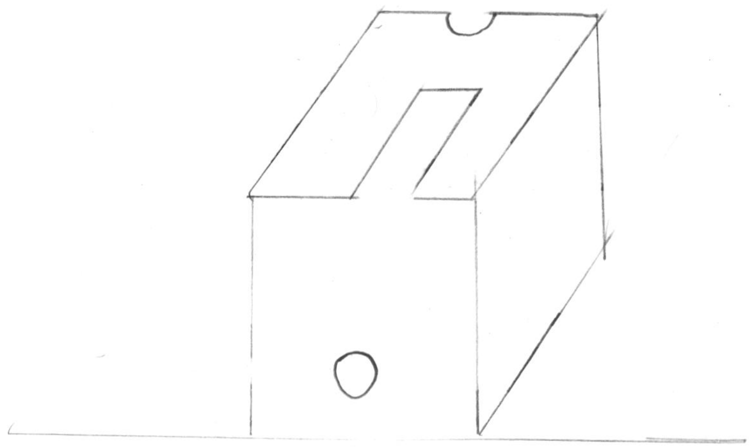

5 Multiview, Auxiliary View, and Sectional View Drawings

MULTIVIEW DRAWINGS

Multiview drawings are two-dimensional representations of three-dimensional objects that are shown in true view. Multiview drawings are necessary to give a complete picture of an object by showing multiple views in detail and without distortion. These drawings use orthographic projection to form views. Each view has a relationship with each of the other views and should be placed on a drawing using specific guidelines.

Orthographic Projection

Orthographic projection is a method of representing the true shape of one view of an object onto a single plane. A *plane of projection* is the plane upon which the object is projected. In orthographic projection, a plane of projection is placed parallel to each plane of an object to show each surface in true view. **See Figure 5-1.**

The *point of view* is the viewing position of the observer when viewing an object. The point of view can be visualized as a point perpendicular to the plane of projection. A *projector* is a line extending from an edge or contour of an object that passes through the plane of projection and the point of view. Projectors run parallel to each other and perpendicular to the plane of projection. All edges and contours of an object are projected onto the plane of projection. Lines of an object that are not visible from the point of view are drawn as hidden lines.

There are six principal planes of projection. Sides and edges of an object that are parallel to one of the principal planes of projection are shown in true view with no foreshortening. Sides and edges that are not parallel to a principal plane of projection are foreshortened. Parallel lines in the object are projected as parallel lines on the projection plane.

View Relationships

The six principal planes of projection form a "glass box" around an object that creates six related views. The panes of the box can be thought of as hinged and can be unfolded to lay the views side-by-side as they would appear in a drawing. The six principal views formed are the front, top, right-side, back, left-side, and bottom views. **See Figure 5-2.**

Multiview Drawings **Media Clip**

TECH TIP

A plane of projection is like a pane of glass that exists between an object and the viewer. The outline of the object on the pane of glass is how the object appears to the viewer. The six principal planes form a "glass box" around an object, and the viewer is always on the outside looking in. An infinite number of projectors can be used to form lines on the pane of glass, outlining every contour of the object and creating a true view of each face.

Orthographic Projection

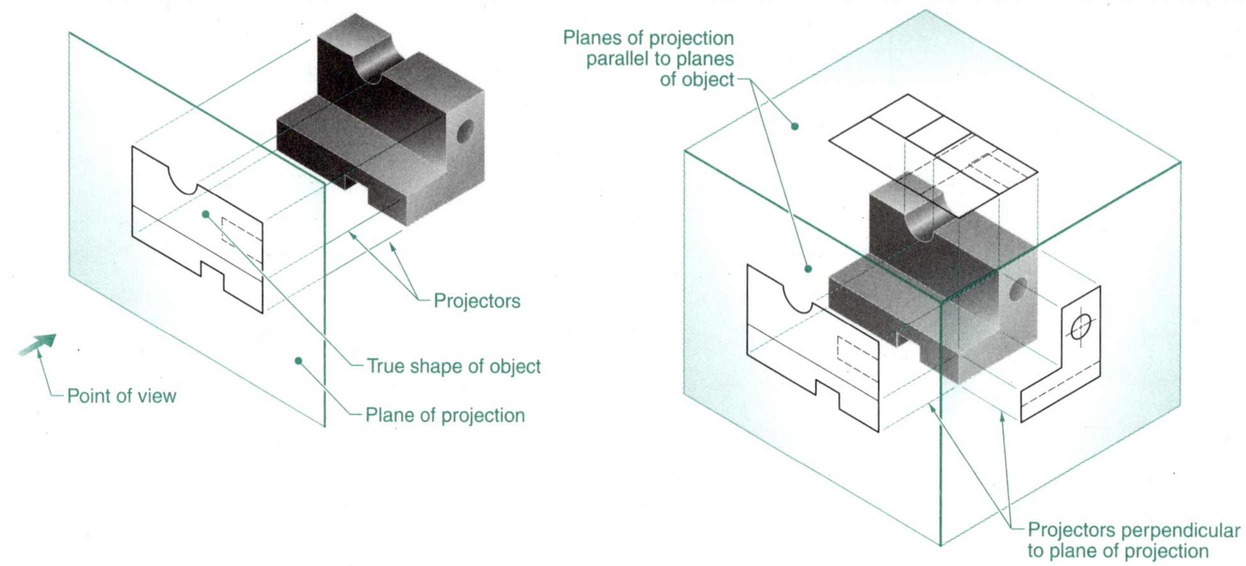

Figure 5-1. Orthographic projection is a method of representing the true shape of one view of an object onto a single plane of projection.

View-Layout Relationships

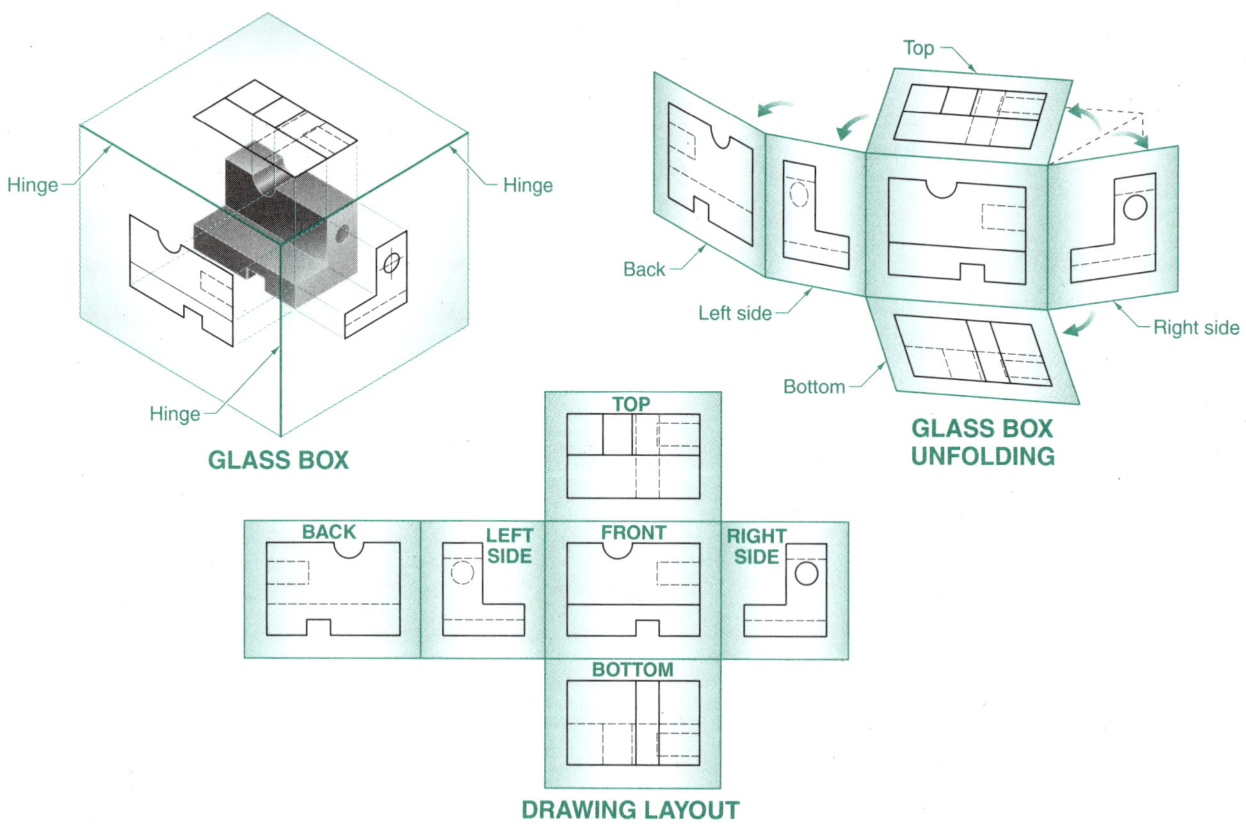

Figure 5-2. The six primary planes of projection create six principal views of an object.

The features and surfaces of each view should be aligned to the same features and surfaces in adjacent views. The six views have direct relationships to each other based on the height, width, and depth dimensions of the object. The back, left-side, front, and right-side views of an object share the same height and line up horizontally. The top, front, and bottom views of an object share the same width and line up vertically. The top, right-side, bottom, and left-side views of an object share the same depth. A 45° miter line can be used with perpendicular lines to form an accurate depth measurement since the views do not line up either horizontally or vertically. **See Figure 5-3.** Views are usually placed on a drawing in the same order to avoid confusion. Views may be presented on a multiview drawing using first-angle projection or third-angle projection. A symbol on the print is often included to indicate the projection method used. **See Figure 5-4.**

First-angle projection is a projection method that places the object between the plane of projection and the observer. First-angle projection shows the top view, front view, and left-side view of an object. First-angle projection is normally used in countries outside the United States and can be identified by the front view being in the upper left-hand corner of the multiview layout.

Projection Methods

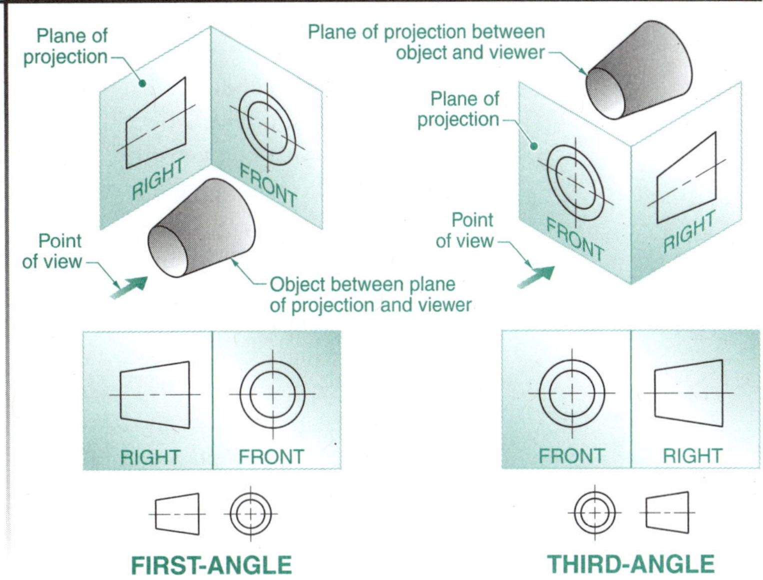

Figure 5-4. Views may be presented on a multiview drawing using first-angle projection or third-angle projection.

View-Size Relationships

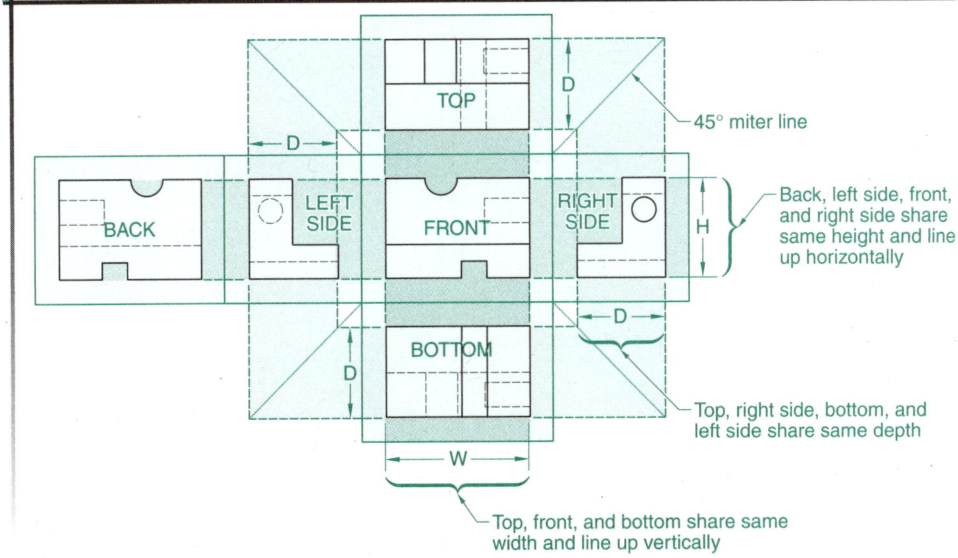

Figure 5-3. The six views of an object have direct relationships to each other based on their height, width, and depth dimensions.

Third-angle projection is a projection method that places the plane of projection between the object and the observer. Third-angle projection shows the top view, front view, and right-side view of an object. Third-angle projection is used in the United States and can be identified by the front view being in the lower left-hand corner of the multiview layout.

In addition to the six principal views, auxiliary views and details can be used to further depict an object. **See Figure 5-5.** An *auxiliary view* is a view that shows the true shape of a plane or feature on an object that is not parallel to one of the six planes of projection. Auxiliary views show the inclined or oblique surface in true view as if the viewing plane is parallel to the auxiliary surface. Auxiliary views are drawn parallel to the surface on the view from which they are projected, or removed and labeled with reference arrows and placed elsewhere on the drawing.

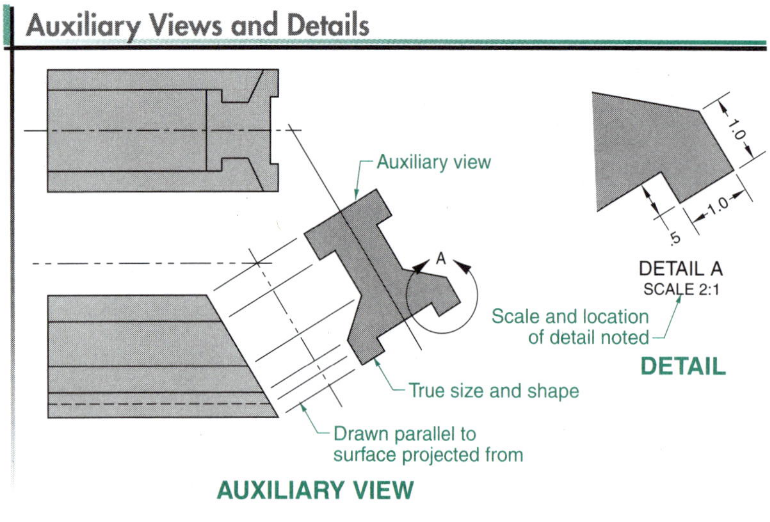

Figure 5-5. Auxiliary views and details can be used to further describe an object.

A *detail* is a portion of a drawing that is enlarged to show features more clearly. Details can be included on a drawing to better illustrate the object. The scale and location on the drawing as a whole should be noted on details.

View Placement

When multiple views of an object are placed on a drawing, the point of view should be identified. Labels, such as front view, top view, or side view, can be used to clearly name each view. View arrangement and reference arrows can also be used to identify views without using labels.

View Arrangement. View arrangement uses standard placement to identify the front, top, and side views for both first-angle and third-angle projections. **See Figure 5-6.** In first-angle projection, the front view is directly above the top view and the left-side view is to the right of the front view. In third-angle projection, the top view is directly above the front view and the right-side view is to the right of the front view.

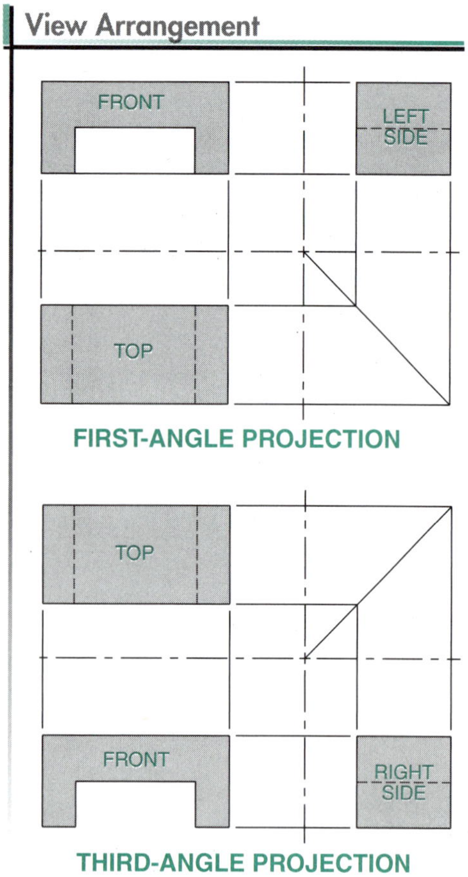

Figure 5-6. View arrangement uses standard placement to identify the front, top, and side views for both first-angle and third-angle projections.

Reference Arrows. Reference arrows indicating the point of view can be used with uppercase letters to identify views. Each individual view is labeled with the letter associated with the proper viewing direction arrow in the multiview. **See Figure 5-7.** This method is useful when space constraints do not allow conventional view arrangement.

Drawing Multiviews

A single view can be used for simple objects, but multiviews are generally drawn with two or more views. Two or more views show the three basic dimensions of height, width, and depth. Most objects are expressed with three views, showing the top, front, and either the left or right side. **See Figure 5-8.** More complex objects may require additional side, bottom, or back views to clearly show all details. Only the views that are unique and that clearly portray the shape of the object should be depicted.

Multiview drawings are based on the conventions described in ANSI/ASME Standard Y14.3, *Multiview and Sectional View Drawings*. Objects can be oriented with any side as the front view. The side that most characteristically shows the shape of the object should be chosen as the front view. The orientation chosen should show the principal features of the object in all selected views with a minimum number of hidden lines.

Most objects require three or more views showing height, width, and depth to accurately depict their shape in a multiview drawing.

Reference Arrows

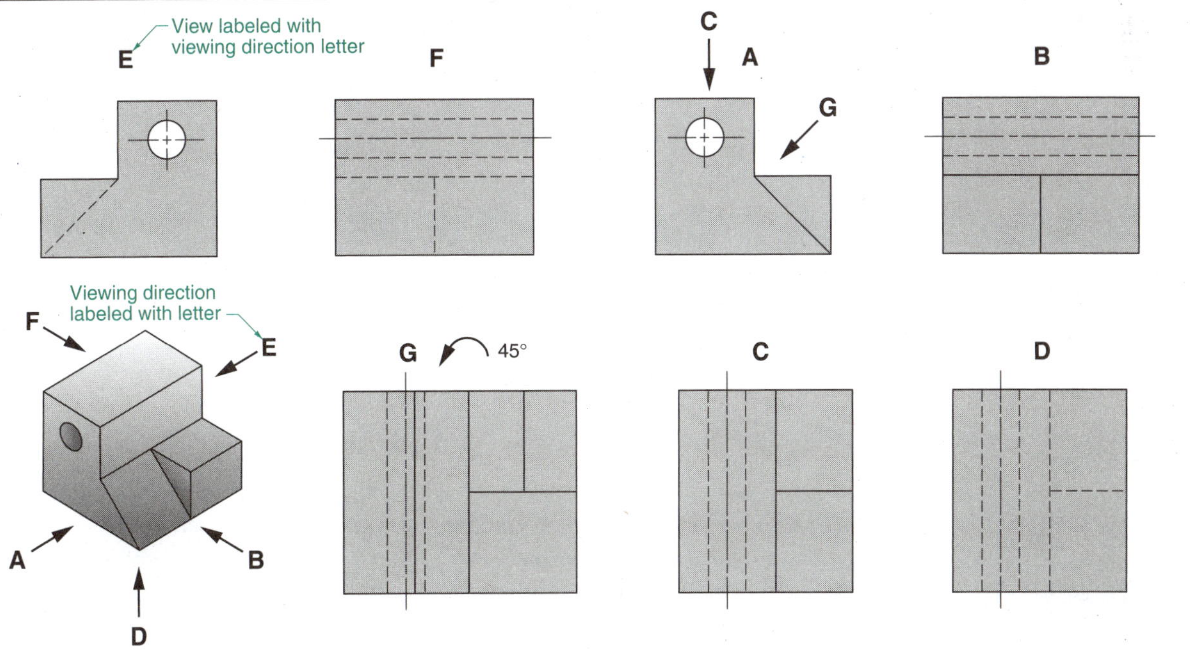

Figure 5-7. Reference arrows indicate points of view and can be used with uppercase letters to identify the different views.

Views

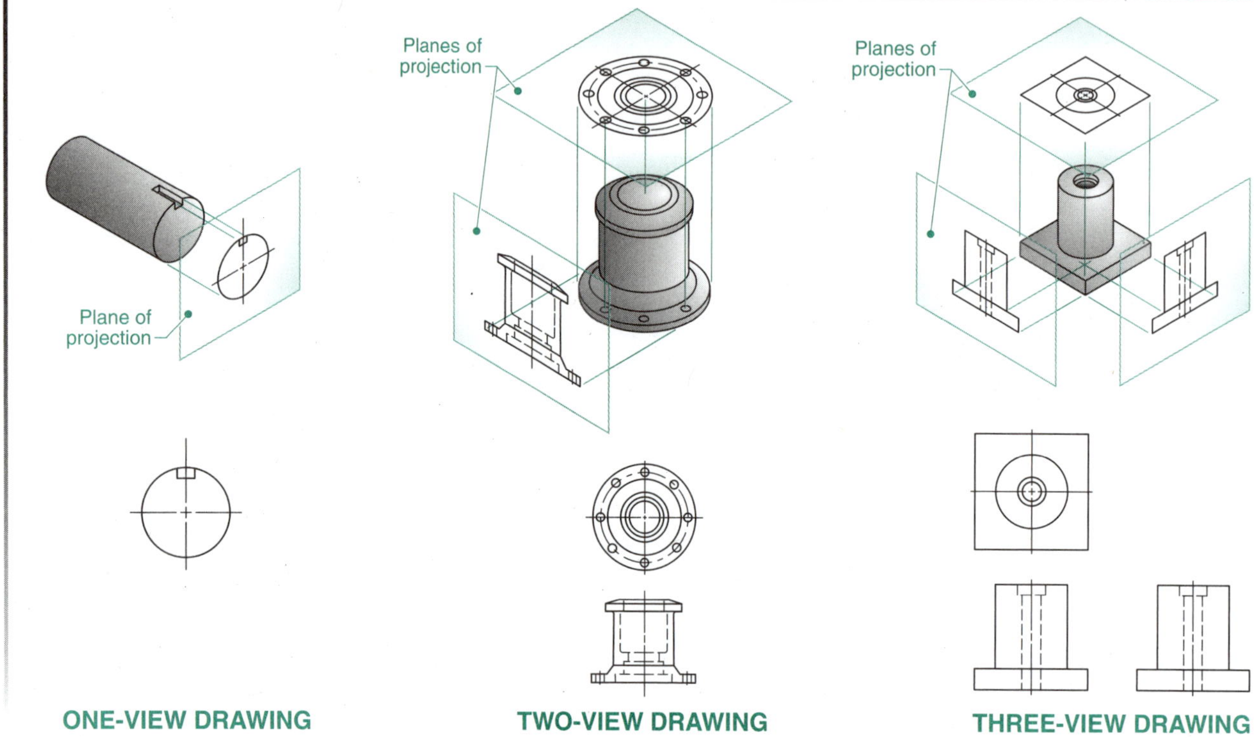

Figure 5-8. Objects are often shown with one view, two views, or three views in multiview drawings.

3.5″ Vise in Print and Drawing Library

Multiviews are often drawn from pictorial drawings. Multiviews that are drawn using first-angle projection are similar to those drawn using third-angle projection. **See Figure 5-9.** To create a multiview drawing using third-angle projection, apply the following procedure:

1. Choose the orientation that best shows the principal features of the object. Block in the width and height dimensions of the front view. *Note:* Place the front view on the drawing paper to allow for even spacing of additional views.
2. Project the width and height dimensions of the front view to the top and right-side views.
3. Establish the depth dimension on the top view. Add a 45° miter line extending from the intersection of the width and height projection lines. From the top view, draw lines extending from the marked depth to the miter line. Use perpendicular lines from the intersection points of the depth lines with the miter line to establish the same depth dimension in the right-side view.
4. Lightly project lines from the remaining features, such as holes or inclined planes, to their appropriate views. Complete the visible lines of the remaining features.
5. Draw hidden lines as required. Add auxiliary views or details as needed.
6. Darken all visible lines to complete the drawing.

Auxiliary Views Media Clip

AUXILIARY VIEW DRAWINGS

Surfaces that are not parallel to the principal planes of projection require auxiliary views to accurately represent the inclined or oblique surface. An auxiliary view has a point of view perpendicular to the auxiliary surface. **See Figure 5-10.** An auxiliary view can be projected from a principal view or from another auxiliary view. The two types of auxiliary views are primary and secondary auxiliary views. Objects in multiview drawings are usually oriented with primary or secondary auxiliary planes bordering the front, top, or side principal plane.

TECH TIP

Objects that have drastically different shapes may have the same true view of one or more sides. Even if an object's shape is shown in perfect outline, it may be difficult to decipher the exact shape of the object. It is important to show all views in order to give a clear indication of the actual shape of the object, thus avoiding confusion.

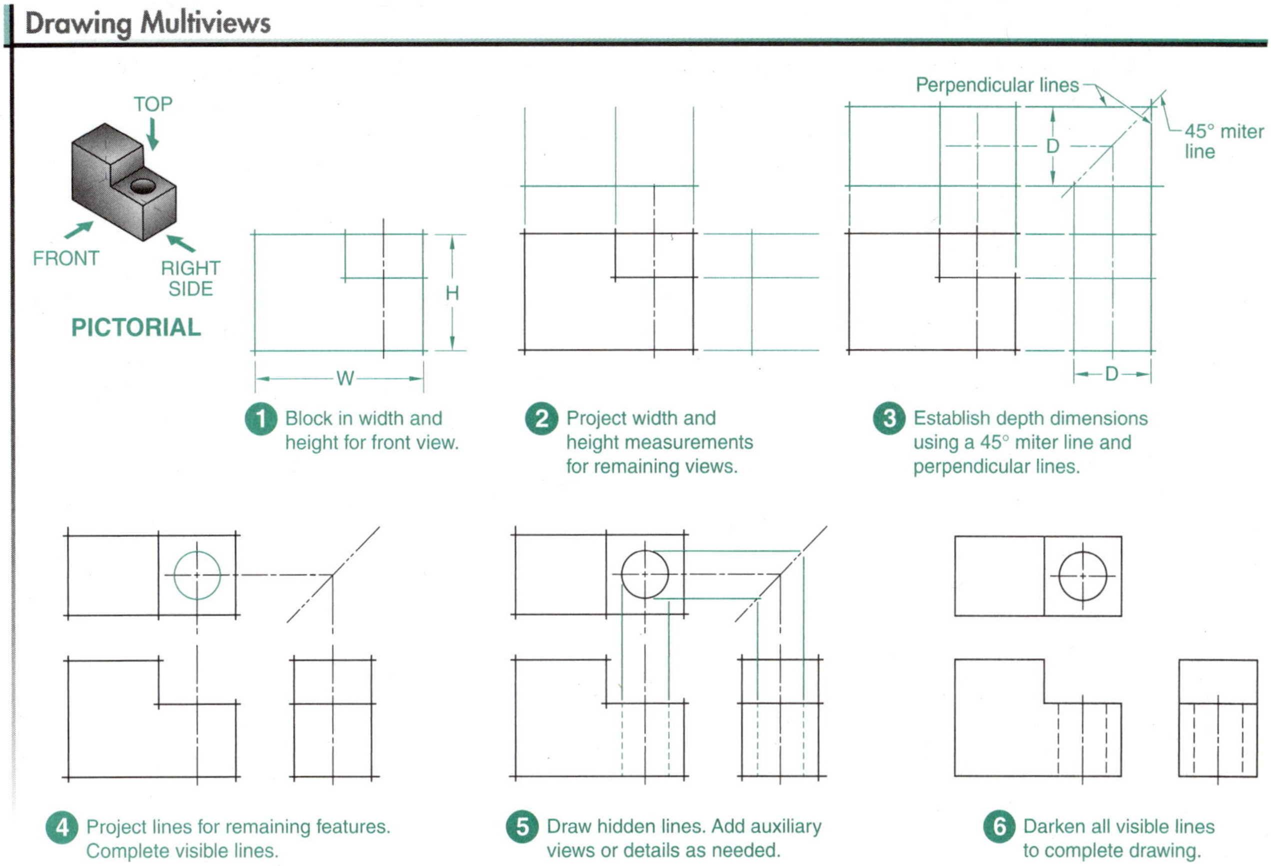

Figure 5-9. After selecting an orientation and the desired views for an object, a multiview can be drawn using the proper procedure.

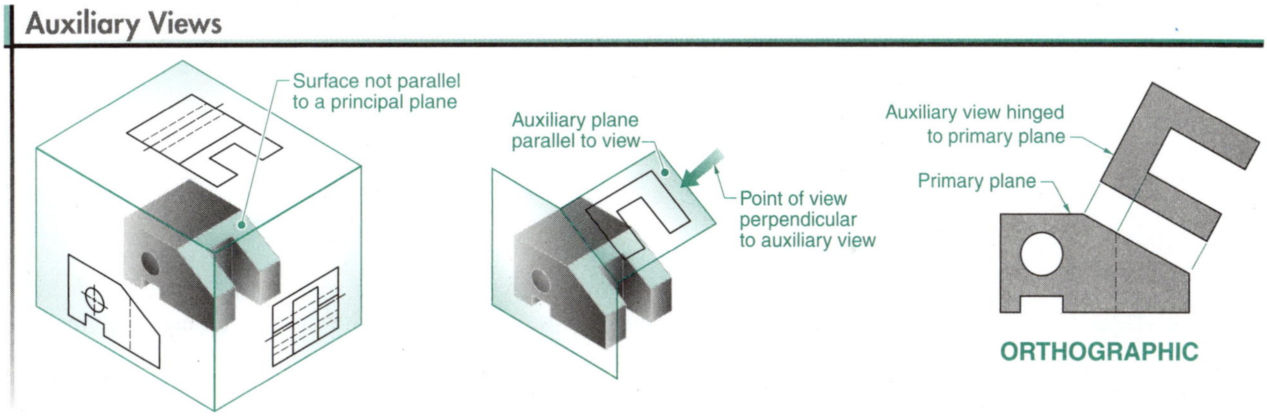

Figure 5-10. Auxiliary views are used to accurately represent an inclined or oblique surface.

Primary Auxiliary Views

A *primary auxiliary view* is a view that is projected to an auxiliary plane that is perpendicular to one of the principal planes but inclined to the remaining planes. The auxiliary plane is parallel to the inclined surface and reflects its true shape. This auxiliary plane is hinged to the principal plane that it is perpendicular to and is unfolded to be seen in true view. A primary auxiliary view is always projected from one of the principal planes in a multiview. **See Figure 5-11.**

Auxiliary views are considered front, top, or side depending on which plane the auxiliary surface is perpendicular to and hinged to. A *front auxiliary view* is a primary auxiliary view that has the inclined surface perpendicular to the front principal plane. A *top auxiliary view* is a primary auxiliary view that has the inclined surface perpendicular to the top principal plane. A *side auxiliary view* is a primary auxiliary view that has the inclined surface perpendicular to the side principal plane.

If the auxiliary view provides a complete understanding of the object and its dimensions, a principal view can be omitted from the multiview drawing. A partial auxiliary view or a symmetrical auxiliary view can also be used to simplify a drawing while still showing the required information. **See Figure 5-12.**

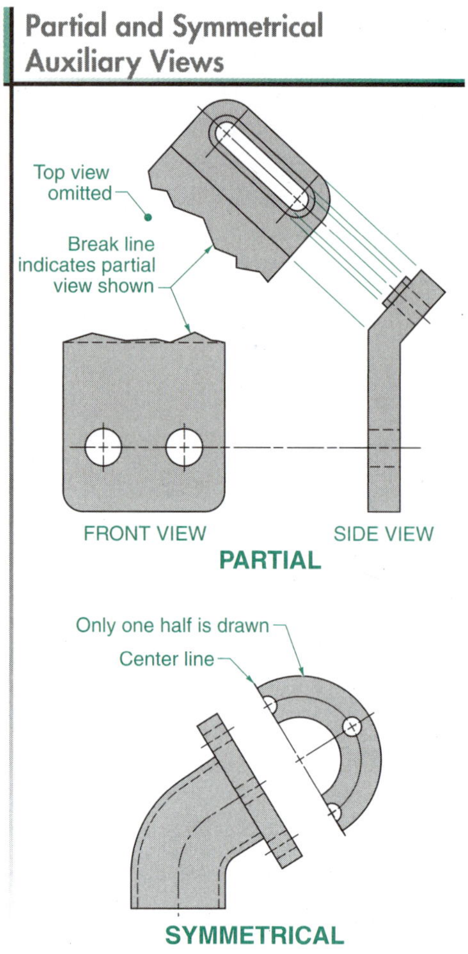

Figure 5-12. A partial or symmetrical auxiliary view can be used to simplify a drawing while still showing the required information.

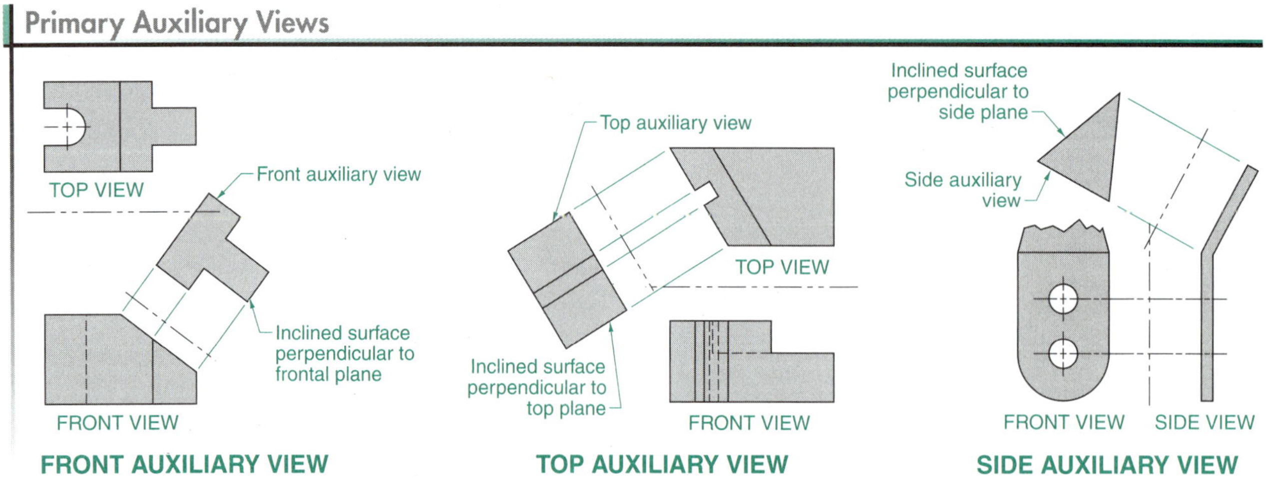

Figure 5-11. A primary auxiliary view is projected to an auxiliary plane that is perpendicular to one of the three principal planes and is inclined to the other two.

Partial Auxiliary Views. A partial auxiliary view shows only the true view of the inclined surface and not the remaining planes as they reflect on the auxiliary plane. The remaining planes are distorted and foreshortened on the auxiliary plane, and their depiction often reduces the clarity of the auxiliary view. A break line can also be used to indicate that only a portion of the auxiliary view from the auxiliary plane is shown. Partial auxiliary views are commonly used.

Symmetrical Auxiliary Views. If an auxiliary view is symmetrical, only half of the auxiliary view needs to be drawn. The auxiliary view is developed from the right or left of the centerline, but often shows the half of the view that is closest to the primary views of the object.

Secondary Auxiliary Views

When the true shape of a surface cannot be shown in a primary auxiliary view, a secondary auxiliary view is necessary. A *secondary (double) auxiliary view* is a view that is projected onto a plane that is oblique to all of the principal planes. A secondary auxiliary view is always projected from a primary auxiliary view. **See Figure 5-13.**

Secondary Auxiliary Views

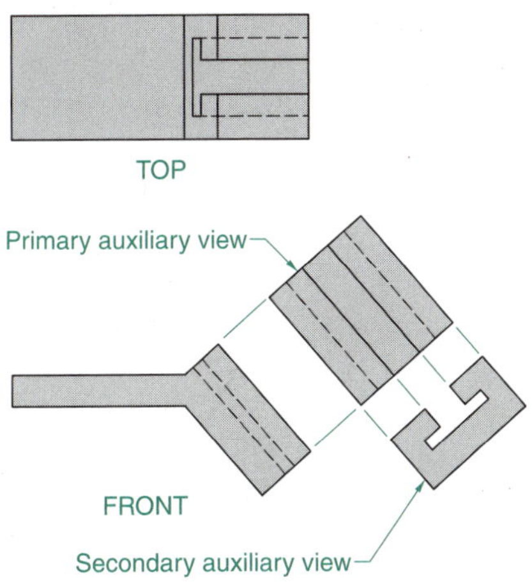

Figure 5-13. A secondary (double) auxiliary view is projected onto a plane that is oblique to all of the principal planes.

Drawing Auxiliary Views

An object requiring an auxiliary view is drawn by starting with the multiview for the object. If the auxiliary view to be drawn will provide the necessary information, all or a portion of a principal view can be omitted. When drawing a primary auxiliary view, projection lines are drawn perpendicular to the inclined surface and the surface is drawn in true view.

An auxiliary view should be placed on a drawing aligned with the principal or auxiliary view it is perpendicular to and hinged to. A partial or symmetrical auxiliary view can be used if desired. If space is an issue, the auxiliary view may be moved to a more suitable place on the drawing if labeled properly.

To draw a primary auxiliary view, complete the following procedure:

1. Draw a multiview of the object. **See Figure 5-14.**
2. Locate the auxiliary plane edge on the principal view that is perpendicular to the auxiliary plane. Draw projection lines perpendicular to the auxiliary plane edge and in the direction of sight. Draw a reference line parallel to the auxiliary plane edge and perpendicular to the projection lines to signify the auxiliary projection plane.
3. Transfer the depth measurement from one of the principal views and mark the depth for the auxiliary view starting from the reference line. Draw the auxiliary view in true shape oriented with the lines of sight and projection plane line. *Note:* Draw only the inclined surface as the auxiliary view.
4. Complete any remaining features, hidden lines, or other elements that are needed to fully portray the object.
5. Darken all visible lines.

TECH TIP

Auxiliary views are always drawn as if the observer is looking straight down on the auxiliary surface from above. This allows the inclined surface to be drawn in true view. Typically, only the inclined surface is projected onto the auxiliary plane and drawn in the auxiliary view. Other views that are not inclined, and therefore not parallel to the auxiliary plane, appear foreshortened and are not in true shape.

Drawing Primary Auxiliary Views

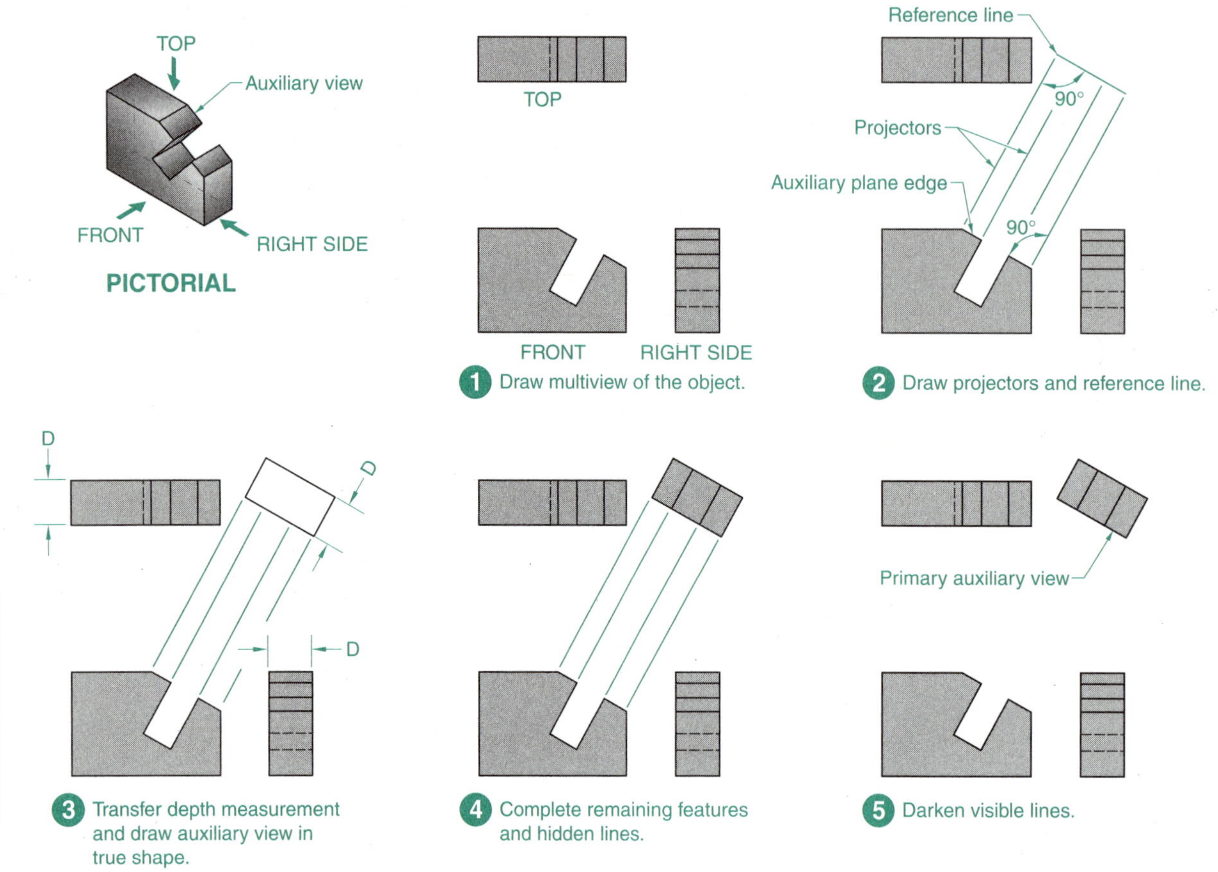

Figure 5-14. Primary auxiliary views are drawn by projecting from the primary plane perpendicular to the auxiliary plane.

A secondary auxiliary view can be drawn if a primary auxiliary view does not depict all surfaces in true view. The primary auxiliary view should be drawn first. The second auxiliary view should be aligned with perpendicular projectors from the primary auxiliary view. Secondary auxiliary views are drawn using the same procedure as primary auxiliary views, but by projecting from the auxiliary view instead of a principal view. **See Figure 5-15.**

SECTIONAL VIEW DRAWINGS

Sectional Views Media Clip

A *sectional view,* or *section,* is a drawing of a cross section of an object. A sectional view is the orthographic representation of a pictorial cutaway drawing. Sectional view drawings can show complex objects that have complicated or internal shapes that would require a large number of hidden lines on a multiview drawing. A sectional view should always be bound by a visible line. Hidden lines are usually omitted from sectional views, unless needed for clarity or dimensioning. **See Figure 5-16.**

A sectional view reveals the internal shape of an object while retaining the exterior outline. A sectional view is created by passing an imaginary cutting plane through an object and removing the cut portion. Cutting-plane lines are used to indicate the location where the cutting plane passes through the object, while section lines are used to indicate the portion of the object that has been cut into parts with the cutting plane.

Drawing Secondary Auxiliary Views

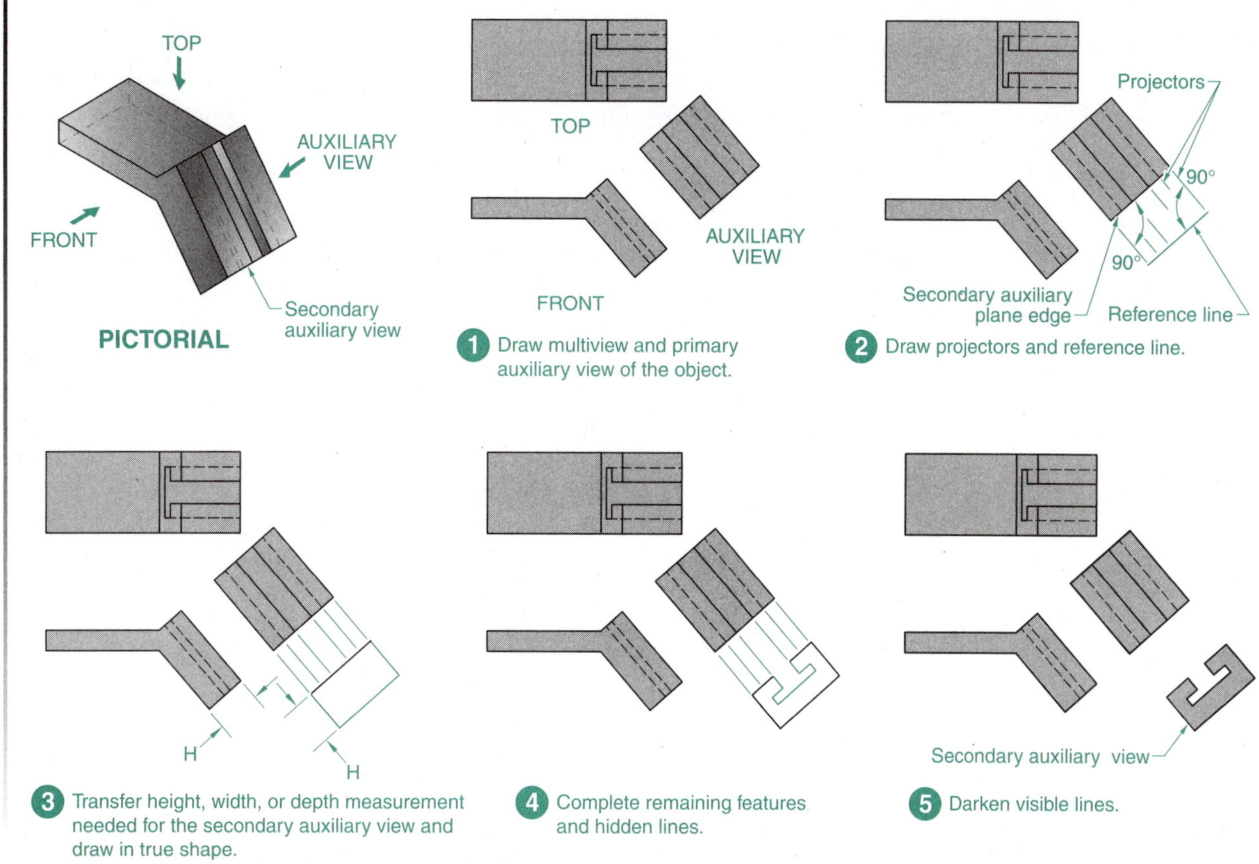

Figure 5-15. Secondary auxiliary views are drawn by projecting from a primary auxiliary view.

Sectional View Drawings

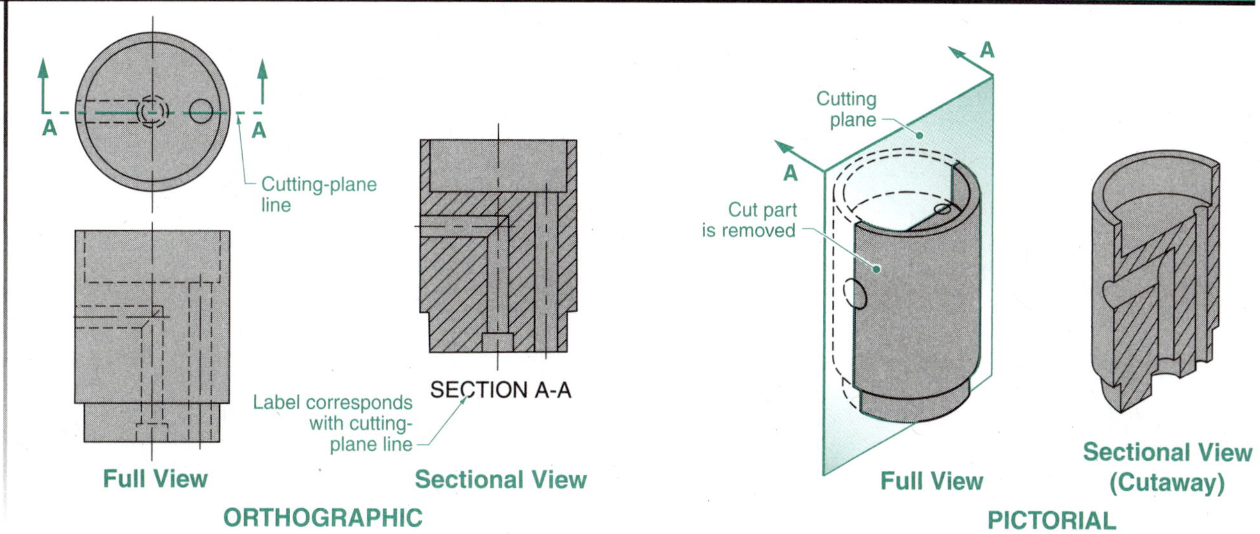

Figure 5-16. A sectional view is an orthographic representation of a cross section of an object cut by a cutting plane.

Sectional in Print and Drawing Library

Cutting-Plane Lines

The cutting plane is shown on the orthographic view with a cutting-plane line. A cutting-plane line is a thick line with ¾″ to 1½″ long dashes and approximately ⅛″ short dashes. All spaces between dashes are approximately 1/16″. Arrowheads point in the direction of sight in which the object is viewed when the sectional view is made. **See Figure 5-17.** If the cutting plane passes through the center of the object or the axis of symmetry, the cutting-plane line can be omitted.

Uppercase letters such as A-A, B-B, C-C, and so on can be used on a drawing to identify the section's location on the object. The letters are placed adjacent to the arrowheads and read horizontally. The section view is marked with the corresponding letter, such as SECTION A-A.

Section letters are used in alphabetical order, omitting the letters I, O, Q, S, X, and Z, which can be confused as numbers. If more than 20 sections are used, the additional sections are indicated by double letters in alphabetical order: AA-AA, BB-BB, and so on. If two or more sections appear on the same sheet, they are generally arranged in alphabetical order from left to right if possible. If only one section is being drawn for an object, letter labels can be omitted.

Section Lines

Section lines are often used to indicate that an object has been sectioned. Section lines consist of thin, evenly spaced diagonal lines spaced at least 1/16″ (1.6 mm) apart. To clearly define the surfaces of a section, thin lines are drawn across the cut area at a 45° angle to the horizontal. **See Figure 5-18.**

If the section consists of two adjacent parts, such as in an assembly drawing, the section lines should run in opposite directions in order to provide contrast. When three parts form the section, the third part adjacent to the first two should have lines sloping at 30° or 60° to the main outline of the view. For additional adjacent parts, the lines may be drawn at any suitable angle where each part stands out separately and clearly.

Cutting-Plane Lines and Labels

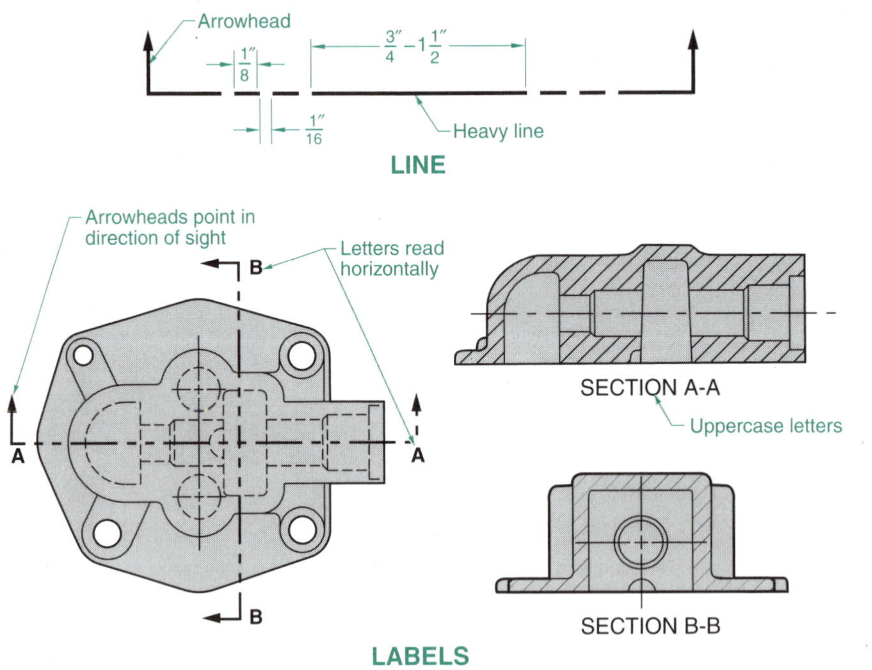

Figure 5-17. The cutting plane is shown with a cutting-plane line and is often labeled with uppercase letters to identify the section's location on the object.

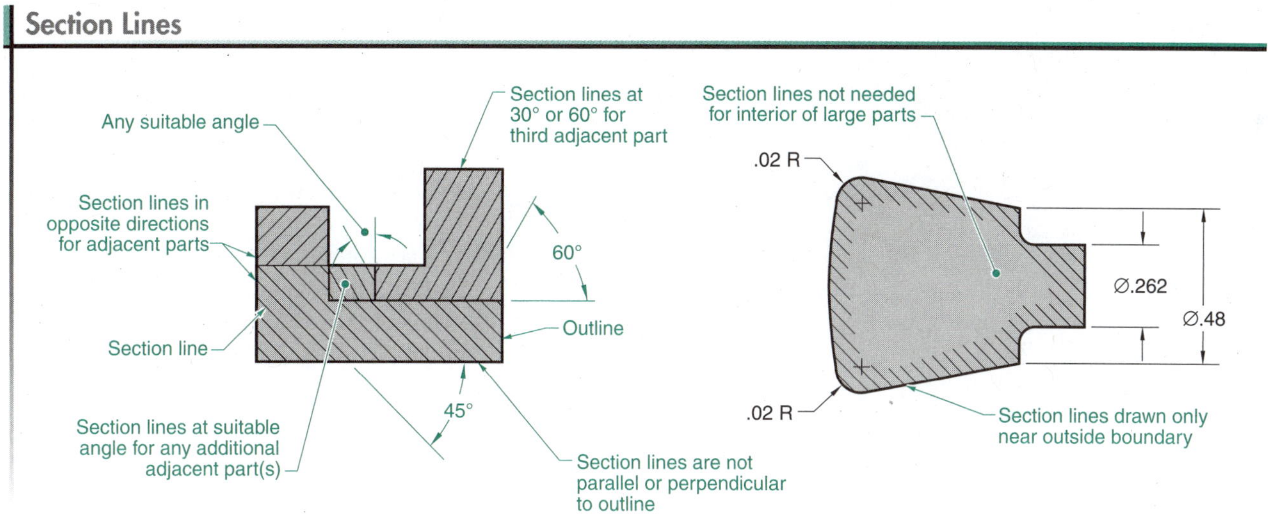

Figure 5-18. Section lines are often used to indicate that an object has been sectioned.

Drawing section lines parallel or nearly parallel to the dominant visible lines of the section should be avoided. All the sectioned areas of a single part should be lined in the same angle of slope. In sectioning very large areas, it is permissible to use section lines only near the adjacent boundary of the sectioned area with the interior portion left clear.

Sectional View Drawing Types

Depending on what is being shown, different types of sectional view drawings can be used. Sectional view drawings can include full sections, half sections, offset and aligned sections, broken-out sections, revolved sections, removed sections, auxiliary sections, and thin sections. Sectional views are drawn based on the conventions described in ANSI/ASME Standard Y14.3, *Multiview and Sectional View Drawings*.

Full Sections. A *full section* is a sectional view in which the cutting plane passes entirely through an object. The half of the object between the observer and the cutting plane is considered removed, and the remaining full section is exposed to view. **See Figure 5-19.** Full sections are commonly used.

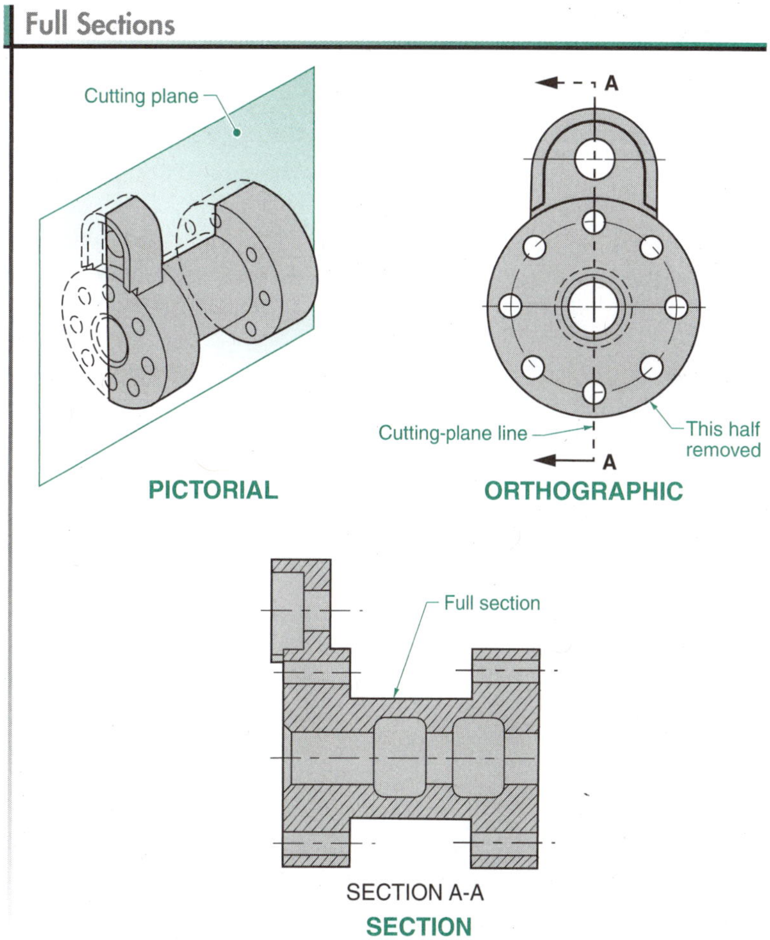

Figure 5-19. A full-section view is cut with a cutting plane that passes entirely through an object.

Boston Gear

Half sections are a convenient way to show both the interior and exterior of an object in a multiview drawing.

Half Sections. A *half section* is a sectional view in which two cutting planes are passed at right angles to each other along center lines or symmetrical axes. The cutting planes permit one-quarter of the object to be removed, and a half section of the interior is exposed to view. **See Figure 5-20.**

A half section has the advantage of showing the interior and exterior shapes of the object at the same time. Half sections are only used with symmetrical objects. A *symmetrical object* is an object in which one half is the mirror image of the other half.

Offset and Aligned Sections. An *offset section* is a sectional view in which the cutting-plane line is offset from the main axis to pass through features that are not located in a straight line. By offsetting the cutting plane in several places, pieces that normally would not be seen with a regular full section can be viewed. Offsets are 90° and are shown in the view indicating the cutting plane. Offsets are not shown in the sectional view. **See Figure 5-21.**

Aligned sections are similar to offset sections. An *aligned section* is a sectional view in which the cutting plane is bent to pass through angled features. These features are then rotated to align in a straight line. Aligned cutting planes should maintain an obtuse angle and be bent less than 90° from a straight line.

Broken-Out Sections. A *broken-out section* is a sectional view in which a small portion is removed and is designated by a freehand break line. By removing only a small portion, it is possible to preserve exterior detail of the object that would otherwise be eliminated in a full or half section. No cutting-plane line is necessary to show a broken-out section. **See Figure 5-22.**

Half Sections

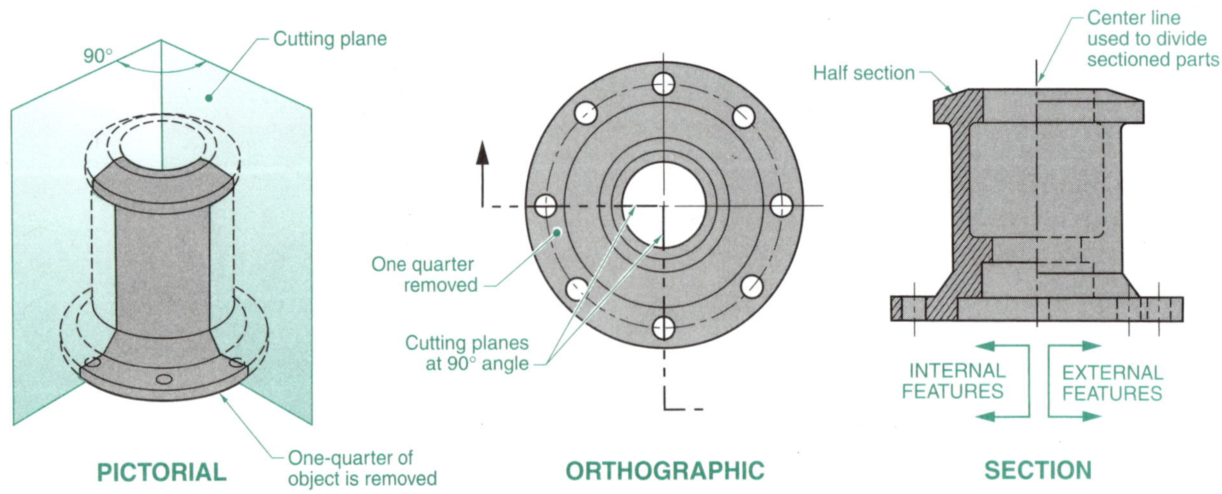

Figure 5-20. A half-section view has two cutting planes that pass at right angles to each other, removing one-quarter of a symmetrical object to show interior and exterior features.

Offset and Aligned Sections

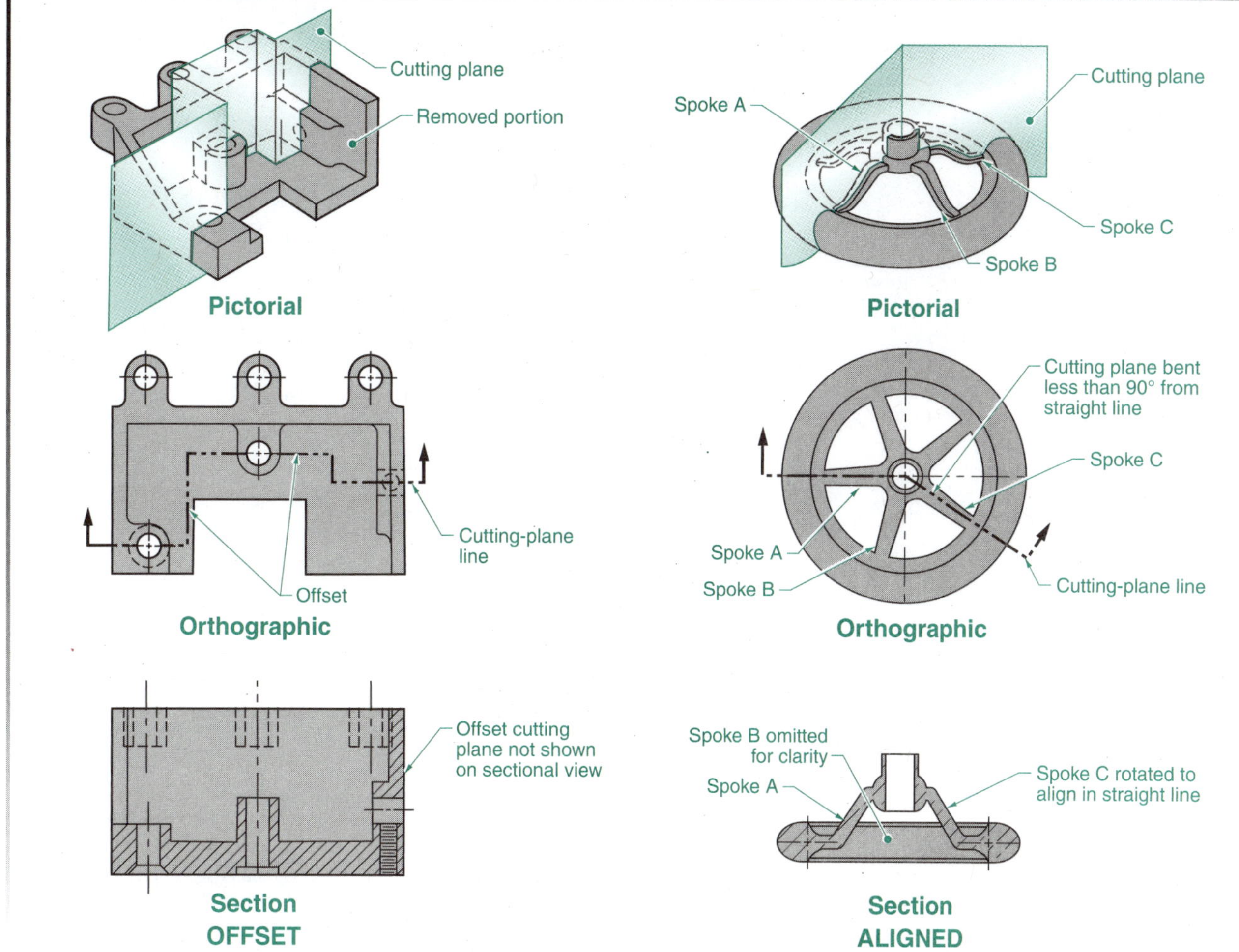

Figure 5-21. Offset and aligned sectional views include a cutting-plane line that is bent to include features that are not located in a straight line.

Revolved Sections. A *revolved section* is a sectional view that shows the cross-sectional shape of elongated parts. The cutting plane is passed perpendicular to the axis of the piece and then revolved in place 90°. No cutting-plane line is necessary to show a revolved section. The true shape of the exposed revolved section should always be retained. The visible lines of nearby views may be removed and broken lines added to clarify the revolved section. **See Figure 5-23.**

TECH TIP

Revolved, or rotated, sections show a portion of an object that is revolved in place. The cutting plane is assumed to be perpendicular to the axis of the feature to be revolved.

Broken-Out Sections

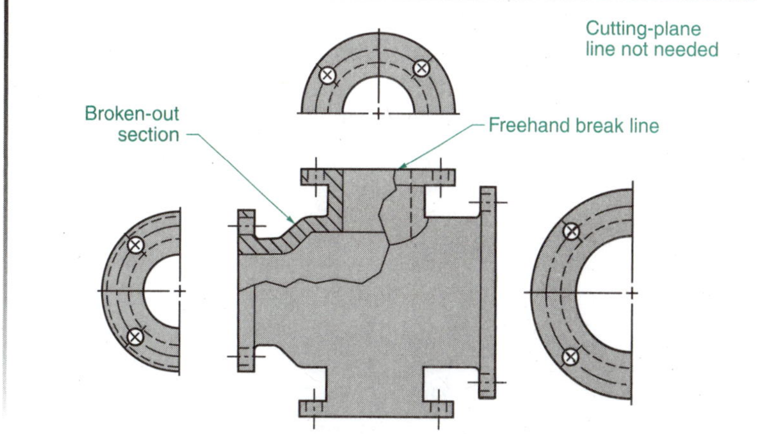

Figure 5-22. A broken-out sectional view has a small portion of the object removed as designated by a freehand break line.

Revolved Sections

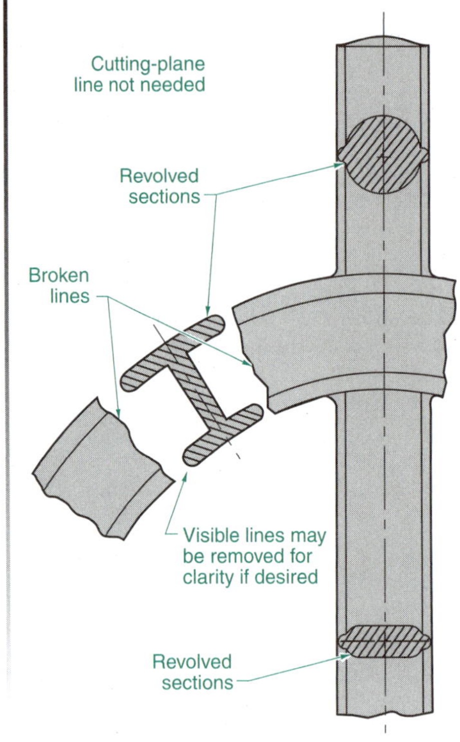

Figure 5-23. A revolved sectional view shows the cross-sectional shape of elongated parts by passing the cutting plane perpendicular to the axis of the piece and revolving it in place 90°.

Removed Sections

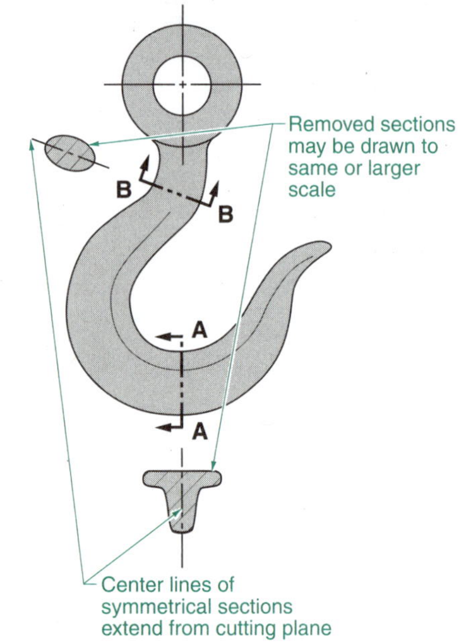

Figure 5-24. A removed sectional view is detached from the projected view and located elsewhere on the sheet.

Removed Sections. A *removed section* is a sectional view that is detached from the projected view and located elsewhere on the sheet. By removing the section, the regular view can be left intact and the removed section can be drawn at a larger scale. If a different scale is used for the removed section, the scale should be noted. A removed section should be labeled with a corresponding letter and cutting-plane line, unless it is clearly identifiable. **See Figure 5-24.**

Auxiliary Sections. An *auxiliary section* is a sectional view that is not one of the principal planes. An auxiliary section may be full, half, broken-out, removed, or revolved. The section should be shown in its normal auxiliary position and clearly identified with a cutting plane and appropriate letters. **See Figure 5-25.**

Auxiliary Sections

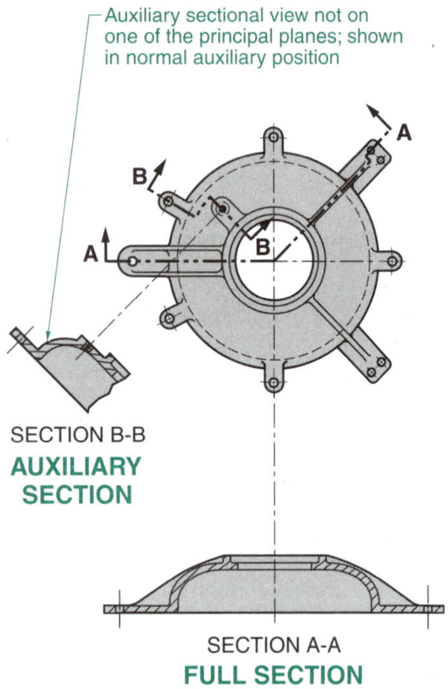

Figure 5-25. An auxiliary sectional view is not shown in one of the principal planes and should be clearly identified.

TECH TIP

Removed sections are similar to revolved sections, except they depict the revolved view elsewhere on the drawing. The cutting plane is indicated and labeled on removed views.

Thin Sections. A *thin section* is a sectional view that is too thin to be shown by the ordinary cross-sectioning convention. **See Figure 5-26.** Cross sections of thin material are drawn in solid black. If two or more pieces are adjacent to each other, a white space is left to separate the parts.

Thin Sections

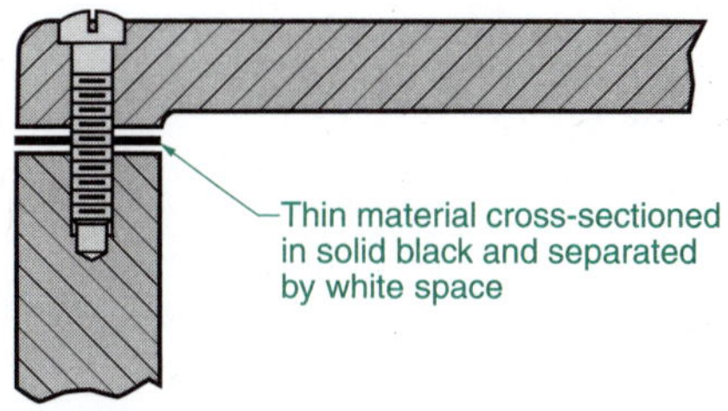

Figure 5-26. A thin sectional view is too thin to be shown by the ordinary cross-sectioning convention and is drawn as solid black.

Chapter 5 Quick Quiz®

Technical Sketching and Drawing

Multiview, Auxiliary View, and Sectional View Drawings

5 Review

Name ______________________________ Date ______________

REVIEW QUESTIONS

______________ **1.** ___ projection is a method of representing the true shape of one view of an object onto a single plane.

T F **2.** A plane of projection is placed perpendicular to each plane of an object to show each surface in true view.

______________ **3.** The ___ is the viewing position of the observer when viewing an object.

______________ **4.** The principal planes of projection form a "glass box" around an object that creates ___ related views.

T F **5.** The back, left-side, front, and right-side views of an object share the same height and line up horizontally.

______________ **6.** A(n) ___° miter line can be used with perpendicular lines to form an accurate depth measurement in multiview drawings.

______________ **7.** ___-angle projection places the object between the plane of projection and the observer.

______________ **8.** ___-angle projection is used in the United States.

______________ **9.** A(n) ___ view shows the true shape of a plane or feature on an object that is not parallel to one of the six planes of projection.

T F **10.** In third-angle projection, the top view is directly above the front view and the right-side view is to the right of the front view.

______________ **11.** In a multiview drawing, ___ or more views show the basic dimensions of height, width, and depth.

______________ **12.** A(n) ___ auxiliary view is always projected from one of the principal planes in a multiview.

T F **13.** A top auxiliary view has the inclined surface perpendicular to the front principal plane.

______________ **14.** A(n) ___ auxiliary view shows only the true view of the inclined surface and not the remaining planes as they reflect on the auxiliary plane.

T F **15.** A secondary auxiliary view is never projected from a primary auxiliary view.

______________ **16.** A(n) ___ view is a drawing of a cross section of an object.

______________ **17.** Sectional views should always be bound by a(n) ___ line.

______________ **18.** A(n) ___ line is used to indicate the location where the cutting plane passes through an object.

T F **19.** If the cutting plane passes through the center of the object or the axis of symmetry, the cutting-plane line can be omitted.

T F **20.** All the sectioned areas of a single part should be lined in the same angle of slope.

IDENTIFICATION – Principal Views

______ **1.** Front

______ **2.** Right side

______ **3.** Left side

______ **4.** Bottom

______ **5.** Top

______ **6.** Back

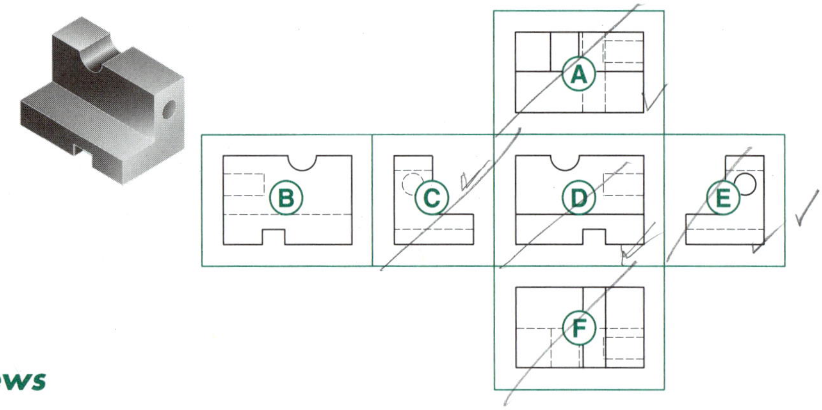

IDENTIFICATION – Sectional Views

______ **1.** Broken-out section

______ **2.** Full section

______ **3.** Removed section

______ **4.** Revolved section

______ **5.** Half section

______ **6.** Thin section

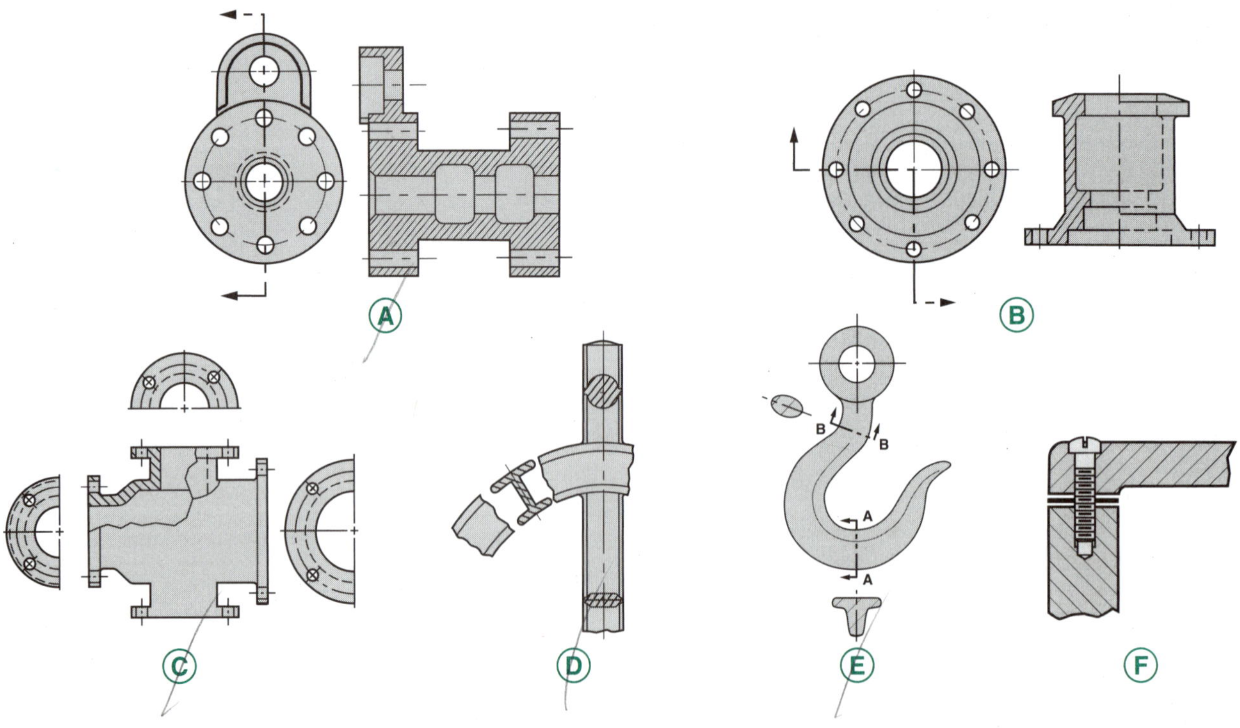

DIMENSIONING

Dimensioning is a method of adding dimensions to a drawing to indicate the geometrical characteristics of an object. A *dimension* is a numerical value that gives the size, form, or location of objects on a drawing. A *size dimension* is a dimension that gives the overall size of an angle or feature. A *location dimension* is a dimension that uses the components of an object to locate an angle or feature on an object. Dimensions are generally shown in combination with lines, symbols, and notes to include all information to fully describe an object. Dimensions should be placed on a drawing so that they allow only one interpretation and include no repetition.

Indicating Dimensions

Lines are used on a drawing to indicate the extent or area of an object being dimensioned. Lines used in dimensioning are thinner than object lines so that they will be visibly different from the lines in the object being drawn. Dimension lines, arrowheads, extension lines, and leaders are used with numbers to apply dimensions to an object. **See Figure 6-1.**

Dimension Lines. A *dimension line* is a line that is used with dimensions to show the size or location. Dimension lines are commonly broken for the placement of numerals giving the measurement. **See Figure 6-2.** If a horizontal dimension line is not broken, the numeral is placed above the dimension line.

Dimensioning Media Clip

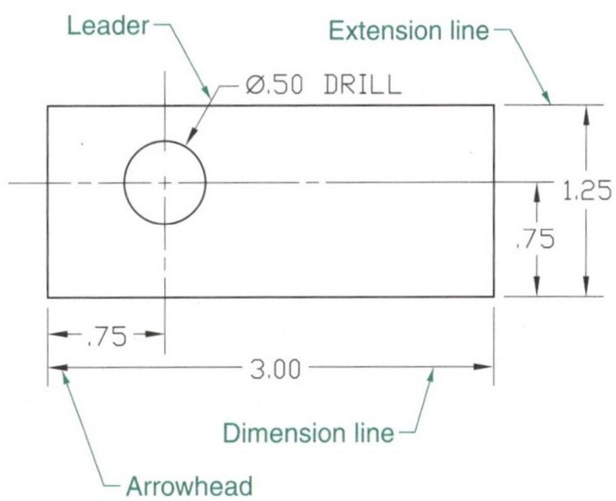

Figure 6-1. Dimension lines, arrowheads, extension lines, and leaders are used with numbers to apply dimensions to an object.

When placing dimension lines, the dimension line nearest the object being dimensioned should be at least ⅜″ (9 mm) from the object. Succeeding dimension lines should be no closer than ¼″ (6 mm) to the previous dimension line. Dimensions are aligned for appearance. When parallel dimension lines are used, the dimensions are staggered to avoid crowding. Dimension lines should not cross other dimension lines unless this is unavoidable.

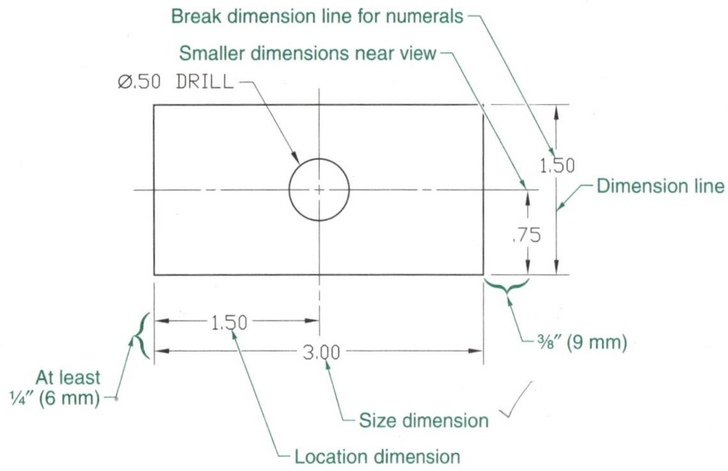

Fig[illegible] A dimension line is a line that is used with dimensions to show size of [illegible] is commonly broken for the placement of numerals.

Arrowheads. An *arrowhead* is a symbol that indicates the extent of a dimension. Arrowheads terminate dimension lines. Arrowheads should be drawn three times as long as they are wide with the width proportional to the thickness of the line. Arrowheads may be open or closed. When there is not adequate space, arrowheads may be placed outside the indicated dimension. **See Figure 6-3.**

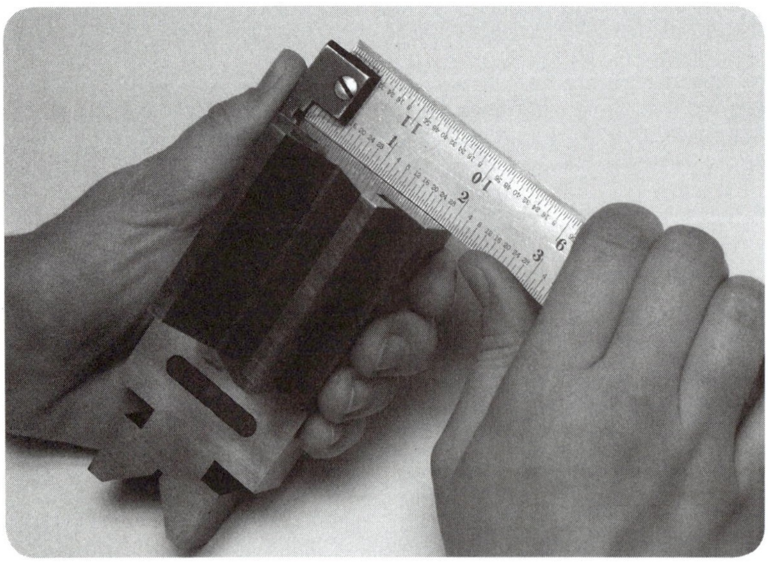

Linear dimensions are used to define an object and may be expressed in decimals or fractions.

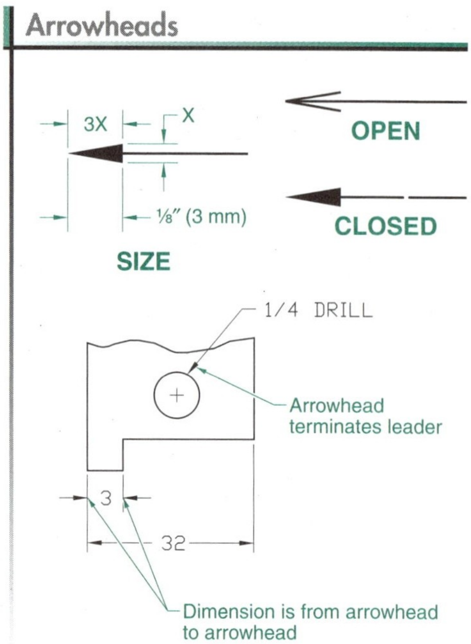

Figure 6-3. An arrowhead is a symbol that indicates the extent of a dimension and terminates dimension lines.

Extension Lines. An *extension line* is a line that extends from a surface feature of an object and ends a dimension line. Extension lines should be placed to leave a gap of approximately 1⁄16″ (1.6 mm) between the extension line and the object or feature from which they are extended. Extension lines extend a short distance beyond the dimension line. Generally, extension lines should not cross other extension lines or dimension lines. **See Figure 6-4.**

Leaders. A *leader,* or *leader line,* is an inclined (slanted) line that connects a dimension, note, or specification with a particular feature of an object. Leaders are drawn at an angle of between 30° and 60° for maximum readability. One end of a leader terminates with an arrowhead or a dot. Arrowheads terminate on a line. Dots are placed within an area and must be at least 1⁄16″ (1.6 mm) in diameter. The other end of the leader is generally connected to a shoulder. A shoulder should be approximately 1⁄4″ (6 mm) and centered on the lettering at the beginning or ending of the dimension, note, or specification. **See Figure 6-5.**

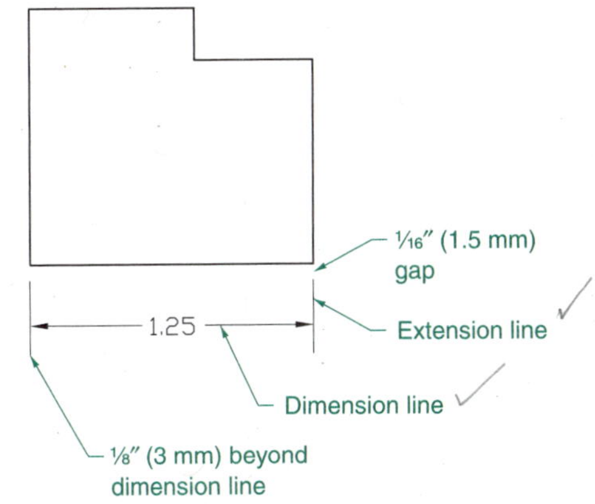

Figure 6-4. An extension line is a line that extends from a surface feature of an object and ends a dimension line.

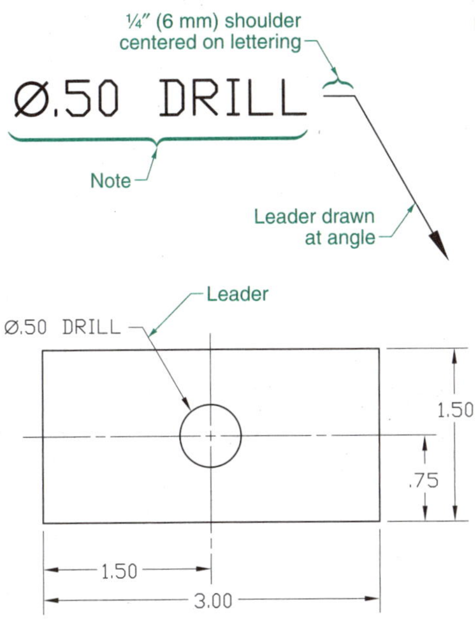

Figure 6-5. A leader, or leader line, is an inclined (slanted) line that connects a dimension, note, or specification with a particular feature of an object.

Dimension Types

Dimensions on a drawing give specific sizes and locations for parts of an object, and they can also contain other pertinent information, including specifications needed to manufacture a finished part or the function of a part in a total assembly. Dimensions are given in the form of linear dimensions or angular dimensions.

Linear Dimensions. A *linear dimension* is a dimension that measures a straight distance and is commonly expressed in English or metric units of measure. If the same linear dimensioning system is used throughout a drawing, unit designations may be omitted. The decimal inch is predominantly used in the United States. A *dual dimension* is a dimension that shows both English and metric dimensions. In a dual dimension, the two numbers are separated by a slash (/) or the second number is enclosed in parentheses or brackets. **See Figure 6-6.**

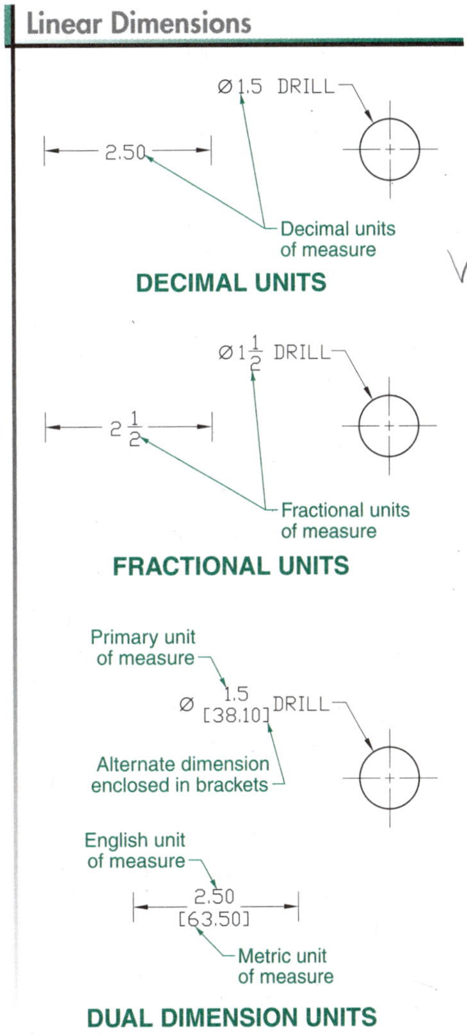

Figure 6-6. A linear dimension is a dimension that measures lines and is commonly expressed in decimal inches, fractional inches, or millimeters.

The use of fractions in linear dimensioning is generally used on drawings not requiring a high degree of precision. Fractional dimensions often have a denominator that is a multiple of two (2, 4, 8, 16).

Angular Dimensions. An *angular dimension* is a dimension that measures the degree of separation between two adjoining lines or points. Angular dimensions are commonly expressed as decimal degrees (32.76°) or as degrees, minutes, and seconds (32°45′36″). Angular dimensions can include a size dimension and a location dimension. See **Figure 6-7.**

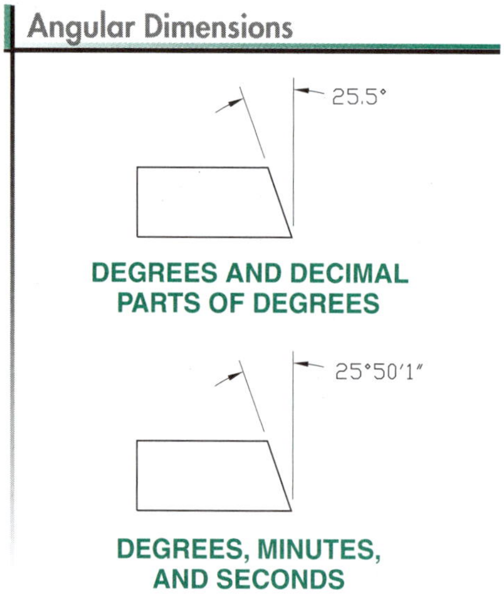

Figure 6-7. An angular dimension is a dimension that measures angles and is commonly expressed in decimal degrees or as degrees, minutes, and seconds.

Units of Measure

If an entire drawing uses one unit of measure, it is unnecessary to use unit labels on the dimensions. A general note on the print is used to specify the measuring system and units used. If any dimensions vary from this system, unit labels are used on those specific dimensions only.

When a dimension is less than 1″, a leading zero is not used before the decimal point. However, a leading zero is always used before the decimal point when a dimension is less than 1 mm. **See Figure 6-8.** For decimal dimensioning, the figures after the decimal point are often rounded to even hundredths. Trailing zeroes can be added to the right of a decimal figure if needed. A 3- or 4-place decimal dimension can be used with drawings requiring more precise figures.

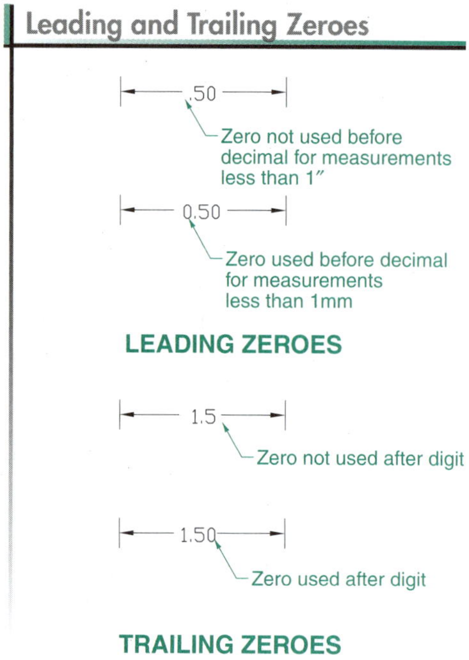

Figure 6-8. Dimensioning units and the amount of precision required in a drawing dictate whether leading or trailing zeroes should be used when dimensioning.

Notes

A *note* is a line (or lines) of explanatory text that is added to a drawing. Notes can be used to supplement direct dimensions. Notes are also used to explain the material required to make a specific object, the number of parts required, the type of finish, and any other data that may be needed to make the drawing complete. Notes should be brief and specific. Notes can be either general or local. A *general note* is a note that applies to the entire drawing. For example, a general note can specify the primary dimensioning system used in a drawing if all dimensions are the same unit of measure. In this case, only dimensions that are exceptions to the primary dimensioning system need to be designated with unit labels. A *local note* is a note that applies only to one or more specific items in a drawing. **See Figure 6-9.**

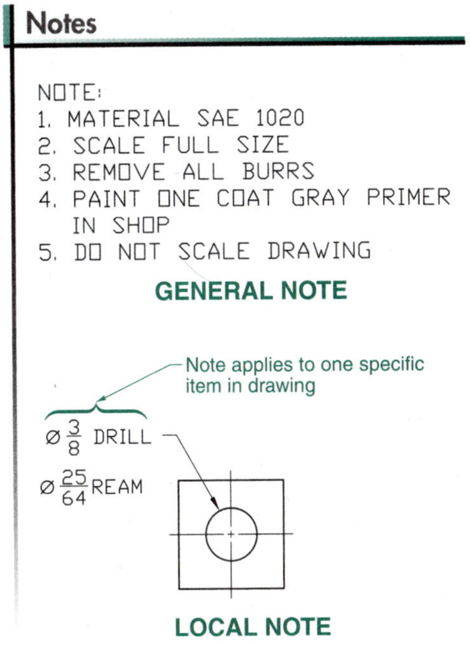

Figure 6-9. A general or local note can be added to a drawing to supplement the direct dimensions.

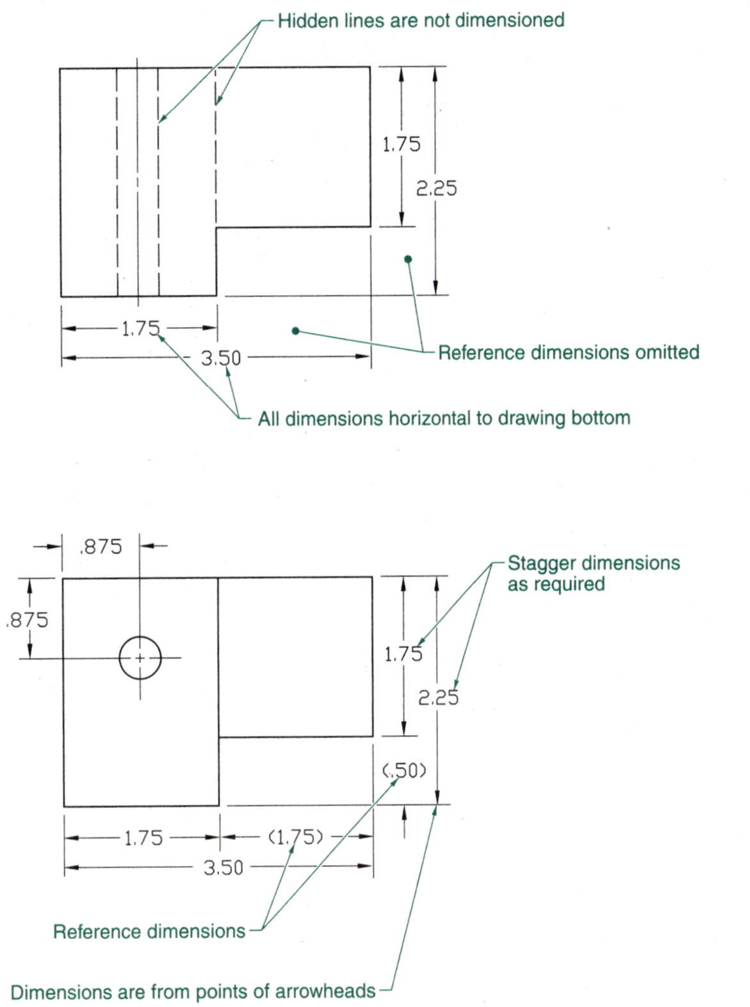

Figure 6-10. Dimension lines are used with numbers to indicate the size or location of different parts of an object.

Dimension Placement

Dimensions should be placed on a drawing so that they allow only one interpretation and include no repetition. The principal view of the object should be selected to show the key dimensions because this view has the most complete representation of the characteristics of the object. Dimensions are placed outside of the view of the object, and they should be grouped uniformly and aligned to read horizontally from the bottom of the drawing. Parallel dimension lines should be at least ¼″ (0.25mm) apart with the numbers staggered for increased readability. Dimensions should not be placed to originate or terminate at hidden lines. **See Figure 6-10.**

Smaller dimensions are placed near the object. Intermediate dimensions are placed outside of smaller dimensions. Overall dimensions are placed the farthest from the object. If an overall dimension is used, one intermediate dimension may be omitted or indicated as a reference dimension. A *reference dimension* is a dimension derived from other dimensions and is indicated by the placement of parentheses around the dimension.

If a drawing being dimensioned is not drawn to scale, the dimensions should be indicated as not to scale (NTS) by underlining the number with a solid straight line or a general note should be included on the drawing.

Dimensioning Circles, Arcs, and Angles. Dimensioning circles, arcs, and angles follows the same general rules as for dimensioning objects. Circles are dimensioned by indicating the diameter or radius, either on the inside of the circle using arrows, or on the outside of the circle using a leader and labeling the number. The number should be labeled with the diameter symbol (∅) for the diameter, or R for the radius. The location of a circle on an object is dimensioned with center lines. A *center line* is a thin line that locates axes or center points of arcs or circles. Center lines can be extended beyond the object and used as extension lines. Circles and holes should always be dimensioned on the view where they appear as circles and not on the view where they are represented by hidden lines. **See Figure 6-11.**

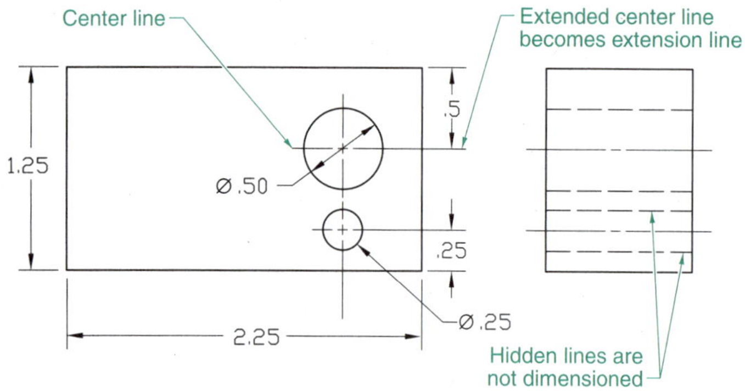

Figure 6-11. A circle is dimensioned by using the diameter or radius to indicate size, and center lines to indicate location.

Only one circle, hole, or raised surface should be dimensioned when dimensioning multiple circles, holes, or raised surfaces that are equal in diameter and spacing. A local note can be added to the dimension to indicate that the dimension is typical and to give the number of holes of equal size and placement. If there are multiple equal circles, holes, or raised surfaces with uneven placement, the dimension is labeled with a number followed by an "x", such as 3x, to indicate that all circles, holes, or raised surfaces are the same size. All dimensions must be indicated if multiple circles, holes, or raised surfaces are sized and spaced unevenly. **See Figure 6-12.**

Dimensioning Multiple Circles

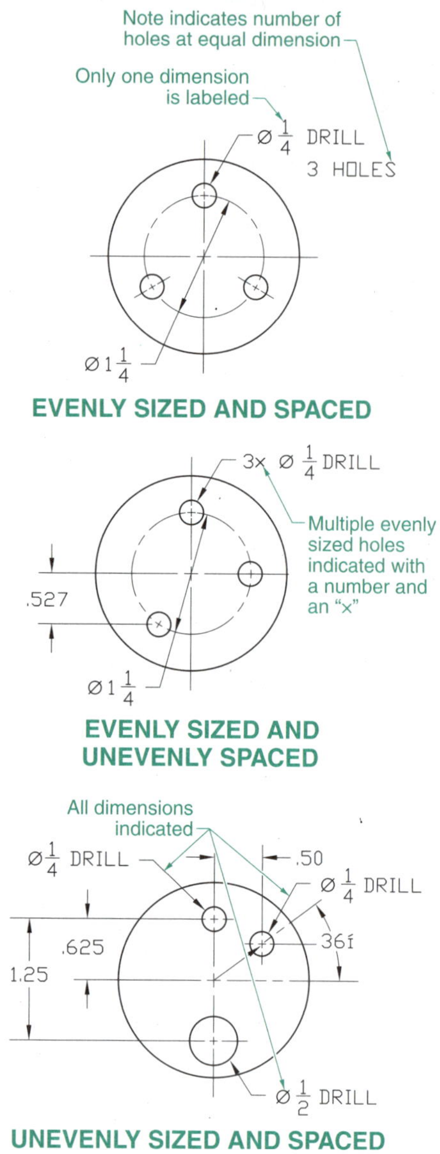

Figure 6-12. Only one circle, hole, or raised surface should be dimensioned when all are equal in diameter. All dimensions must be indicated if multiple circles, holes, or raised surfaces are sized and spaced unevenly.

TECH TIP

Arcs can be dimensioned using the arc's linear distance or radius. When dimensioning an arc using the radius, the center point of the arc should be indicated by two short lines that are perpendicular to each other and form a small cross. Arcs in a drawing that are not being dimensioned should not have their center points indicated.

The linear distance of an arc is dimensioned with the dimension line following the curve of the arc. An *arc* is a curved line that is part of the circumference of a circle. When dimensioning an arc, the extension lines are always either vertical or horizontal. When indicating the linear distance of an arc, an arc symbol (⌒) is placed above the number. Arcs can also be dimensioned using a center point to find the radius. When dimensioning a radius, only one arrowhead extending from the center point to the center of the arc should be used. If space allows, the dimension number is placed between the center point and arrowhead, or the dimension may be placed outside of the arc using a leader. The dimension is labeled with the letter R to indicate that it is a radius dimension. **See Figure 6-13.**

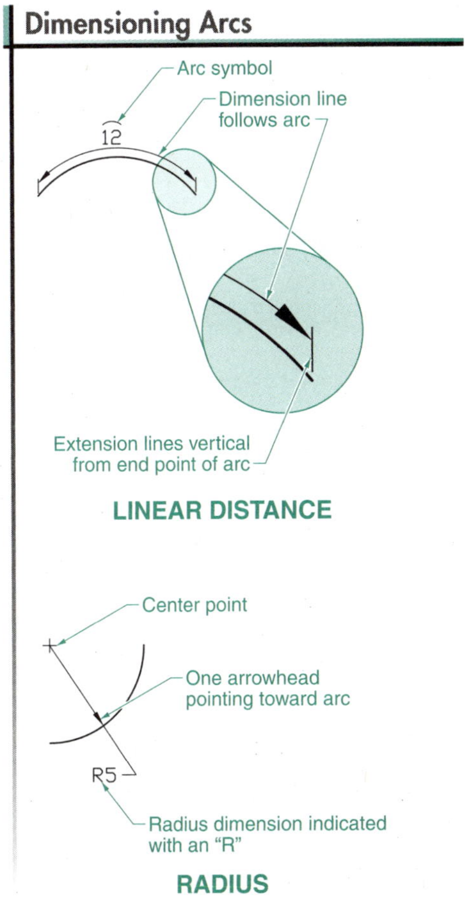

Figure 6-13. Arcs can be dimensioned to show the linear distance or the radius.

An angle is the intersection of two lines. Angles are dimensioned with the dimension line following the curve of the arc created within the angle. When dimensioning an angle, the extension lines should extend from the lines that form the angle. Angles may also be dimensioned within the angle without the use of extension lines. **See Figure 6-14.**

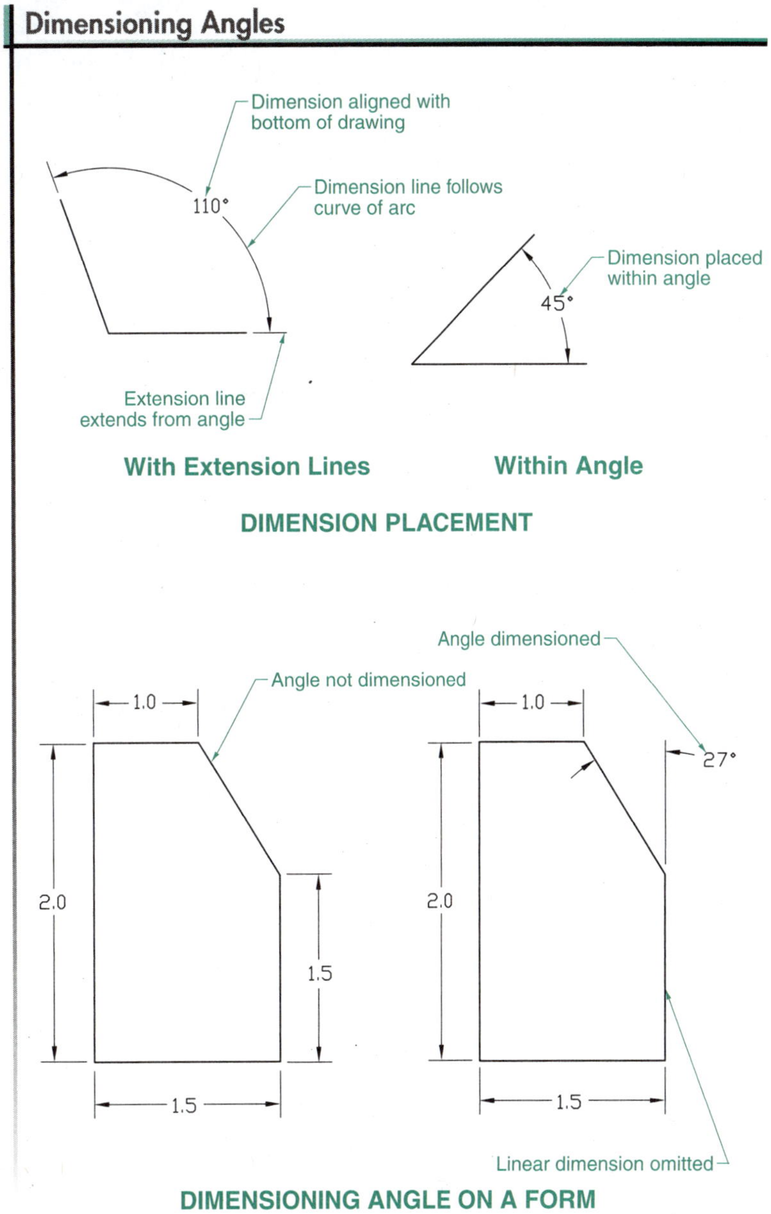

Figure 6-14. Angles are dimensioned with the dimension line following the curve of the arc created with the angle.

Tolerances are required when manufacturing precision parts to ensure an accurate fit.

TOLERANCING

Tolerance is the amount of measured variation allowed in a given dimension. Tolerances should be included in a drawing because the variations in size and shape among manufactured objects can affect the interchangeability of parts. *Direct tolerancing* is the practice of specifying a dimension's permissible range directly within the dimensioning lines. The two common methods of direct tolerancing are limit dimensioning and plus and minus tolerancing.

Limit Dimensioning

Limit dimensioning is a direct tolerancing method that includes only the maximum and minimum values of a dimension on a drawing. The two limits can be expressed in a single line containing the low limit and the high limit separated by a dash, or in two lines showing the high limit above the low limit. **See Figure 6-15.** The high and low limits should be displayed with the same number of decimal places for uniformity. Zeroes can be added to limits for uniformity if necessary.

Plus and Minus Tolerancing

Plus and minus tolerancing is a direct tolerancing method that provides an ideal dimension and includes the allowable deviations in the positive and negative directions. The tolerances are placed after the dimension with the positive tolerance always shown above the negative tolerance. The dimension and tolerances should be displayed with the same number of decimal places. If necessary, zeroes can be added to the dimensions and tolerances for uniformity. When the numerical tolerance is the same in both directions, the plus and minus values can be combined into a single expression using the ± symbol, such as 15.0 ±1.

Direct Tolerancing

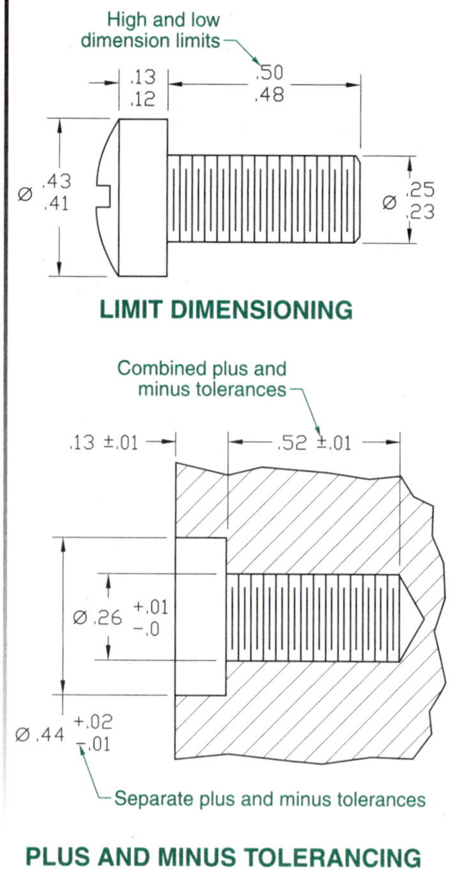

Figure 6-15. Direct tolerancing specifies a dimension's permissible range directly within the dimensioning lines and can be shown using one of two methods: limit dimensioning or plus and minus tolerancing.

If there is a large number of dimensions on a drawing that require the same degree of tolerance, a general note may be added to indicate the common tolerance for these dimensions. Only tolerances that differ from the general values need to be included on the drawing.

Dimensioning Methods

Dimensions are added to a drawing to communicate object information in a clear and concise manner. The four dimensioning methods that can be used to dimension a drawing include point-to-point, rectangular coordinate, tabular, and polar coordinate. The point-to-point and rectangular coordinate methods are the most common dimensioning methods used.

Point-to-Point Dimensioning. *Point-to-point dimensioning* is a dimensioning method that specifies sizes along an object from one feature to another. Since each dimension is based on the previous dimension, tolerances for each successive dimension are added together. The tolerances continue to increase for each successive feature. Because the tolerances increase with each dimension indicated, this method should not be used for objects that require a high degree of precision. **See Figure 6-16.**

Point-to-Point Dimensioning

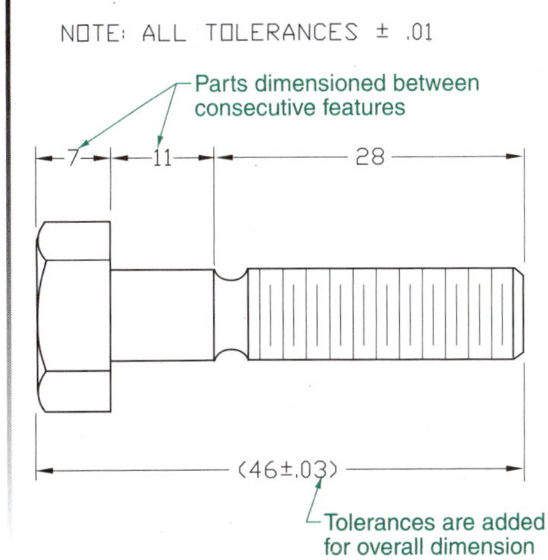

Figure 6-16. Point-to-point dimensioning specifies sizes along an object from one feature to another and where tolerances are cumulative.

Rectangular Coordinate Dimensioning. *Rectangular coordinate dimensioning* is a dimensioning method that specifies all dimensions from a common baseline. The advantage of rectangular coordinate dimensioning is that tolerances do not increase with each dimension because the dimension of each feature is not dependent on prior dimensions. **See Figure 6-17.** Even though rectangular coordinate dimensioning can require more space on a drawing than point-to-point dimensioning, it gives more precise dimensions than point-to-point dimensioning.

Rectangular coordinate dimensioning can be drawn in a more compact manner by removing the dimension lines and placing the numerical dimensions at the ends of the extension lines. Each baseline is labeled as zero and each extension line is dimensioned from the zero baseline.

Tabular Dimensioning. *Tabular dimensioning* is a dimensioning method that assembles all numerical feature dimensions in a table. Features are identified using letters or symbols, and all dimensions are determined from a zero baseline using the rectangular coordinate method. The feature symbols and the associated dimensions are listed together in a table. Only overall dimensions and baseline indicators are typically included on the drawing itself. **See Figure 6-18.**

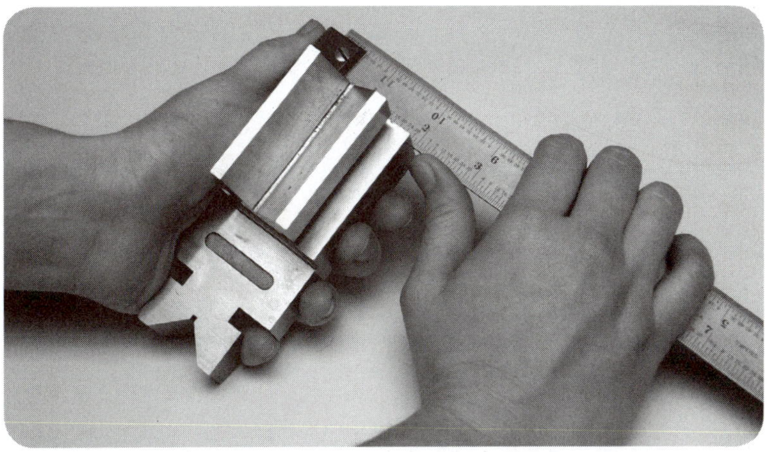

During the measurement and dimensioning of an object, tolerances are not cumulative if all measurements are taken from a common baseline.

Rectangular Coordinate Dimensioning

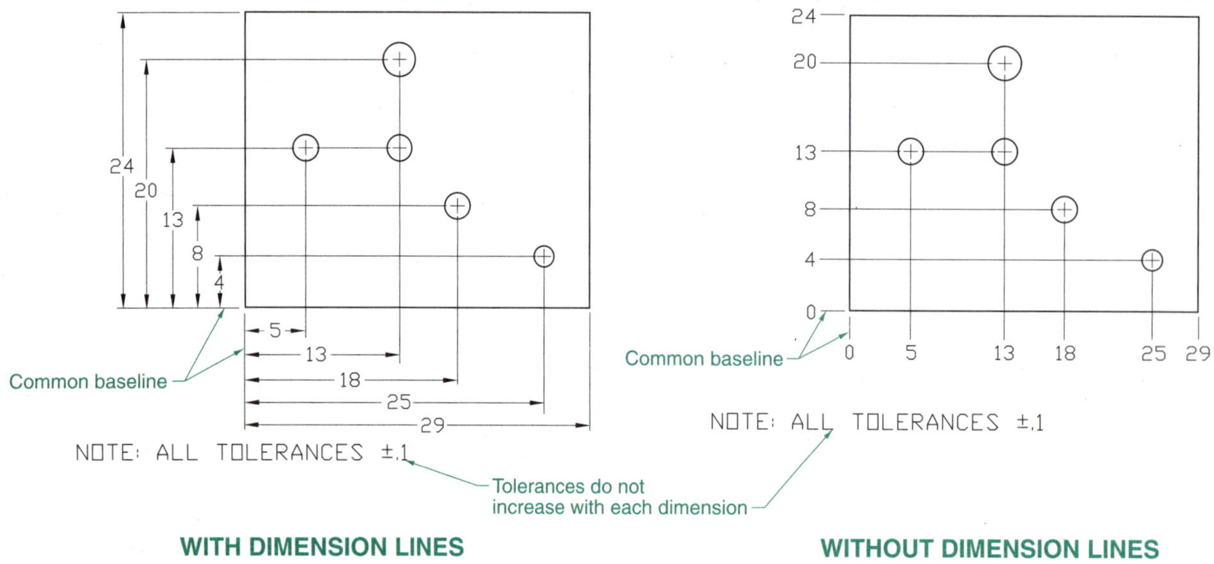

Figure 6-17. Rectangular coordinate dimensioning specifies all dimensions from a common baseline with no cumulative tolerances.

Tabular Dimensioning

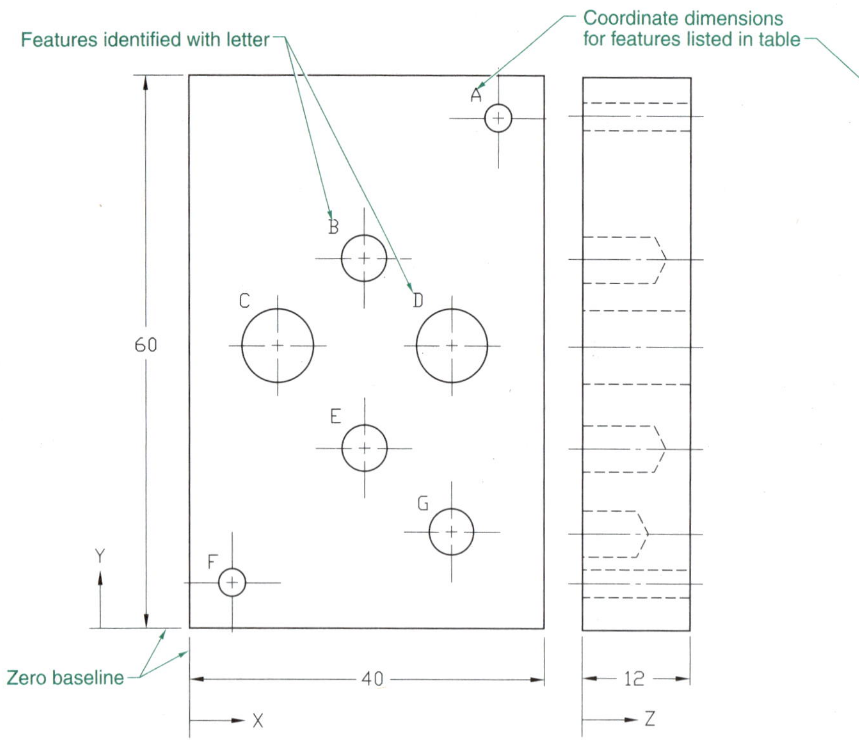

HOLE	DIA.	X	Y	Z
A	3	35	55	THRU
B	5	20	40	8
C	8	10	30	THRU
D	8	30	30	THRU
E	5	20	20	8
F	3	5	5	THRU
G	5	30	10	6

Figure 6-18. Tabular dimensioning assembles all numerical feature dimensions in a table by identifying features using letters or symbols and measuring from a common baseline.

TECH TIP

For precision parts, it is recommended that each dimension be based on only one other dimension; otherwise, the tolerances may add together and make the overall dimensions less accurate.

Polar Coordinate Dimensioning. *Polar coordinate dimensioning* is a dimensioning method that determines location using angle and radius dimensions. The polar coordinate dimensioning method is particularly useful for parts with features arranged in a circle. When features are arranged in a circle, it is usually more convenient to indicate the circle's radius or diameter and the angular distance of each feature from the circle's center lines. **See Figure 6-19.**

GEOMETRIC DIMENSIONING AND TOLERANCING

Geometric dimensioning and tolerancing is a method of specifying the size, shape, and location of features and their allowable tolerances. Geometric dimensioning and tolerancing is necessary in order to make an object or part as exact as possible, which in turn allows the part to fit and function within an assembly as needed. Accepted specifications for dimensioning types, methods, and practices are described in ANSI/ASME Standard Y14.5M, *Dimensioning and Tolerancing.*

Geometric Dimensioning

Dimensions can be used to describe the specific size and shape of parts to be manufactured. Most dimensions shown on a drawing have general or specific tolerances. However, basic and reference dimensions are two types of dimensions that do not include tolerances.

Basic Dimensions. A *basic dimension* is a dimension that is used to describe a theoretical exact size, shape, or location of a feature on an object. In other words, a basic dimension is the ideal size, not accounting for any variations or tolerances and serves as the basis for establishing permissible tolerances on other dimensions. A basic dimension is indicated by enclosing it in a rectangular frame. **See Figure 6-20.**

Polar Coordinate Dimensioning

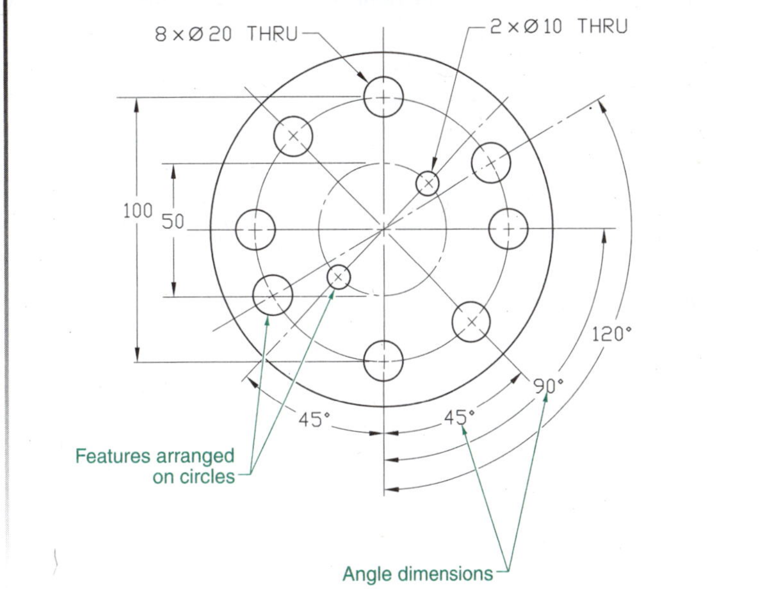

Figure 6-19. Polar coordinate dimensioning determines location using angle and radius dimensions.

Basic Dimensions

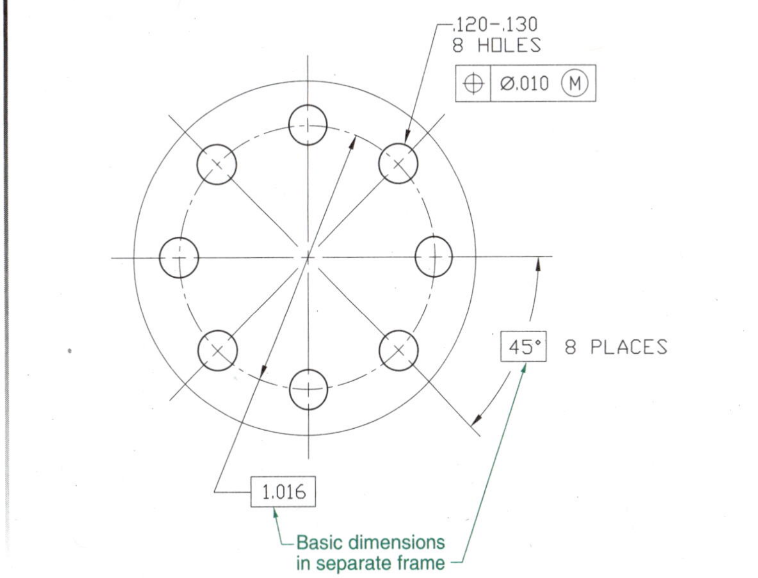

Figure 6-20. A basic dimension describes a theoretical exact size, shape, or location of a feature on an object and is enclosed in a rectangular frame.

TECH TIP

The manufacture of high-quality products requires that parts be interchangeable. Machining parts to an exact size can be difficult and expensive. By adding allowable variations during the production of parts, the manufacturer can produce the assembly on an automated assembly line, making the price of the finished assembly more affordable.

Reference Dimensions. A reference dimension is derived from other values on the drawing and is used for informational purposes only. Reference dimensions are enclosed in parentheses or are followed by the abbreviation REF. **See Figure 6-21.**

Reference Dimensions

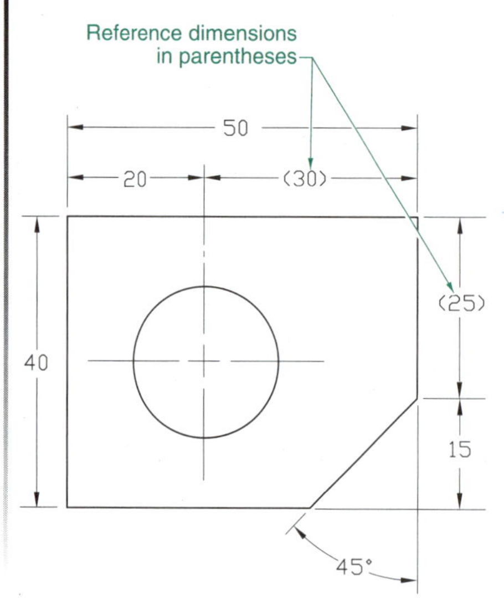

Figure 6-21. A reference dimension is derived from other values on the drawing and is enclosed in parentheses.

Datums

A *datum* is a theoretical exact point, axis, or plane that serves as the origin for the location or geometric characteristics of object features. Datums are used to identify related features. Datums are labeled on a drawing in the primary orthographic view, or in any alternate view that is needed to clearly establish the required relationships between features. The two types of datums are datum features and datum targets. Each datum type is signified on a drawing with a specific symbol.

Datum Features. A *datum feature* is a datum that is an actual feature of an object that is used to establish a datum. Datums are identified using uppercase letters enclosed in a square frame (datum frame). The frame is connected to the datum feature with a leader that is terminated with a triangle, which may be filled or unfilled. Datum frames can be connected directly to the datum's object line, extension line, or to a feature control frame already in place. **See Figure 6-22.** A *feature control frame* is a rectangular frame divided into sections that encloses characteristic symbols, numerical tolerances, datum references, and modifier symbols.

Letters are assigned to each datum feature beginning with A and continuing through the alphabet, excluding I, O, and Q. If more than 23 datums are required, AA through AZ, BA through BZ, and so on may be used.

Datum Targets. A *datum target* is a datum established by a point, line, or area designated on an object. Unlike a datum feature, a datum target is not an actual part feature; however, a datum target is associated with a datum feature. Datum targets can be points, lines, or areas. Points are marked with a large × symbol. Lines are drawn using phantom lines and appear as a point when seen on an edge. Areas are filled with general-purpose section lining. The location of a target is given with basic dimensions.

A datum target symbol is a circular frame divided into two halves. The bottom half contains a letter identifying the associated datum feature and a serial number. The top half contains the size of the target when identifying a target area, or is left blank when a target area is not provided.

Feature Control

Feature control is a method of identifying and controlling features on an object. A *feature* is any surface, angle, hole, round, or other characteristic of an object that can be dimensioned and controlled. Features can be individual or related. An *individual feature* is a feature that is an independent characteristic of a part. A *related feature* is a feature that is associated with a specific datum or datums. Relationships with datums determine whether a feature is individual or related.

Datum Features and Datum Targets

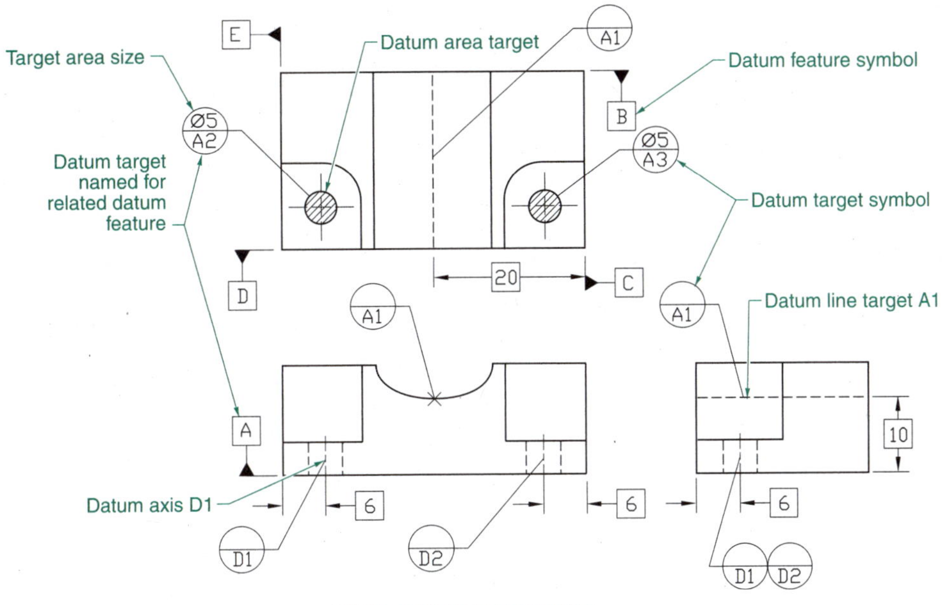

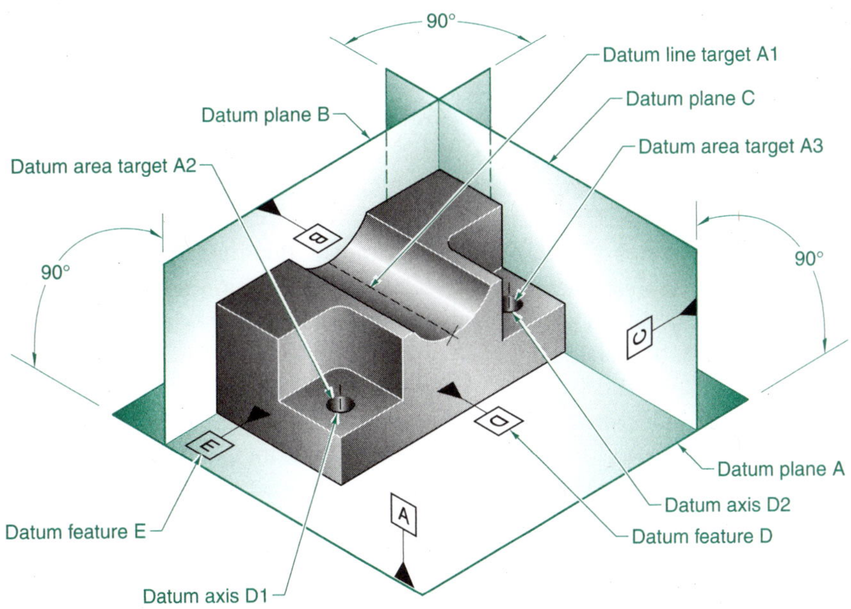

Figure 6-22. A datum feature is an actual feature on an object. A datum target is established by a point, line, or area designated on an object.

Geometric dimensioning and tolerancing uses five types of controls with features, including form, profile, orientation, location, and runout. Characteristics involving the form are controlled using individual features. Characteristics involving the profile are controlled using either individual or related features. Characteristics involving the location, orientation, or runout of an object are controlled using related features.

Individual Features. Tolerances on individual features stipulate how much the part may vary from a theoretical ideal geometric form. The applicable geometric forms include straightness, flatness, circularity, and cylindricity. For example, a piece of square bar stock may have a straightness tolerance of .002″.

Related Features. Tolerances on related features specify the allowable deviation from a relationship with one or more datums. The relationships include position, concentricity, symmetry, angularity, perpendicularity, parallelism, circular runout, and total runout. For example, a hole may have a position tolerance of 1 mm in relation to the part's edge.

The control of profile tolerances may be applied to either individual or related features. This means that profiles may or may not require reference to a datum.

Feature Control Frames. The ANSI standard defines a system of symbols and references for applying tolerances to features. This information, which includes characteristic symbols, numerical tolerances, datum references, and modifier symbols, is enclosed within a rectangular feature control frame. **See Figure 6-23.**

A characteristic symbol is a symbol that stands for one of the feature characteristics defined by ANSI. Individual features, related features, and modifiers each have a unique symbol that should be used to identify the characteristics in the feature control frame. **See Appendix.**

A numerical tolerance is applied to the characteristic represented by the symbol in the frame. For example, a drawing may require that a surface be flat to a tolerance of .08″. The feature control frame for that surface would include the symbol for flatness and the tolerance of .08.

Chapter 6 Quick Quiz®

Feature Control Frames

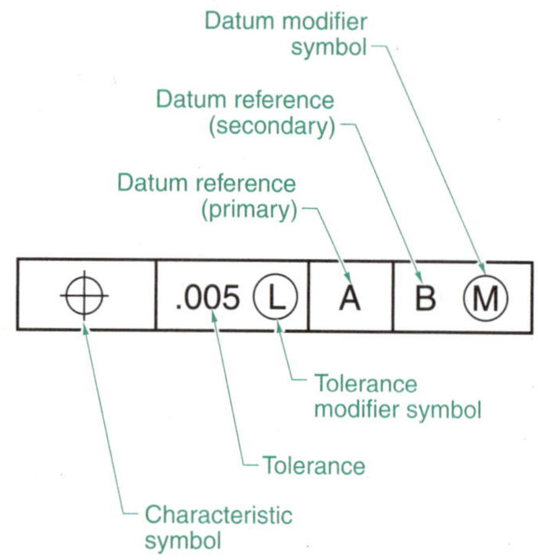

Figure 6-23. A feature control frame is a rectangular frame divided into sections that encloses characteristic symbols, numerical tolerances, datum references, and modifier symbols.

Datum references are included in the feature control frame for related features. Multiple datums can be referenced and are typically added individually in their order of importance for the feature. There may be primary, secondary, and tertiary datums. Multiple datums may be identified as primary. This could occur when multiple datums are used to determine a feature. An example of this is when two datum planes intersect and create a line that can be used as the center axis of a cylinder.

A *modifier symbol* is a symbol that represents a specific condition of the part or feature when the tolerance is applied. For instance, one tolerance may require that it be applied when another feature is at a maximum tolerance condition. Modifiers may not be necessary at all or they may appear multiple times within a feature control frame.

Technical Sketching and Drawing

Dimensioning and Tolerancing

6

Review

Name ____________________ Date ____________________

REVIEW QUESTIONS

____________ 1. ___ is a method of adding dimensions to a drawing to indicate the geometrical characteristics of an object.

T F 2. A location dimension gives the size of an angle or feature.

____________ 3. Dimension lines, arrowheads, extension lines, and ___ are used with numbers to apply dimensions to an object.

____________ 4. A(n) ___ is a symbol that indicates the extent of a dimension that can be open or closed.

____________ 5. A(n) ___ dimension measures a straight distance and is commonly expressed in English or metric units of measure.

T F 6. If the same linear dimensioning system is used throughout a drawing, unit designations may be omitted.

____________ 7. A(n) ___ dimension shows both English and metric dimensions.

____________ 8. A(n) ___ is a line (or lines) of explanatory text that is added to a drawing to supplement direct dimensions.

T F 9. Dimensions should be placed on a drawing so that they allow for multiple interpretations and include repetition.

T F 10. The principal view of an object should be selected to show key dimensions because it has the most complete representation of the characteristics of the object.

____________ 11. Dimensions should be grouped uniformly and aligned to read horizontally from the ___ of the drawing.

T F 12. Dimensions should be placed to originate and terminate at hidden lines.

T F 13. When dimensioning an angle, the extension lines should extend from the lines that form the angle.

____________ 14. ___ should be included on drawings because the variations in size and shape among manufactured objects can affect the interchangeability of parts.

____________ 15. Basic and ___ dimensions are two types of dimensions that do not include tolerances.

T F 16. Point-to-point dimensioning specifies sizes along an object from one feature to another and has tolerances that add together.

________________ **17.** A(n) ___ is a theoretical exact point, axis, or plane that serves as the origin for the location or geometric characteristics of object features.

T F **18.** A feature control frame is a circular frame divided into sections that encloses characteristic symbols, numerical tolerances, datum references, and modifier symbols.

________________ **19.** A(n) ___ is any surface, angle, hole, round, or other characteristic of an object that can be dimensioned and controlled.

________________ **20.** A(n) ___ symbol represents a specific condition of the part or feature when the tolerance is applied.

IDENTIFICATION — Lines

c **1.** Dimension line

d **2.** Arrowhead

b **3.** Extension line

a **4.** Leader

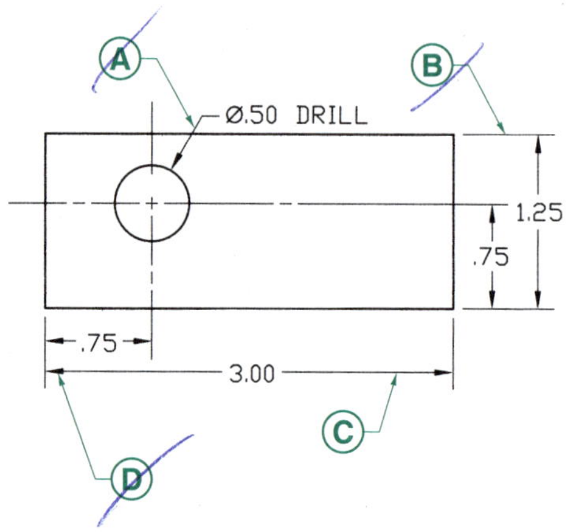

IDENTIFICATION — Feature Control Frames

f **1.** Characteristic symbol

a **2.** Datum reference (primary)

d **3.** Tolerance modifier symbol

b **4.** Datum reference (secondary)

c **5.** Datum modifier symbol

e **6.** Tolerance

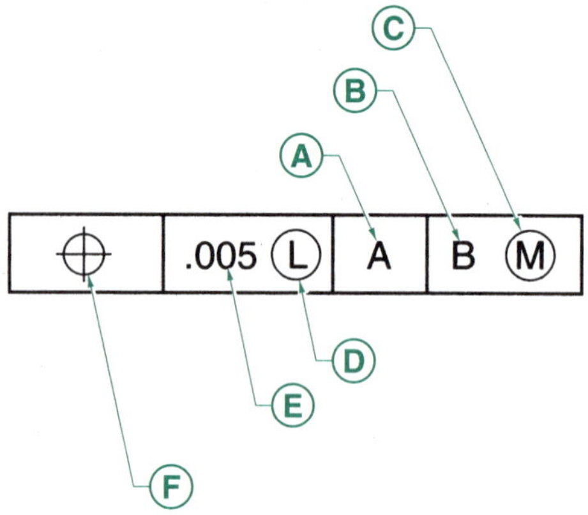

IDENTIFICATION — Datums

D ____ **1.** Datum line target

a ____ **2.** Datum plane

d ____ **3.** Datum axis

e ____ **4.** Datum feature

c ____ **5.** Datum area target

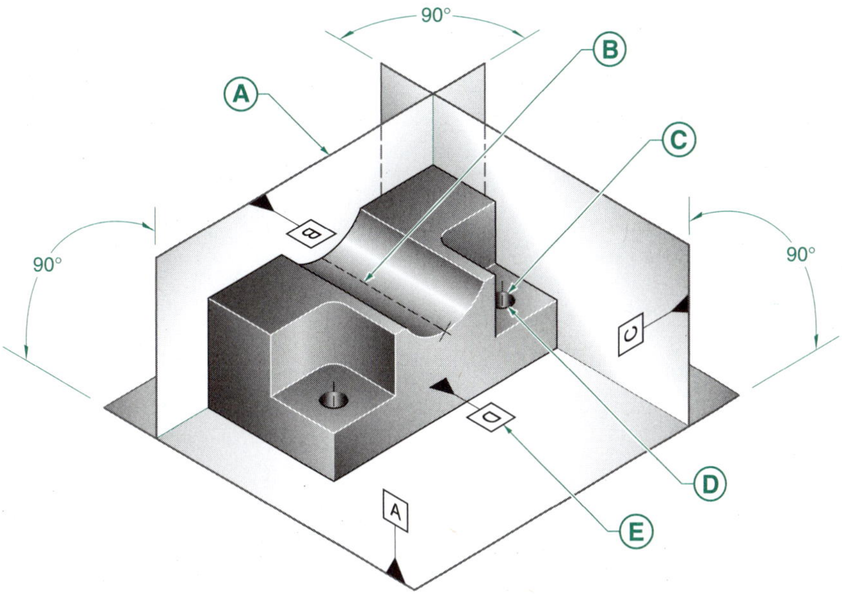

EXERCISES

Use the grid provided to add dimensions to the following drawings. Each grid square equals 1".

1.

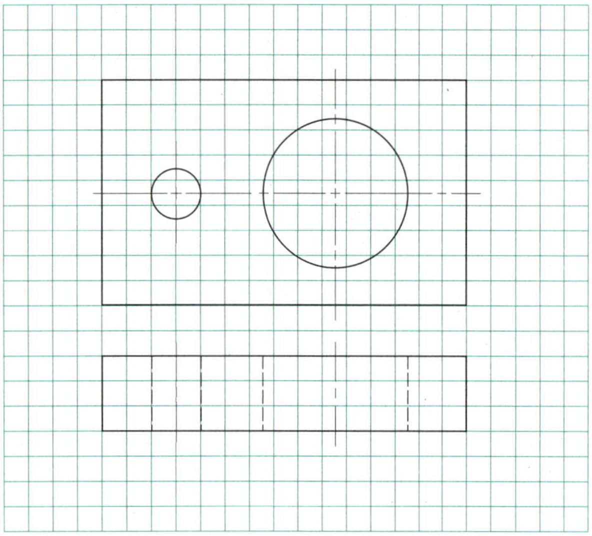

2.

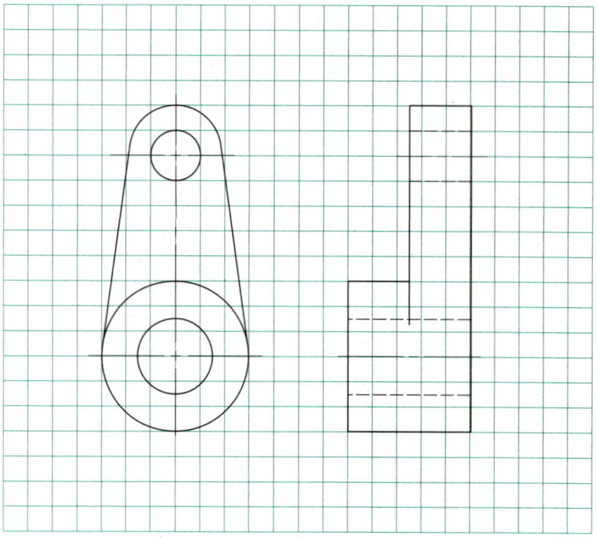

3.

4.

Use the scale provided and a divider to measure and find the dimensions of the following drawings. Include all necessary dimensions.

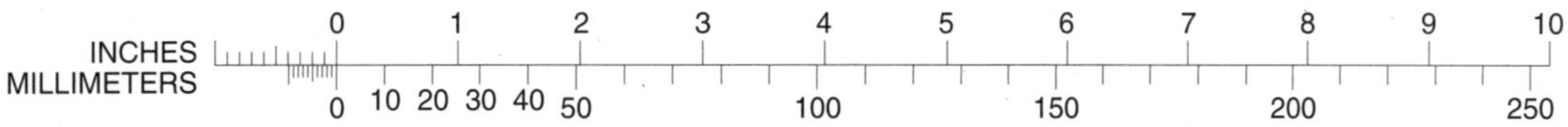

5. Add dimensions in inches.

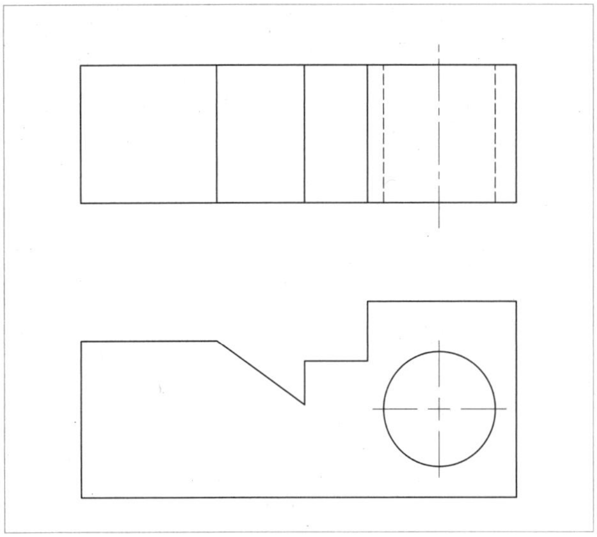

6. Add dimensions in millimeters.

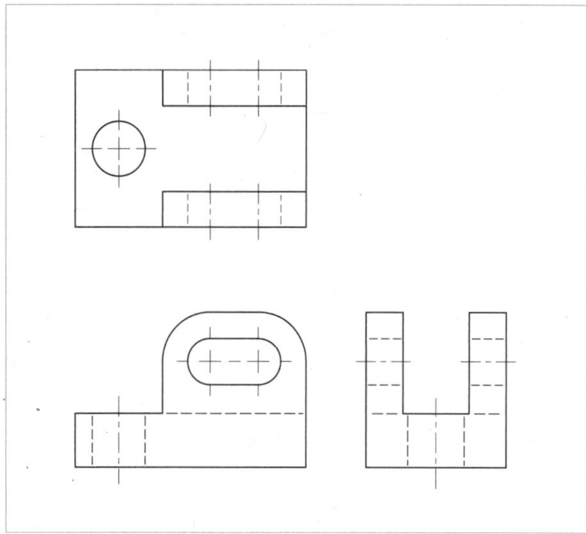

7. Add dimensions in inches.

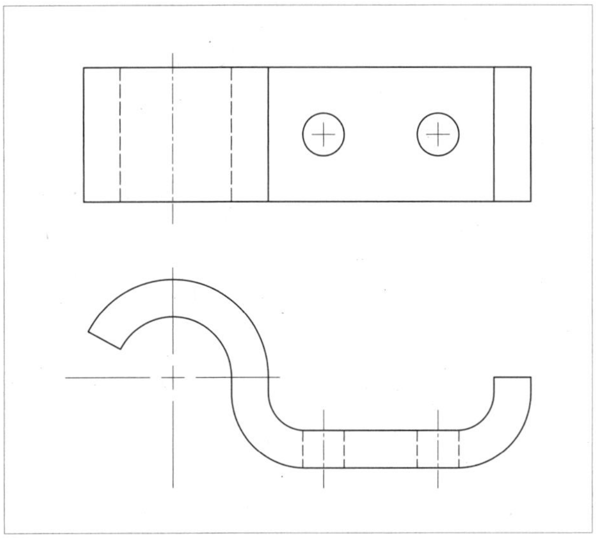

8. Add dimensions in millimeters.

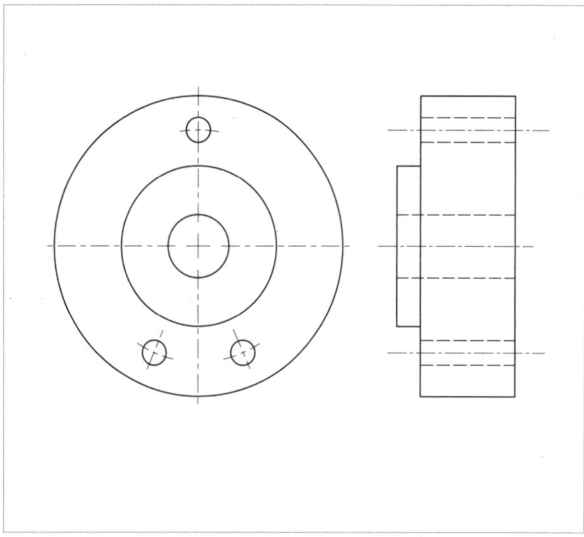

9. Create a multiview drawing from the drawing provided. Include only the views that are necessary. Include all necessary dimensions.

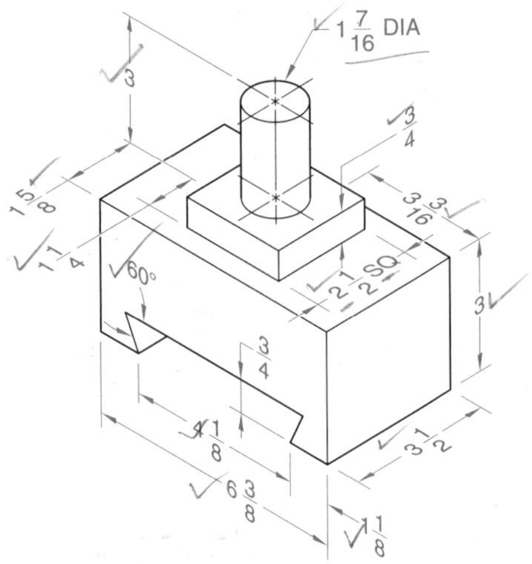

10. Create a multiview drawing from the drawing provided. Include only the views that are necessary. Convert all fractional dimensions to decimal dimensions. Include all necessary dimensions.

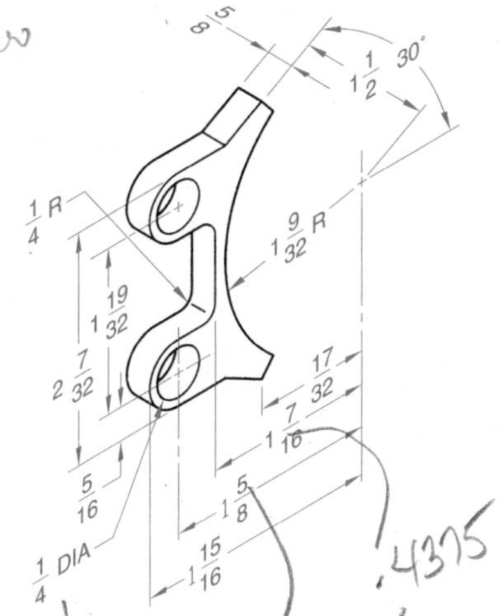

11. Create a multiview drawing from the drawing provided. Include only the views that are necessary. Include all necessary dimensions using the rectangular coordinate dimensioning method.

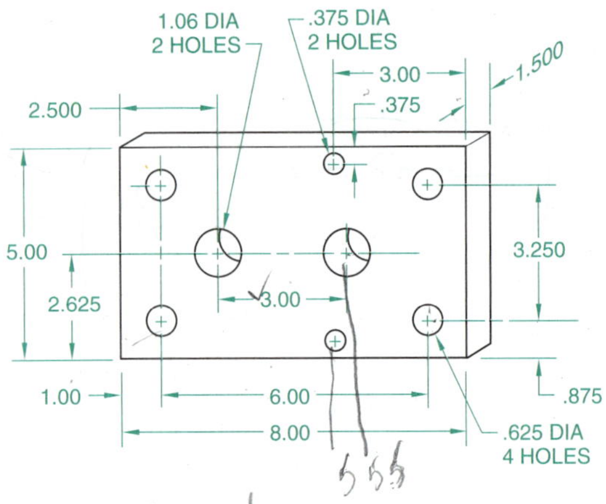

Technical Sketching and Drawing

7 Prints

PRINTS

Many professions require a technical artist to create, read, and understand prints. A *print* is a drawing that shows the structure and related characteristics of an object, building, or site. **See Figure 7-1.** Prints are used in various trades and can display a wide variety of information required to complete a project.

Prints used in the construction trades are labeled with a capital letter prefix and are often divided into the following five main divisions:

A— Architectural prints provide general information about a building's location, structure, and utilities.

S— Structural prints provide information about sizes, styles, and placement for foundations, columns, beams, joists, and other framing and load-bearing members.

M—Mechanical prints provide information about the plumbing, heating, ventilation, and air conditioning systems of a building, including ductwork, piping, and fire protection systems.

Prints

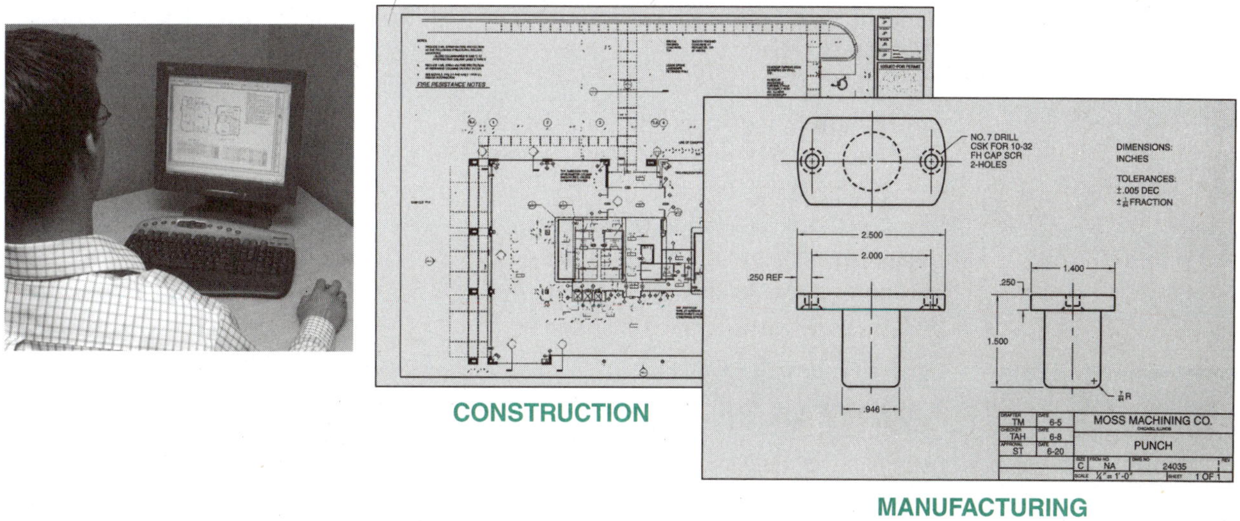

Figure 7-1. Prints show the structure and related characteristics of an object, building, or site.

E— Electrical prints indicate the capacity and placement of power plant systems, lighting, wiring, switches, and other electrical installations.

C— Civil prints provide information about the overall placement of a building on the site, parking lots and paving, drainage, grading, elevations, topographical information, and landscaping.

In addition to letter labels, prints are numbered in ascending order within each division beginning with the number 1. Print numbers can be subdivided with secondary numbers when several pages apply to the same elements. **See Figure 7-2.**

Print Divisions

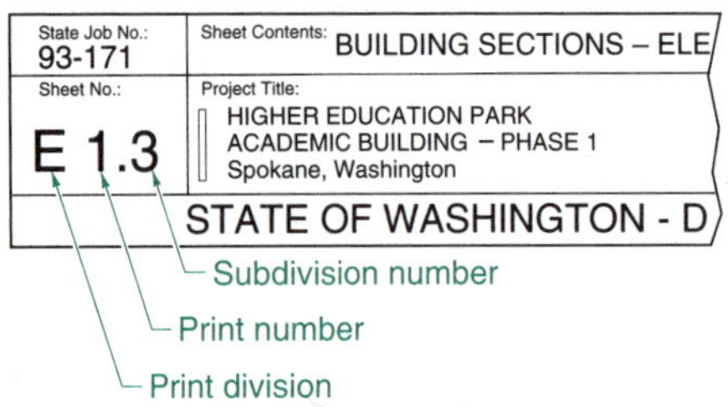

PRINT DIVISIONS
A = ARCHITECTURAL
S = STRUCTURAL
M = MECHANICAL
E = ELECTRICAL
C = CIVIL

Figure 7-2. Prints are numbered in ascending order within each division beginning with the number 1.

Manufacturing prints often contain specialized manufacturing-related information. Objects and parts that are to be fabricated need specific drawings to outline their shape, manufacturing processes and specifications, and assembly.

Elevations in Print and Drawing Library

PRINT FORMAT AND FEATURES

Prints have standard formats and features in order to present information in a clear and concise manner. Prints are drawn as a plan or an elevation. **See Figure 7-3.**

A *plan* is an orthographic drawing of an object viewed from above and contains a horizontal cutting plane that passes through the object. Plans are a type of sectional drawing because the upper portion of the object is removed to reveal interior parts. An *elevation* is an orthographic view of an object with the viewing plane parallel to one side of the object. Elevations generally depict the front, back, left side, or right side of an object and are depicted in true shape with no perspective.

Plans and Elevations

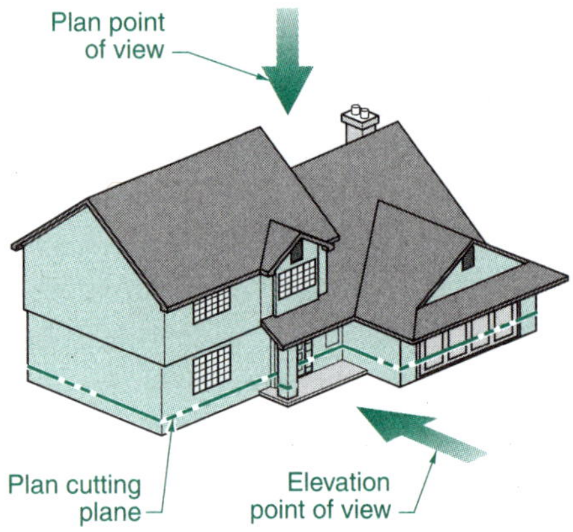

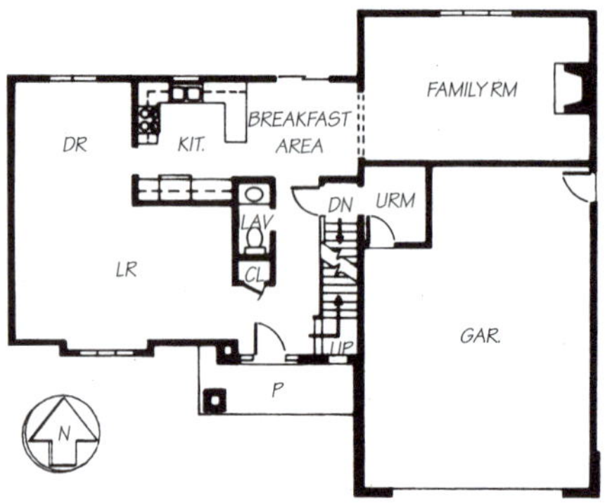

PLAN

ELEVATION

Figure 7-3. Prints are drawn as a plan or an elevation.

Prints use standardized abbreviations and symbols, and can include a title block, schedules, zones, notes, and specifications. Prints can also include section and detail drawings.

Abbreviations and Symbols

Abbreviations and symbols are used on prints to conserve space and promote consistency of term use from print to print. An *abbreviation* is a letter or group of letters that represents a term or phrase. Abbreviations use only uppercase letters, and any abbreviations that form a word are followed by a period to avoid confusion. Some abbreviations are universal, while others have meanings that are specific to the type of print they are found on. **See Figure 7-4.**

A *symbol* is a graphic representation of an object, quantity, or unit. Symbols can also be used to show the type of material from which an object is made. Abbreviations and symbols should appear on a legend at the beginning of a set of prints.

Title Blocks

A *title block* is the area of a print that contains identifying information about the contents of the print. The title block is located in the lower right-hand corner of the drawing sheet. While title blocks can be customized, they usually include the company name and address, drawing title, drawing number, paper size, scale, and sheet number. **See Figure 7-5.** A separate revision block can also be included with the description, date, and approvals of changes made to a print.

Schedules

A *schedule* is a chart used to conserve space and display information in a concise and organized format. Schedules resemble spreadsheets with detailed information displayed in columns and rows. **See Figure 7-6.** A wide variety of schedules are found on different types of prints. A schedule can also be called a parts list or bill of materials if the given information relates to material components needed for construction.

Abbreviations and Symbols

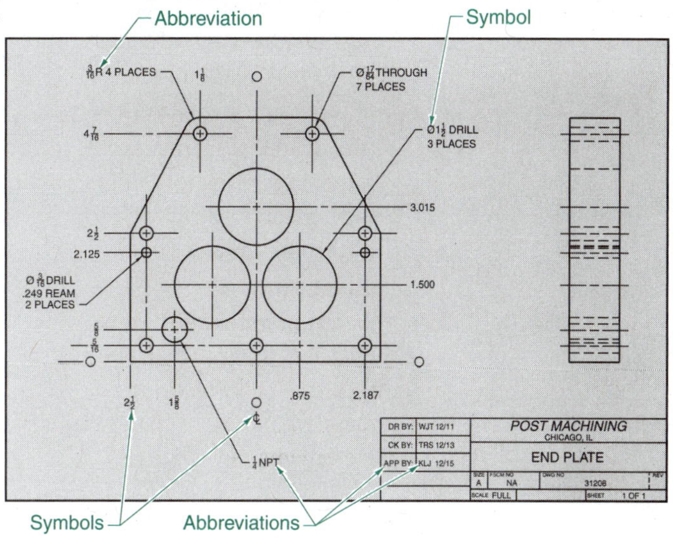

Figure 7-4. Abbreviations and symbols are used on prints to conserve space and promote consistency.

Title and Revision Blocks

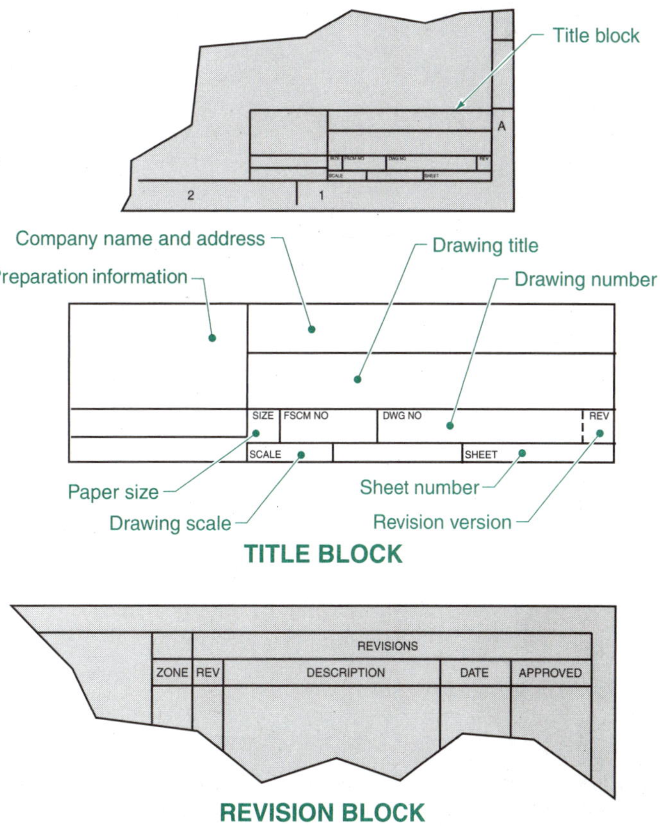

Figure 7-5. A title block and revision block give identifying information about the contents of a print and its subsequent revisions.

Schedules

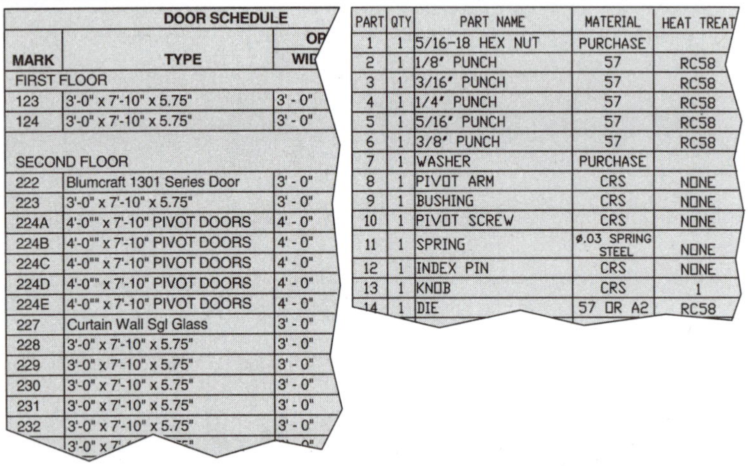

DOOR SCHEDULE		
MARK	TYPE	OP WID
FIRST FLOOR		
123	3'-0" x 7'-10" x 5.75"	3' - 0"
124	3'-0" x 7'-10" x 5.75"	3' - 0"
SECOND FLOOR		
222	Blumcraft 1301 Series Door	3' - 0"
223	3'-0" x 7'-10" x 5.75"	3' - 0"
224A	4'-0"" x 7'-10" PIVOT DOORS	4' - 0"
224B	4'-0"" x 7'-10" PIVOT DOORS	4' - 0"
224C	4'-0"" x 7'-10" PIVOT DOORS	4' - 0"
224D	4'-0"" x 7'-10" PIVOT DOORS	4' - 0"
224E	4'-0"" x 7'-10" PIVOT DOORS	4' - 0"
227	Curtain Wall Sgl Glass	3' - 0"
228	3'-0" x 7'-10" x 5.75"	3' - 0"
229	3'-0" x 7'-10" x 5.75"	3' - 0"
230	3'-0" x 7'-10" x 5.75"	3' - 0"
231	3'-0" x 7'-10" x 5.75"	3' - 0"
232	3'-0" x 7'-10" x 5.75"	3' - 0"

PART	QTY	PART NAME	MATERIAL	HEAT TREAT
1	1	5/16-18 HEX NUT	PURCHASE	
2	1	1/8' PUNCH	57	RC58
3	1	3/16' PUNCH	57	RC58
4	1	1/4' PUNCH	57	RC58
5	1	5/16' PUNCH	57	RC58
6	1	3/8' PUNCH	57	RC58
7	1	WASHER	PURCHASE	
8	1	PIVOT ARM	CRS	NONE
9	1	BUSHING	CRS	NONE
10	1	PIVOT SCREW	CRS	NONE
11	1	SPRING	ø.03 SPRING STEEL	NONE
12	1	INDEX PIN	CRS	NONE
13	1	KNOB	CRS	1
14	1	DIE	57 OR A2	RC58

Figure 7-6. Schedules resemble spreadsheets with detailed information displayed in columns and rows and are used to display information in an organized format.

Schedules can show the required information for a building's windows, lighting fixtures, and exterior finish specifications.

Schedules in Print and Drawing Library

Zones

A *zone* is an area on a print designated by horizontal and vertical labels along the margin. Zones are used to reference a specific location on a print and can vary in size from print to print. A print is separated into a regular grid with letter and number labels in the margins. Each area on the grid can be identified by a coordinate of the letter and number where it appears. Commonly, the print's base margin is labeled with numbers, while the right and/or left margins are labeled with letters. **See Figure 7-7.**

Zones

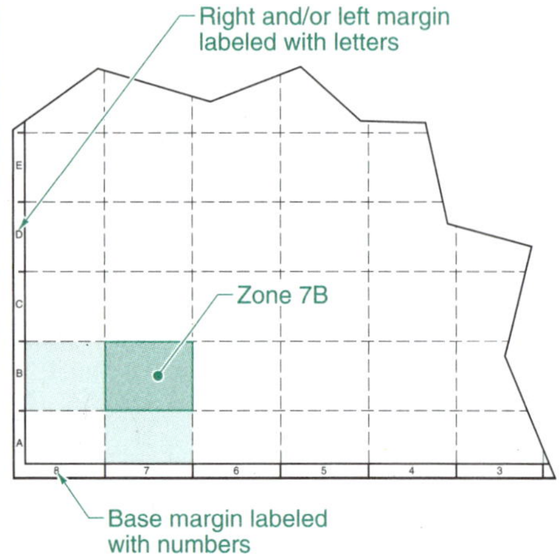

Figure 7-7. A zone, which is used to reference a specific location, is an area on a print designated by horizontal and vertical labels along the margin.

Notes

Prints can include general notes and sheet notes to incorporate additional information. General notes apply to the entire drawing and can cover broad or general topics. A *sheet note* is a note that provides information for a specific item on the print on which it appears. Sheet notes are identified by a symbol consisting of a number enclosed within a circle. The symbol may be located adjacent to the specific item or connected with a leader line. **See Figure 7-8.**

Specifications

A *specification* is a supplemental document with written instructions and information that is not necessarily shown on the print. Depending on the size of the job, the specifications may consist of a few notations on the prints or a separate book with hundreds of pages. Specifications can include information on legal responsibilities, insurance, and other necessary details. **See Figure 7-9.**

Notes

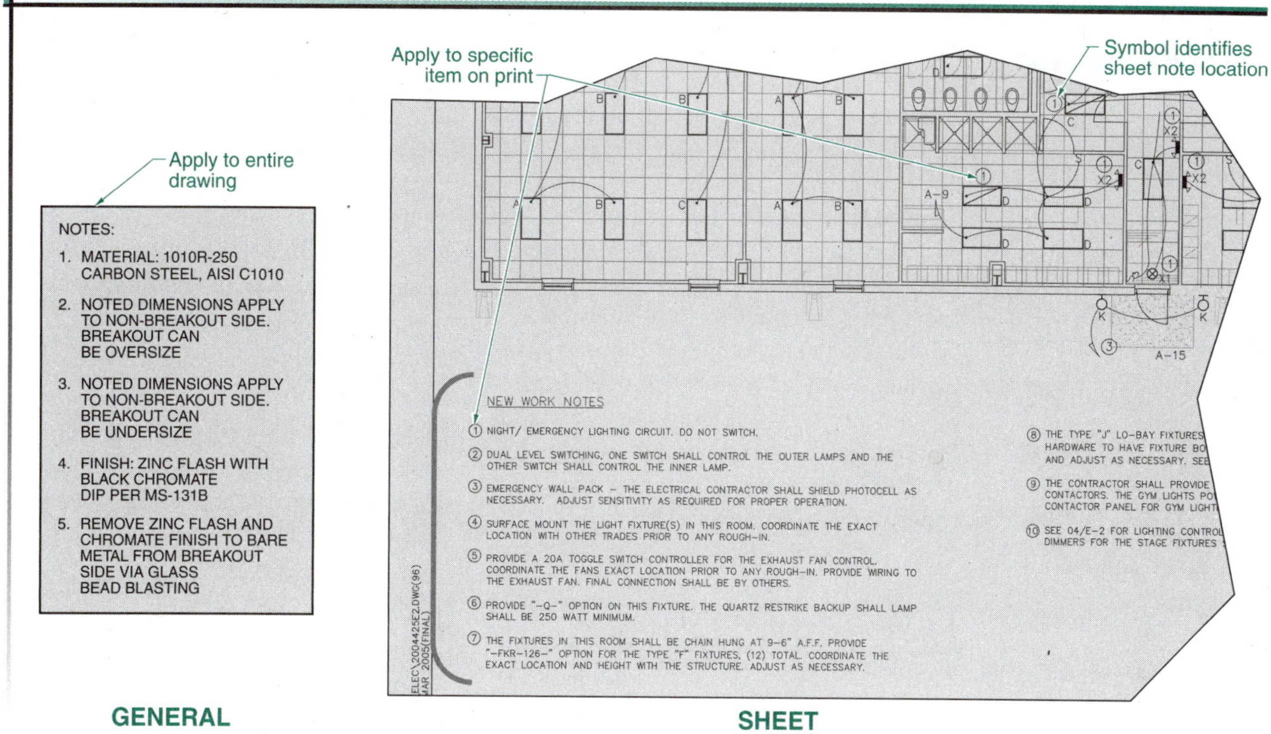

Figure 7-8. Prints can include general notes and sheet notes.

Specifications

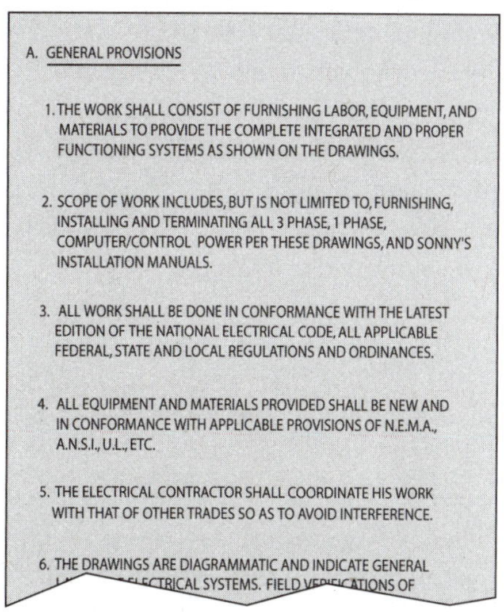
A. GENERAL PROVISIONS

1. THE WORK SHALL CONSIST OF FURNISHING LABOR, EQUIPMENT, AND MATERIALS TO PROVIDE THE COMPLETE INTEGRATED AND PROPER FUNCTIONING SYSTEMS AS SHOWN ON THE DRAWINGS.

2. SCOPE OF WORK INCLUDES, BUT IS NOT LIMITED TO, FURNISHING, INSTALLING AND TERMINATING ALL 3 PHASE, 1 PHASE, COMPUTER/CONTROL POWER PER THESE DRAWINGS, AND SONNY'S INSTALLATION MANUALS.

3. ALL WORK SHALL BE DONE IN CONFORMANCE WITH THE LATEST EDITION OF THE NATIONAL ELECTRICAL CODE, ALL APPLICABLE FEDERAL, STATE AND LOCAL REGULATIONS AND ORDINANCES.

4. ALL EQUIPMENT AND MATERIALS PROVIDED SHALL BE NEW AND IN CONFORMANCE WITH APPLICABLE PROVISIONS OF N.E.M.A., A.N.S.I., U.L., ETC.

5. THE ELECTRICAL CONTRACTOR SHALL COORDINATE HIS WORK WITH THAT OF OTHER TRADES SO AS TO AVOID INTERFERENCE.

6. THE DRAWINGS ARE DIAGRAMMATIC AND INDICATE GENERAL ... ELECTRICAL SYSTEMS. FIELD ... ATIONS OF

Figure 7-9. A specification is a document that supplements a print with written instructions that give additional information.

Notes and specifications on prints should be referenced during a construction job.

Sections in Print and Drawing Library

Section and Detail Drawings

Section and detail drawings can further describe what is being depicted on a print. Symbols are used to identify the location of a section or detail drawing on the whole print. The section or detail drawing header symbol usually contains a circle divided into two parts. The top part denotes what the view or drawing is named. The bottom part denotes the sheet where the section or detail drawing originated. In a section drawing, an arrow can also be added to the symbol to denote the direction of viewing. Section and detail drawings can be drawn as separate drawings or combined into one drawing if desired. **See Figure 7-10.**

Section and Detail Drawings

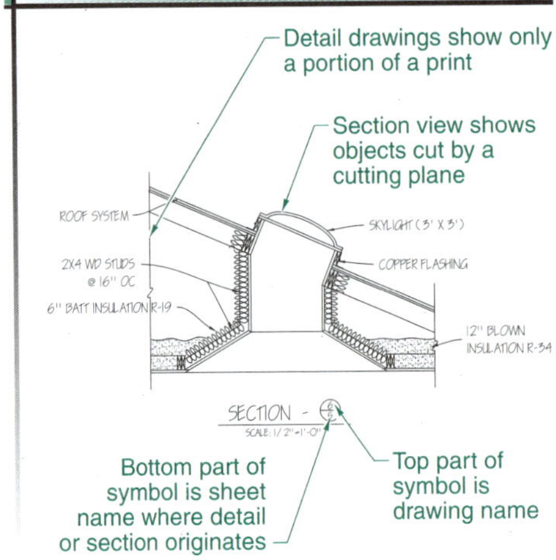

Figure 7-10. Section and detail drawings can be included on prints to further illustrate the subject.

Plot Plan in Print and Drawing Library

Floor Plan in Print and Drawing Library

ARCHITECTURAL PRINTS

Architectural prints contain various drawings to fully describe a building and its location, structure, and utilities. Architectural prints can also contain elements that are included on other types of prints, such as structural and mechanical components, but usually in less detail. Architectural prints generally include plans and elevations.

Architectural Plans

Architectural plans show various types of information needed to complete a project. Some plan types can be combined on the same print. **See Figure 7-11.** Plan types include the following:

- plat plans—give an overview of the entire area in which the building lot is located, including utility lines and nearby properties and buildings
- plot plans (site plans)—give details of the specific building and lot within the plat
- floor plans—show the specific features, such as walls and doorways, of one floor of a building with a separate floor plan for each floor
- foundation plans—show a building's foundations and structural supports
- structural plans—show the type, placement, and fabrication of materials used for the structure of a building
- utility plans—indicate the locations of utilities on the site

Architectural Elevations

Architectural elevations may be interior or exterior elevations. **See Figure 7-12.** Interior elevations show the shape and size of interior walls and partitions of a building and indicate any special interior finishes or fixtures. Exterior elevations give the shape and size of the exterior walls or roof of a building and include information on exterior finish materials; size and elevation dimensions; exterior door, window, and vent openings; and roof type.

TECH TIP

Architectural prints use plans to show a cross-section of room arrangements, partitions, walls, cabinets, and fixtures. Plans are viewed as if the viewer is looking down on the structure from above. A horizontal cutting plane passes through the structure at approximately 5′ above the floor of the level being shown. Objects below the cutting plane are shown using visible lines, while objects above the cutting plane are shown using hidden lines.

Architectural Plans

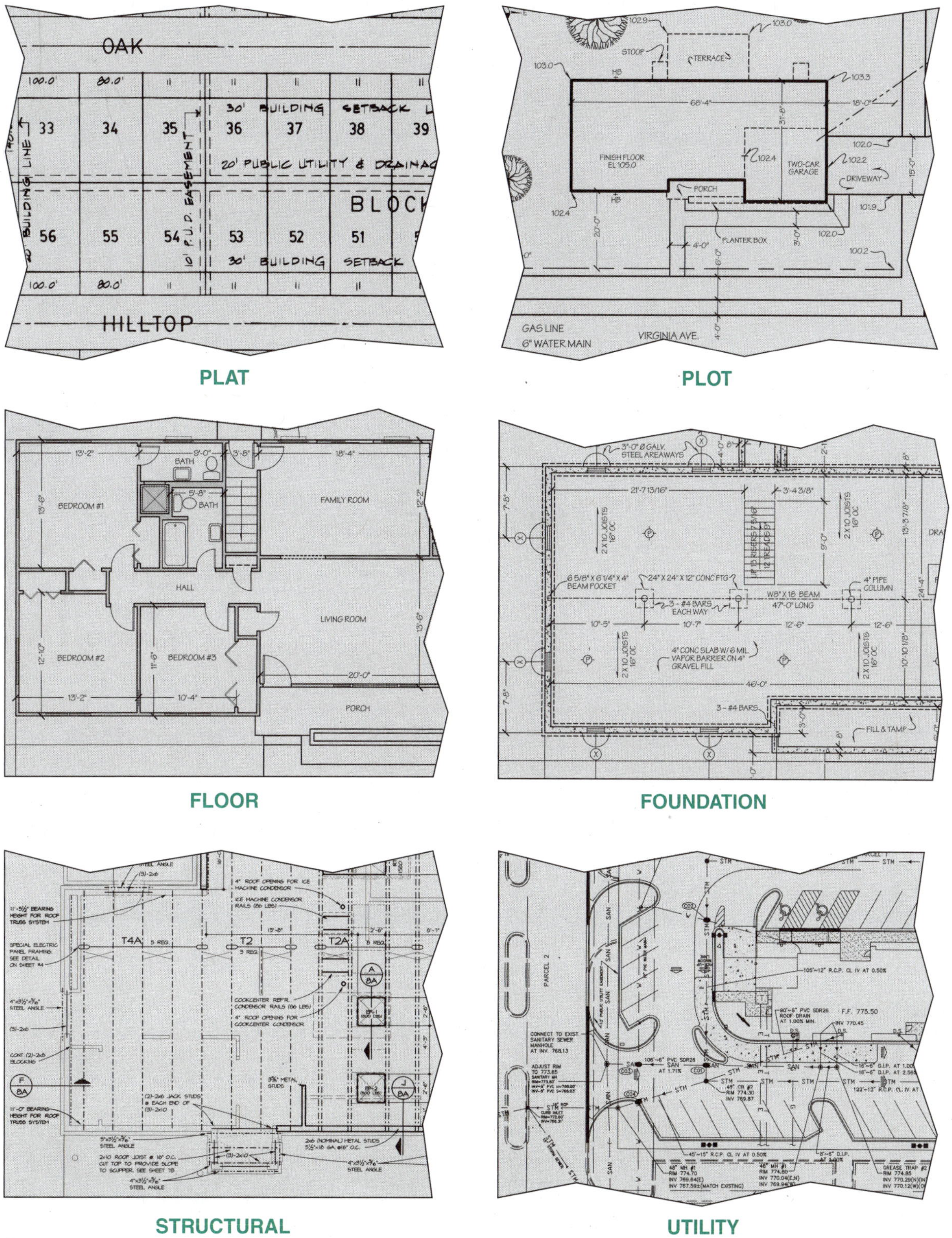

Figure 7-11. Architectural plans can include plat plans, plot plans, floor plans, foundation plans, structural plans, and utility plans.

Elevations

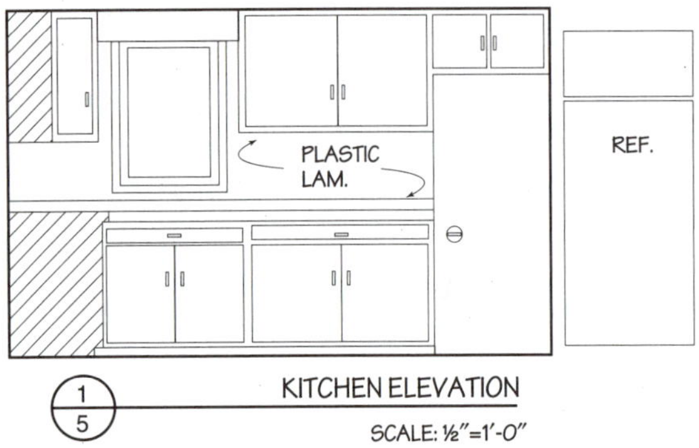

INTERIOR

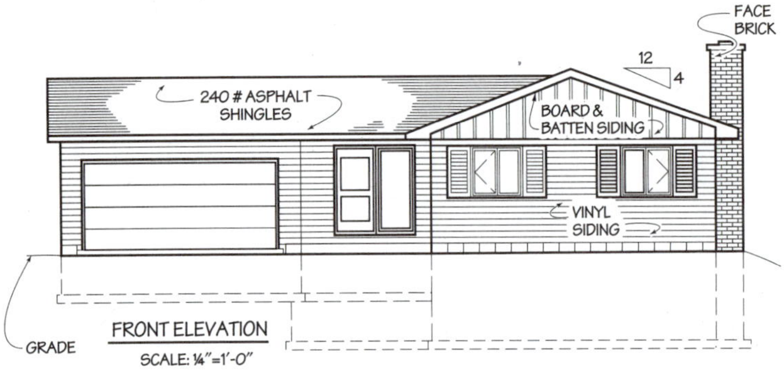

EXTERIOR

Figure 7-12. Elevations shown in architectural prints generally include interior and exterior elevations.

Structural Print Types

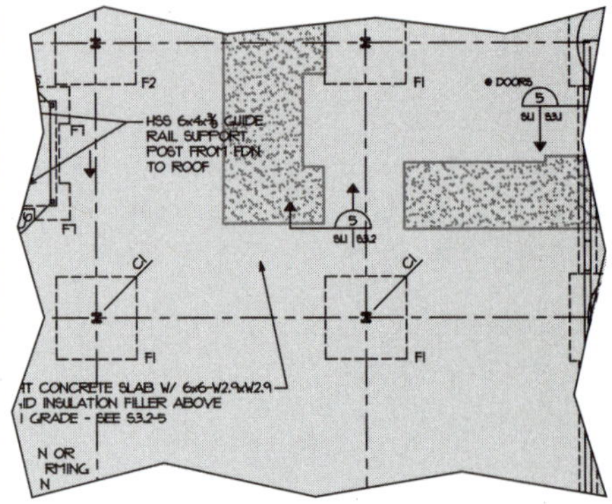

FOUNDATION PLAN

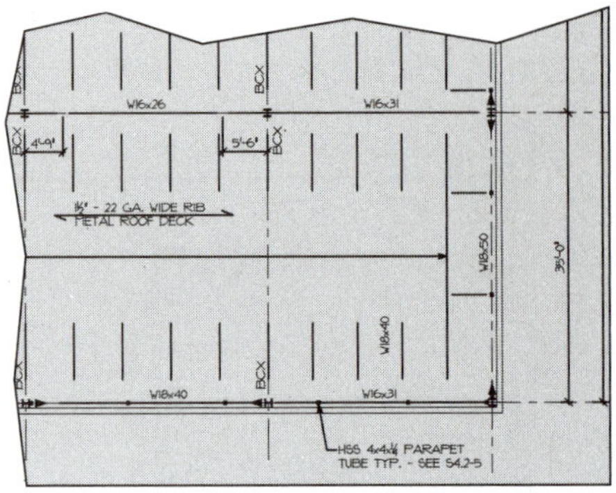

ROOF FRAMING PLAN

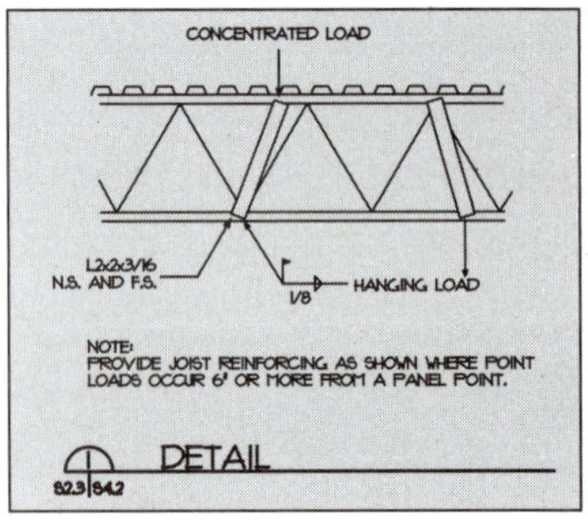

LOAD AND FASTENING INFORMATION

Figure 7-13. Structural prints show the framing and support systems of a structure, as well as foundation plans, load information, and erection sequences.

Foundation Plan in Print and Drawing Library

Roof Plan in Print and Drawing Library

STRUCTURAL PRINTS

Structural prints show the framing and support systems of a structure, as well as foundation plans, load information, and erection sequences. **See Figure 7-13.** Structural supports include columns, beams, joists, and girders. Columns are vertical support members, while beams, joists, and girders are horizontal support members. Fastening methods, such as bolting and welding, are also indicated on structural prints. Structural prints give the specific types and sizes of products required for a project and how they will be used, as well as detailed drawings outlining how each individual part is fabricated.

MECHANICAL PRINTS

Mechanical prints include information on the mechanical systems that circulate fluids, gases, and air for a building.

The main types of mechanical prints are plumbing prints; heating, ventilating, and air conditioning (HVAC) prints; and fire protection system prints. Information related to mechanical systems can also be found on other types of prints, such as architectural or civil prints.

Plumbing Prints

Plumbing prints include information on drainage systems, piping systems, and water supplies. Plumbing prints use symbols to depict the specific fixtures, fittings, and valves, and generally use piping diagrams to convey information. **See Figure 7-14.**

Piping Diagrams. Piping diagrams show the placement and relationship of pipes and fixtures in a structure. While piping can be depicted in plan view, piping is often drawn orthographically using a single line for piping and graphic symbols for fixtures. Two common types of piping diagrams are schematic and isometric. **See Figure 7-15.**

A *schematic piping diagram* is an orthographic diagram of a plumbing system without regard to scale and exact location of the fixtures shown. Pipes are usually drawn at horizontal and vertical axes with joints at right angles. A single line is generally used to depict pipes, but pipes can also be shown pictorially using two lines.

Plumbing Prints

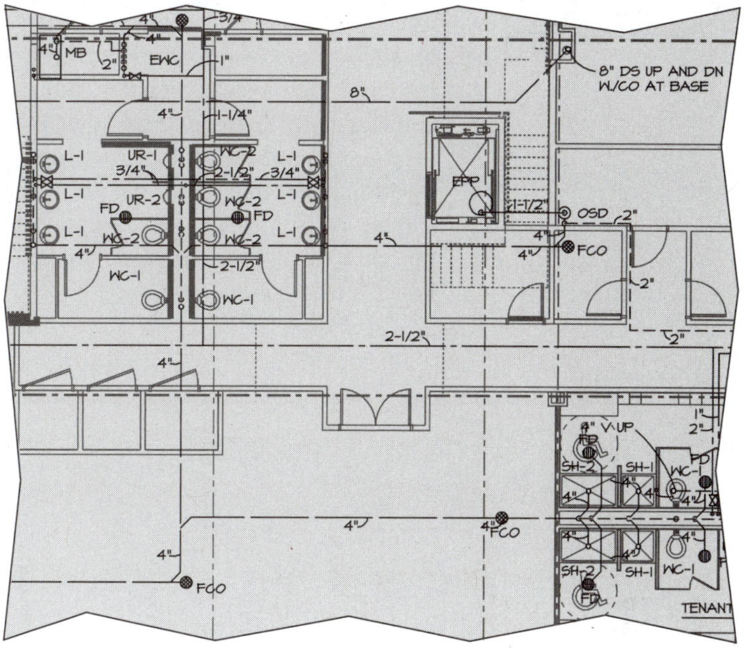

Figure 7-14. Plumbing prints include information on drainage and piping systems for a structure or site.

An *isometric piping diagram* is an isometric diagram of a plumbing system. All horizontal pipes are drawn as 30° lines and all vertical pipes are drawn as vertical lines.

Piping Diagram in Print and Drawing Library

Piping Diagrams

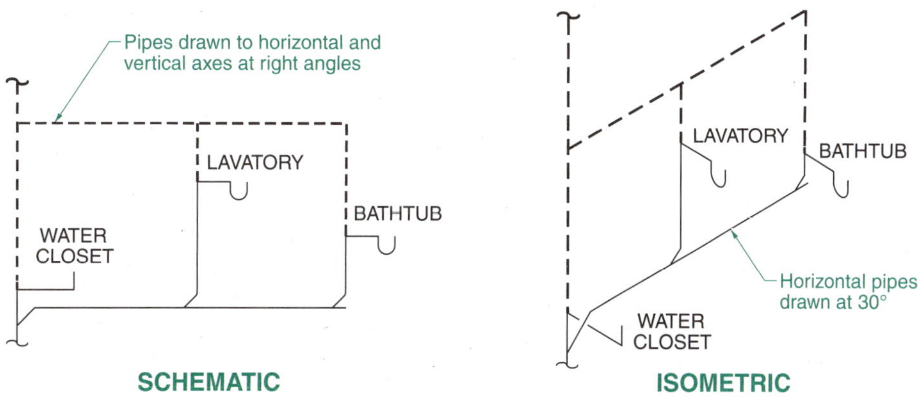

Figure 7-15. Common types of piping diagrams are schematic and isometric, which show pipe placement and the relationships between plumbing pipes and fixtures in a structure.

Mechanical Plan in Print and Drawing Library

HVAC Prints

HVAC prints show systems of ductwork used for heating and cooling a structure, along with the locations of heating/cooling units, thermostats, registers, and vents, and ductwork sizes and airflow. HVAC prints use symbols to show the type and location of equipment for the project. **See Figure 7-16.** Any specific information on the equipment or system is included in the notes, specifications, or schedules.

Fire Protection System Prints

Fire protection system prints show fire protection equipment and systems. These systems detect and/or suppress fire and include devices that sense heat, smoke, or flame, and activate a fire alarm or suppression system. Fire protection system prints indicate locations of equipment and devices such as pipes, sprinkler heads, valves, and alarm boxes that are related to a fire protection system. **See Figure 7-17.**

HVAC Prints

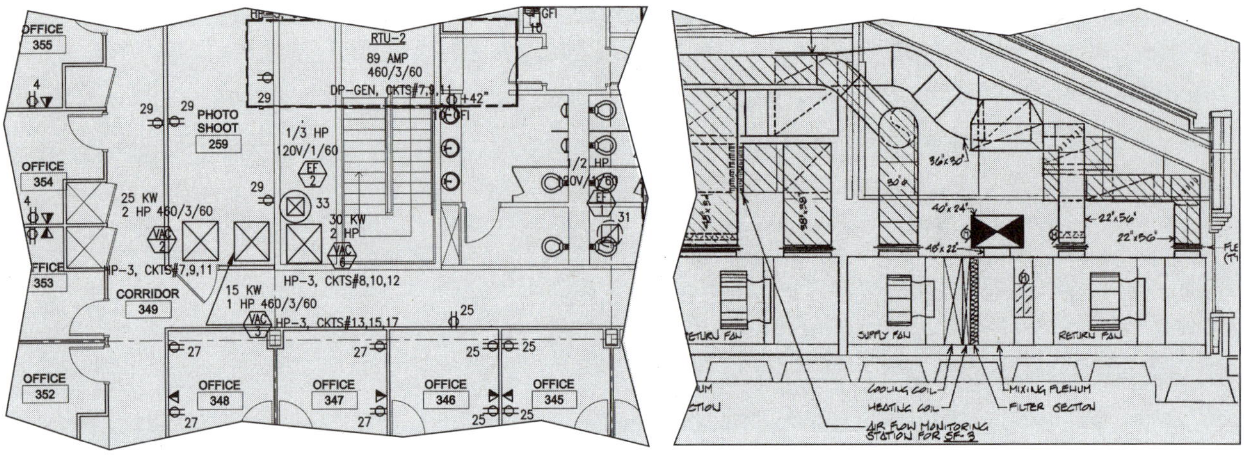

Figure 7-16. HVAC prints show systems of ductwork and equipment used for heating and cooling a structure.

Fire Protection System Prints

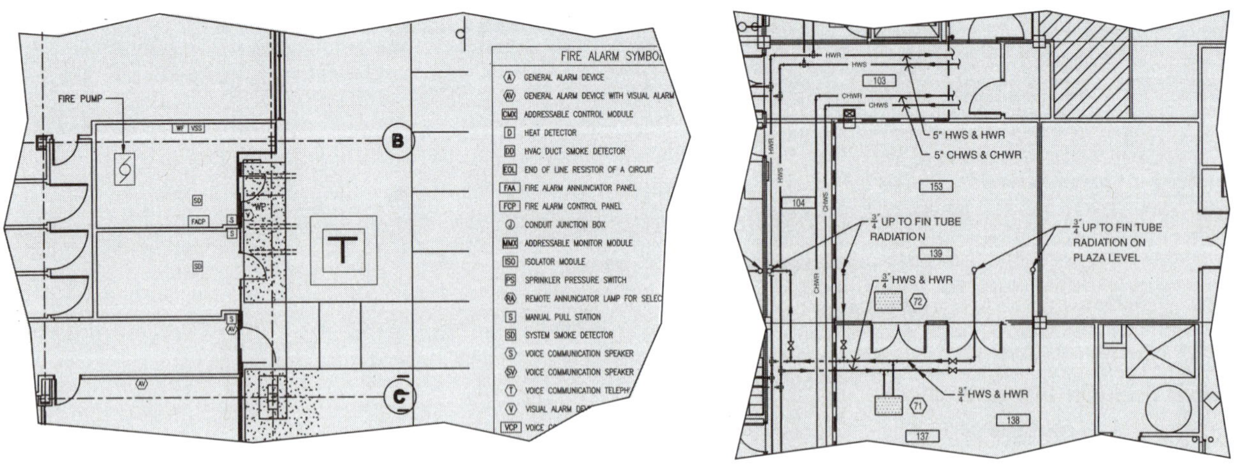

Figure 7-17. Fire protection system prints show fire protection equipment and systems, which detect and/or suppress fire in a structure.

ELECTRICAL PRINTS

Electrical prints show the locations and connections of electrical systems including connections to power supplies. These include wiring and lighting fixtures, power outlets, installation of attachments and power for electrical equipment, and wiring and finish information for communication, security, and fire alarm systems. **See Figure 7-18.**

Electrical information can be shown on the site plan, floor plan, or in a separate diagram. Electrical diagrams and schematic diagrams are two common types of drawings that are included on an electrical print.

Electrical Diagrams

Electrical diagrams use symbols to show the location of electrical elements in a simple and graphic way. **See Figure 7-19.** Electrical diagrams can include the following:

- one-line diagrams—use a single line and symbols to show a basic overview of the power flow
- wiring diagrams—give the general location of devices and components and how they are wired in the circuit
- ladder (line) diagrams—use horizontal and vertical lines with symbols to show the logic of a control circuit; connections are shown, but not in actual circuit positions

Schematic Diagrams

A *schematic diagram* is a drawing that shows the electrical connections and functions of a circuit using graphic symbols. A schematic diagram shows the relationship between devices and components without regard to physical location, size, or shape of items in a circuit. **See Figure 7-20.**

CIVIL PRINTS

Civil prints give measurements and other technical information pertaining to a building site. *Surveying* is the process of measuring land by recording elevations, noting directions and lengths of property lines, and noting sizes of structures. Surveying uses known reference points and surveying stations on a property as points of reference to mark the accurate locations of a property's major characteristics.

Electrical Plan in Print and Drawing Library

Electrical Prints

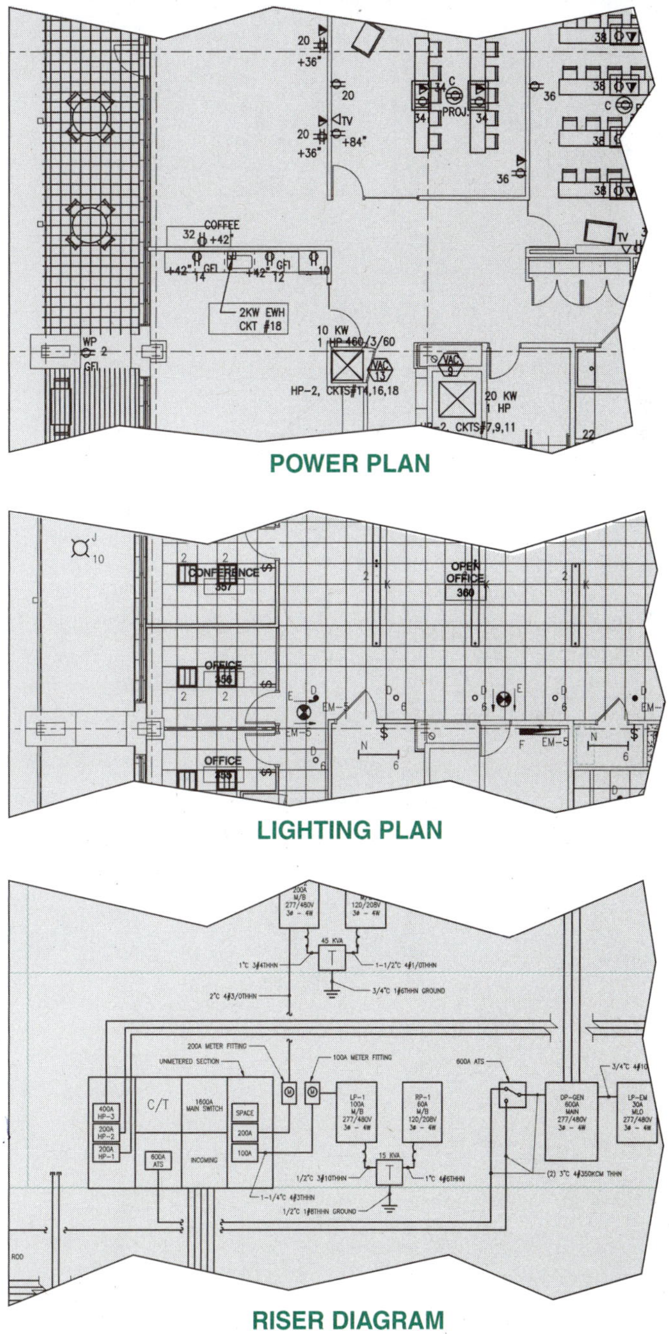

Figure 7-18. Electrical prints use lines and symbols to show the locations and connections of various electrical systems, such as power and lighting.

Electrical Diagrams

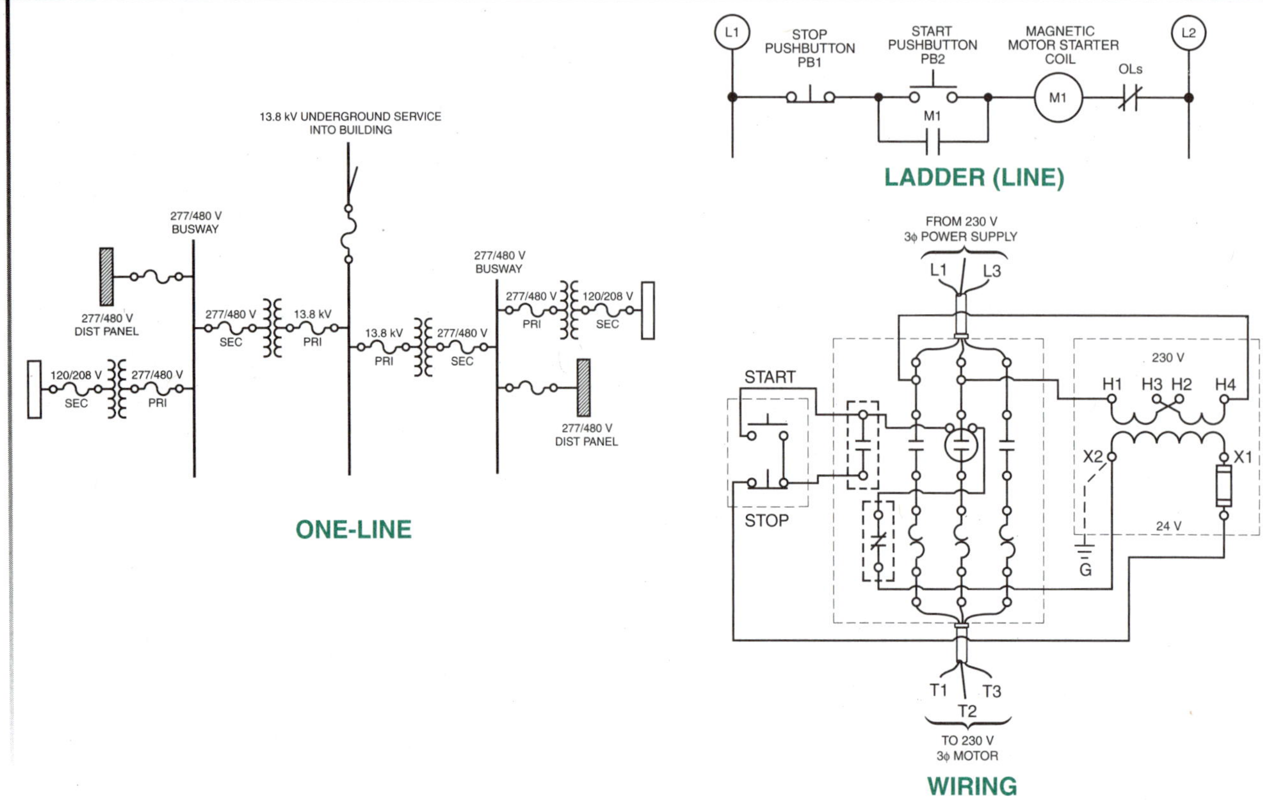

Figure 7-19. Electrical diagrams use symbols to show the location of electrical elements and include a one-line diagram, ladder (line) diagram, and wiring diagram.

Schematic Diagrams

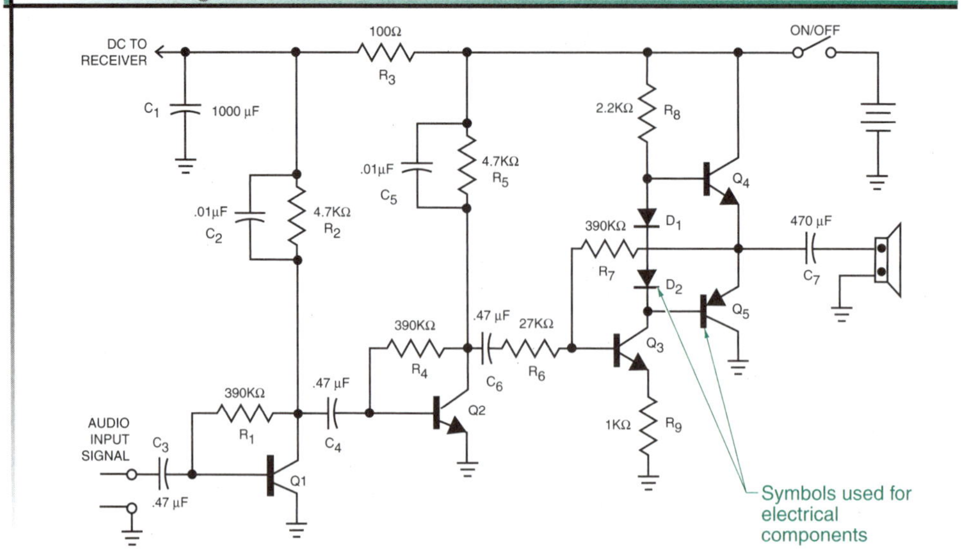

Figure 7-20. A schematic diagram shows the electrical connections and functions of a circuit using graphic symbols and without regard to physical location, size, or shape of items in a circuit.

Property or reference lines are indicated using compass directions and distance from reference points or a radius and the length of the curvature. Lines indicating property elevations and utility placement, connections, piping, and cables on the building site are found on civil prints. **See Figure 7-21.**

Civil Prints

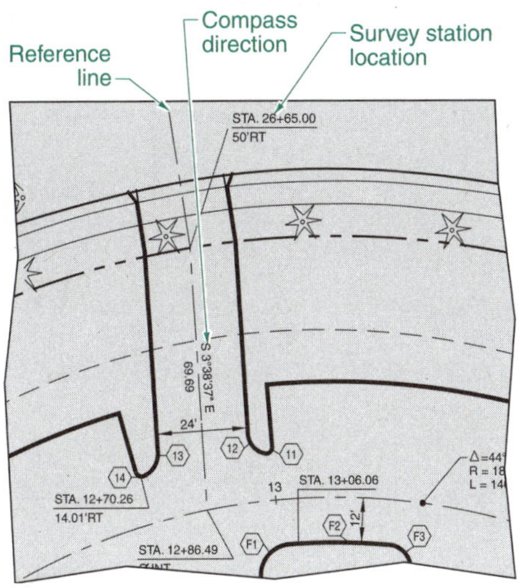

Figure 7-21. Civil prints give measurements and other technical information pertaining to a building site.

Civil prints also provide information on the soil and topographical characteristics, drainage systems, site layout, driveway and parking areas, and construction-related needs, such as landscape removal and equipment storage. Landscaping and plant and soil treatments of the area are also included in civil prints, or can be included in a separate division, generally labeled "L".

MANUFACTURING PRINTS

Manufacturing prints show detailed information needed to manufacture a part, such as required welds, and any specifics needed to cast, forge, stamp, or machine the part. Manufacturing prints often contain a parts list (bill of materials) describing necessary parts for manufacture.

Manufacturing prints can include detail and assembly prints. Detail prints may show various views, as well as the necessary dimensions, tolerances, notes, or additional information needed to manufacture an object. **See Figure 7-22.**

3.5" Vise in Print and Drawing Library

Detail Prints

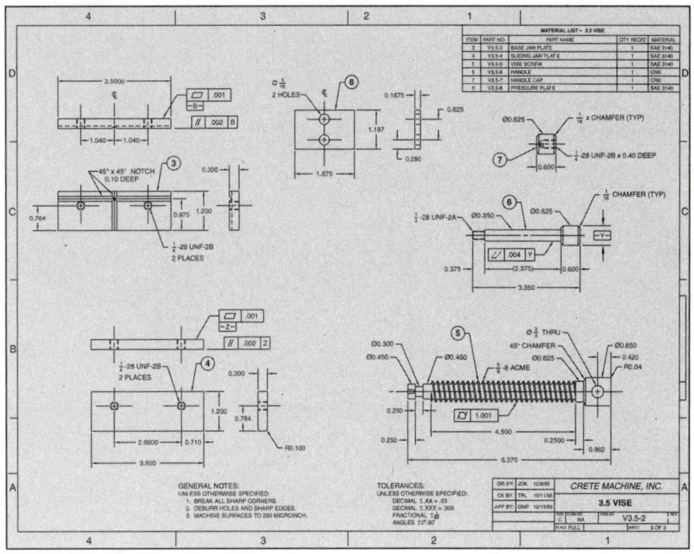

Figure 7-22. Detail prints show various views, dimensions, tolerances, notes, and any additional information needed to manufacture an object.

An *assembly print* is a print that shows how two or more parts fit together. All manufactured and purchased parts required for the assembly are identified on the print using a leader with an encircled number that matches the parts list. **See Figure 7-23.**

Assembly prints can be either orthographic or schematic representations of the finished assembly. Schematic assembly prints often show circuits and layouts of systems such as piping or ventilation ducts. The are drawn in plan or pictorial view and are not drawn to scale. Detail and assembly prints can be combined in the same drawing if desired.

> **TECH TIP**
> *Assembly prints show how parts fit together and often include the overall height, width, and depth measurements of the finished assembly. The print can include conventional views, sectional views, detail views, exploded views, or any other drawing type needed to portray the finished assembly. Assembly prints can also portray installation specifications, such as clearance requirements and dimensions for mounting and minimum space.*

Assembly Prints

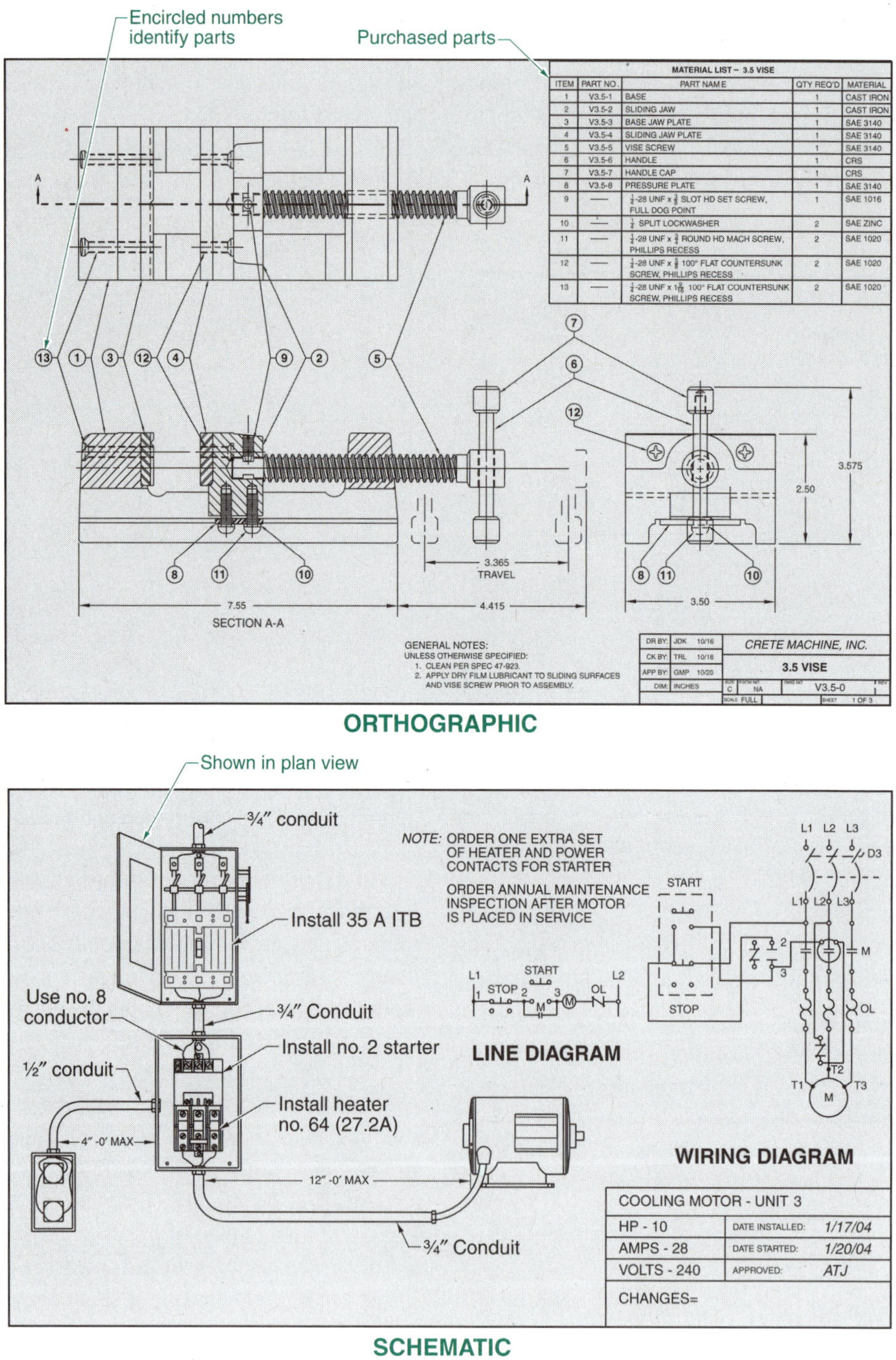

Figure 7-23. Assembly prints can be either orthographic or pictorial (schematic) representations of the finished assembly.

Chapter 7 Quick Quiz®

Prints

Review 7

Name ______________________ Date ______________

______________ **1.** A(n) ___ is a drawing that shows the structure and related characteristics of an object, building, or site.

T F **2.** Print divisions are labeled with a lowercase letter prefix.

______________ **3.** Print divisions include ___, structural, mechanical, electrical, and civil.

______________ **4.** A(n) ___ is an orthographic drawing of an object viewed from above and contains a horizontal cutting plane that passes through the object.

______________ **5.** ___ generally depict the front, back, left side, or right side of an object and are depicted in true shape with no perspective.

______________ **6.** A(n) ___ is a letter or group of letters that represents a term or phrase.

T F **7.** The title block is located in the lower right-hand corner of the drawing sheet.

______________ **8.** A(n) ___ is a chart used to conserve space and display information in a concise and organized format.

______________ **9.** A(n) ___ note provides information for a specific item on the print on which it appears.

______________ **10.** Architectural prints generally include ___ and elevations.

T F **11.** Manufacturing prints show the framing and support systems of a structure, as well as foundation plans, load information, and erection sequences.

______________ **12.** The three main types of mechanical prints are ___ prints, HVAC prints, and fire protection system prints.

T F **13.** A schematic piping diagram is drawn to scale and has horizontal pipes drawn as 30° lines.

______________ **14.** Electrical diagrams can include one-line diagrams, ___ diagrams, and ladder (line) diagrams.

______________ **15.** A(n) ___ diagram shows the relationship between electrical devices and components without regard to the physical location, size, or shape of items in a circuit.

T F **16.** Civil prints give measurements and other technical information pertaining to a building site.

______________ **17.** ___ is the process of measuring land by recording elevations, noting directions and lengths of property lines, and noting sizes of structures.

______________ **18.** ___ prints show detailed information needed to manufacture a part, such as required welds, and any specifics needed to cast, forge, stamp, or machine the part.

T F **19.** A detail print shows how two or more parts fit together.

________________ **20.** ___ prints can be either orthographic or pictorial representations of the finished assembly.

IDENTIFICATION—Electrical Drawings

______d______ **1.** Schematic

______a______ **2.** One-line

______c______ **3.** Wiring

______b______ **4.** Ladder (line)

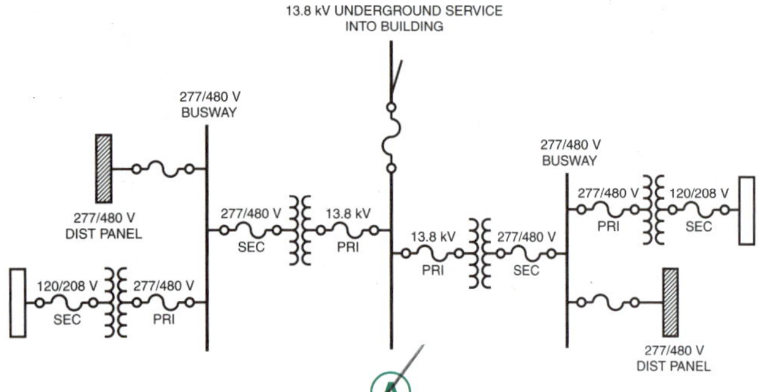

A

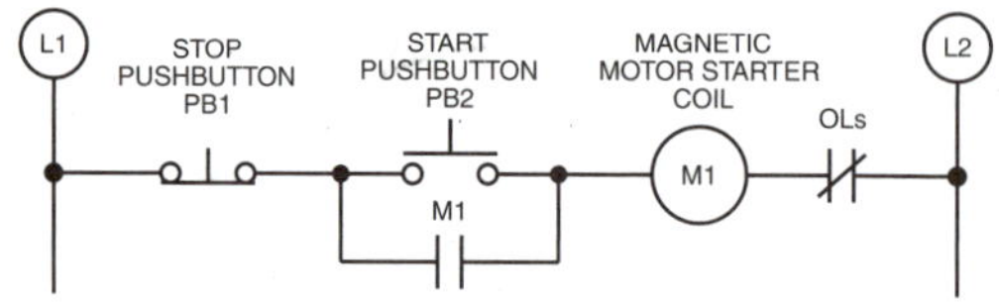

B

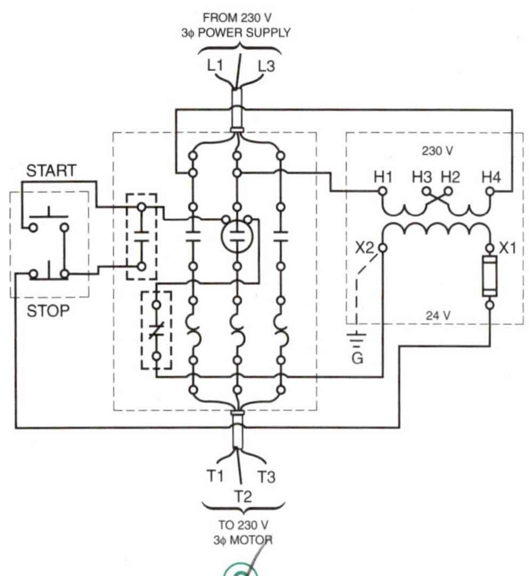

C

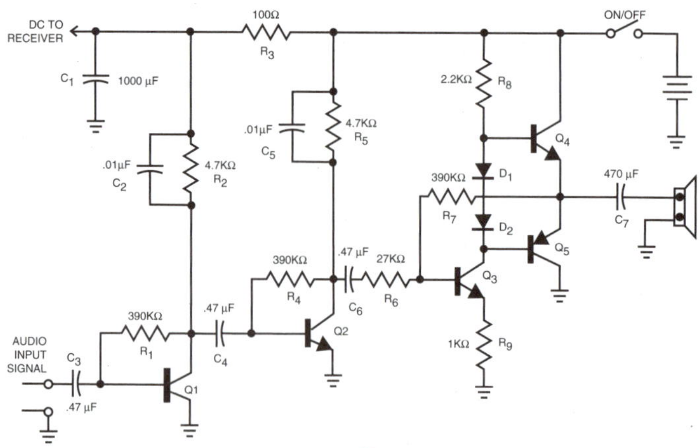

D

IDENTIFICATION—Architectural Plans

______b______ **1.** Plot

______a______ **2.** Foundation

_______________ **3.** Utility

_______________ **4.** Floor

_______________ **5.** Structural

_______________ **6.** Plat

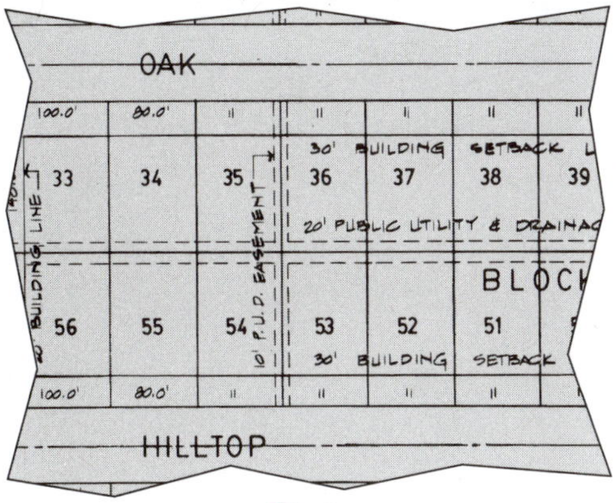

A

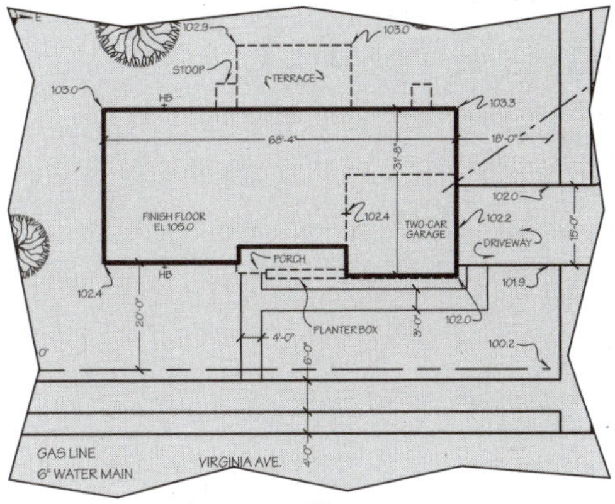

B

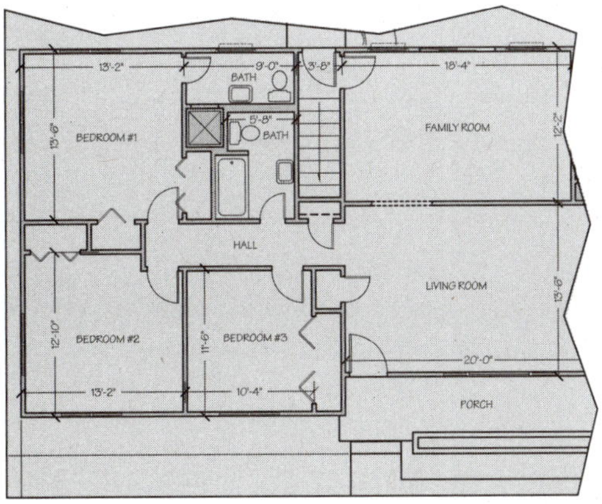

C

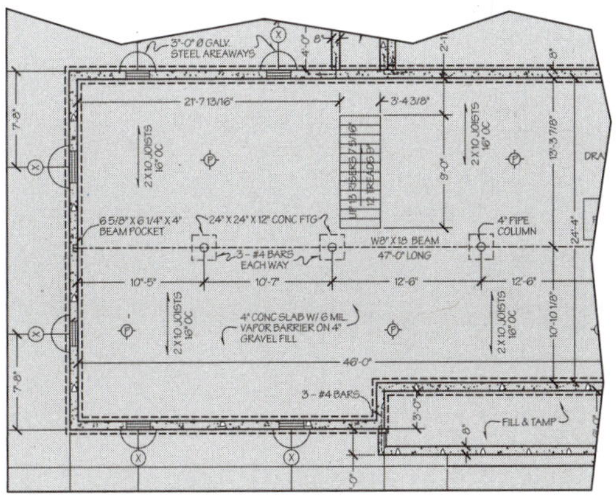

D

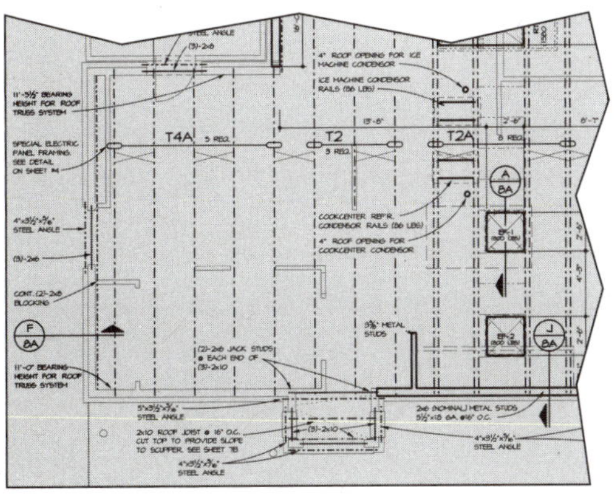

E

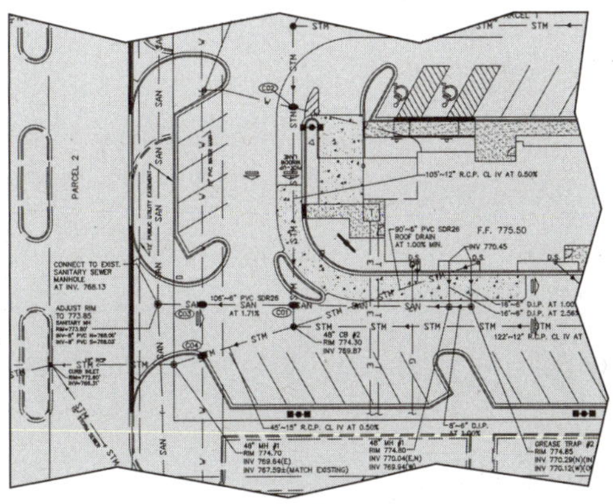

F

EXERCISES

Use the pictorial drawings provided to identify the correct elevation view.

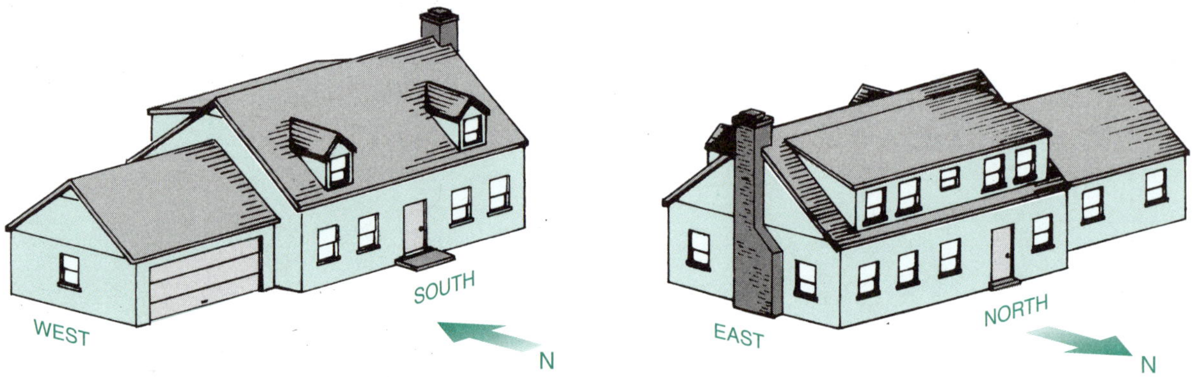

_______ 1. West elevation

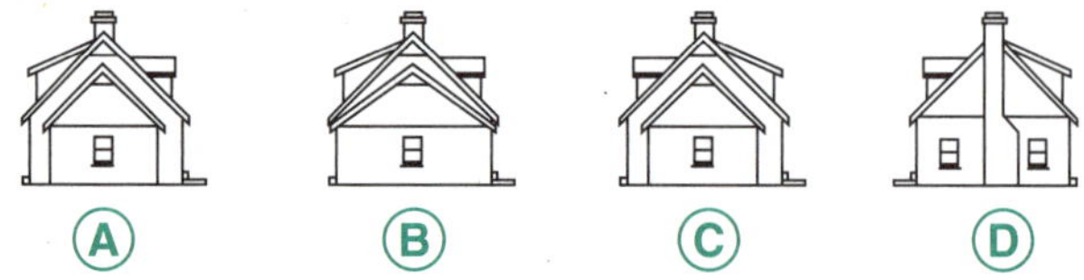

_______ 2. South elevation

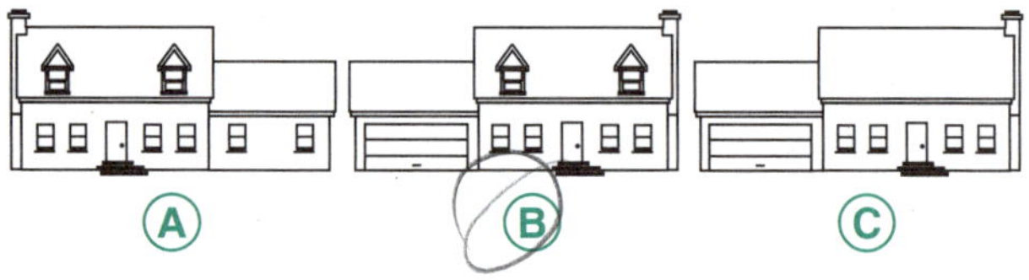

_______ 3. East elevation

_______ 4. North elevation

Use the pictorial drawings and floor plan provided to answer the following questions.

_______________ **5.** A person standing at X would see the ___ elevation.

_______________ **6.** A person standing at Y would see the ___ elevation.

_______________ **7.** Wall section A borders what room area based on the floor plan?

_______________ **8.** Wall D borders what room area based on the floor plan?

_______________ **9.** Wall E borders what room area based on the floor plan?

_______________ **10.** Wall F borders what room area based on the floor plan?

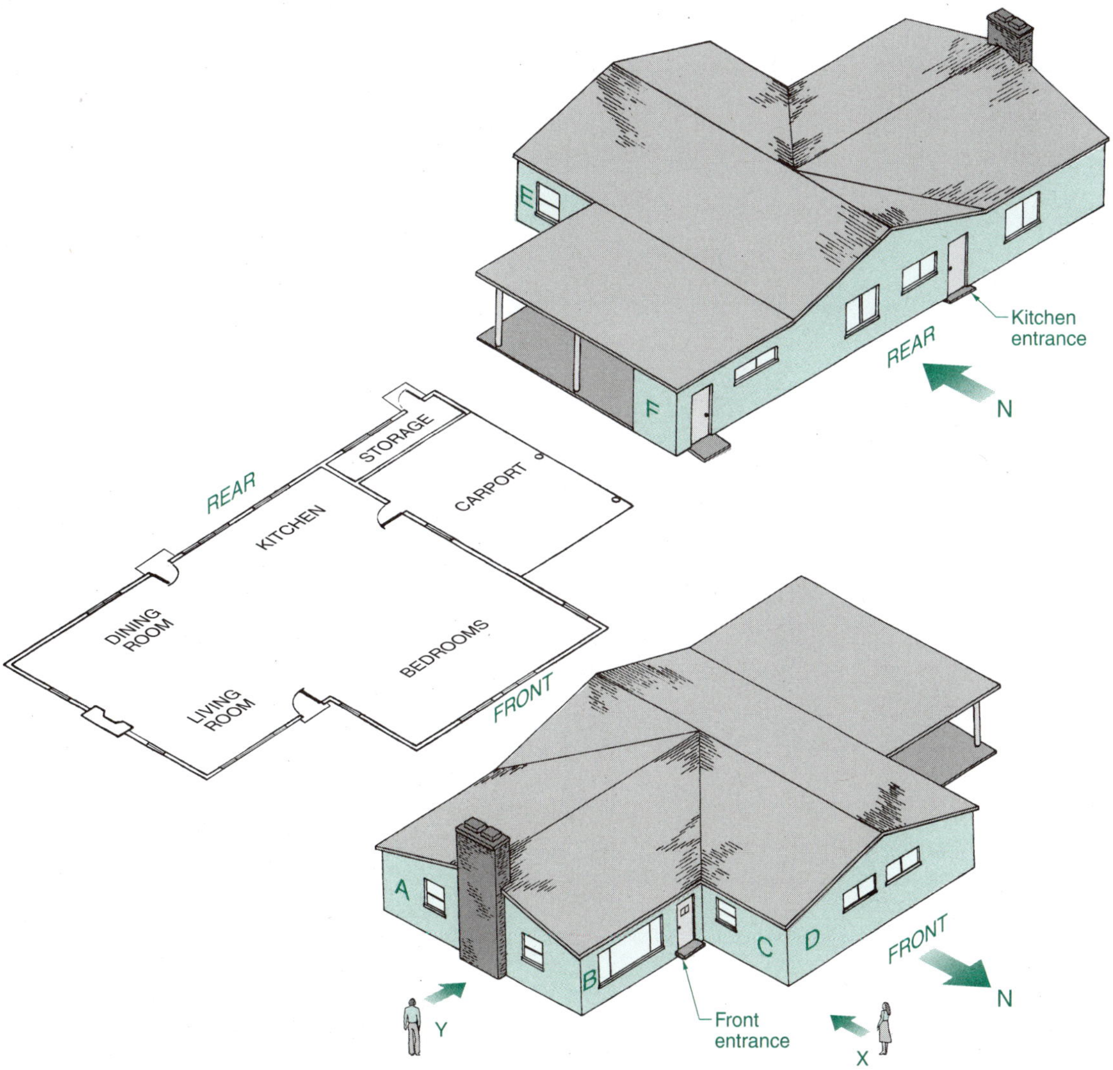

Use the print provided to answer the following questions.

____________ **11.** The radius of the curve at A is ___″.

____________ **12.** How many holes are indicated at B with a diameter of $17/64$″?

____________ **13.** What is the diameter of the hole indicated at C?

____________ **14.** The hidden line at D is the edge of one of the ___″ diameter holes.

____________ **15.** What view of the object is shown at E?

____________ **16.** What type of line is at F?

____________ **17.** How far from the center line of the object is the line at G?

____________ **18.** Where does the dimension at H originate from?

____________ **19.** How far apart is the line indicated at I and the center of the hole at J?

____________ **20.** How far apart are the centers of the holes at B and K?

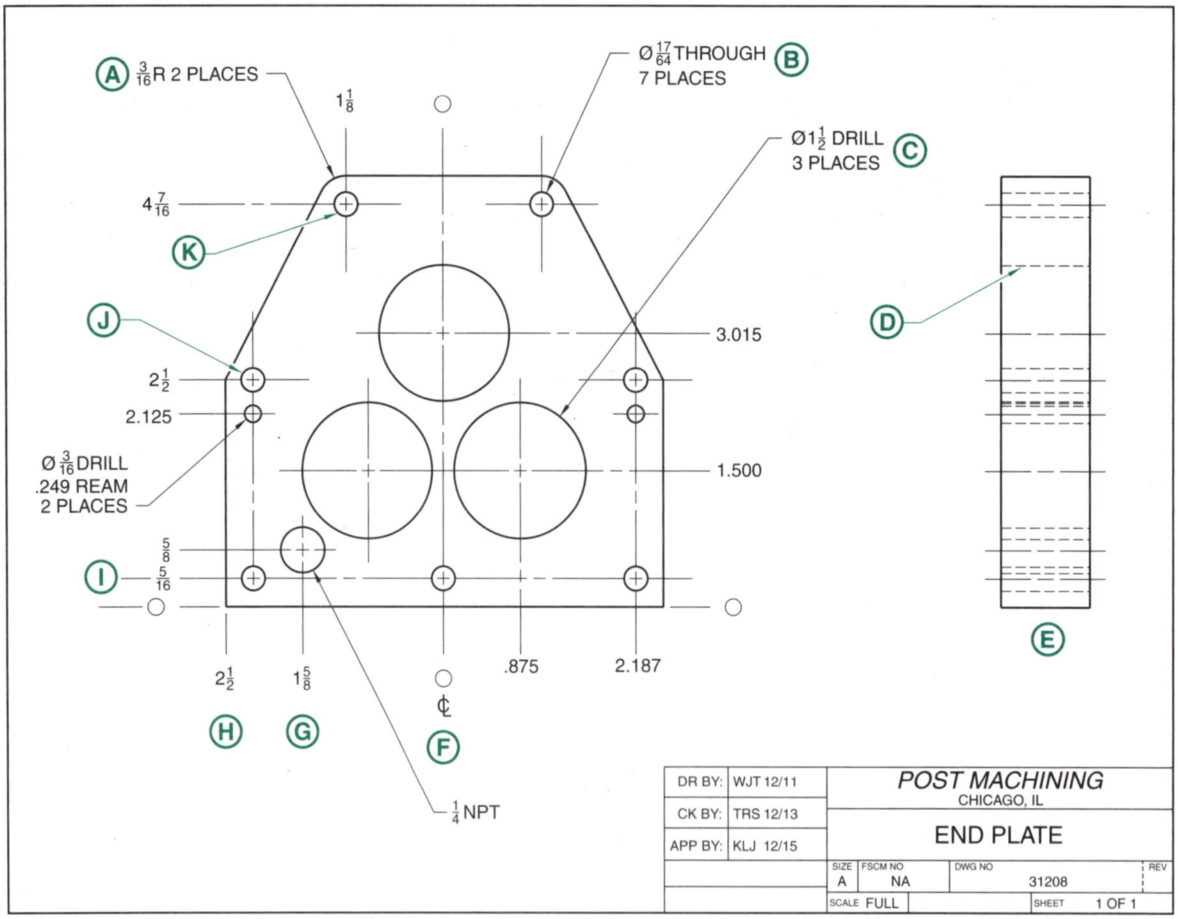

Use the assembly print provided to answer the following questions.

____________ **21.** How many parts does the final assembly contain?

____________ **22.** Part A is ___ mm long.

____________ **23.** Part B is ___″ long.

______ **24.** How many SPACER B-803-28533 parts are required for the assembly?

______ **25.** How many HEX HD. SCR. #10-24 × ¾ LG. (STN. STL.) parts are required for the assembly?

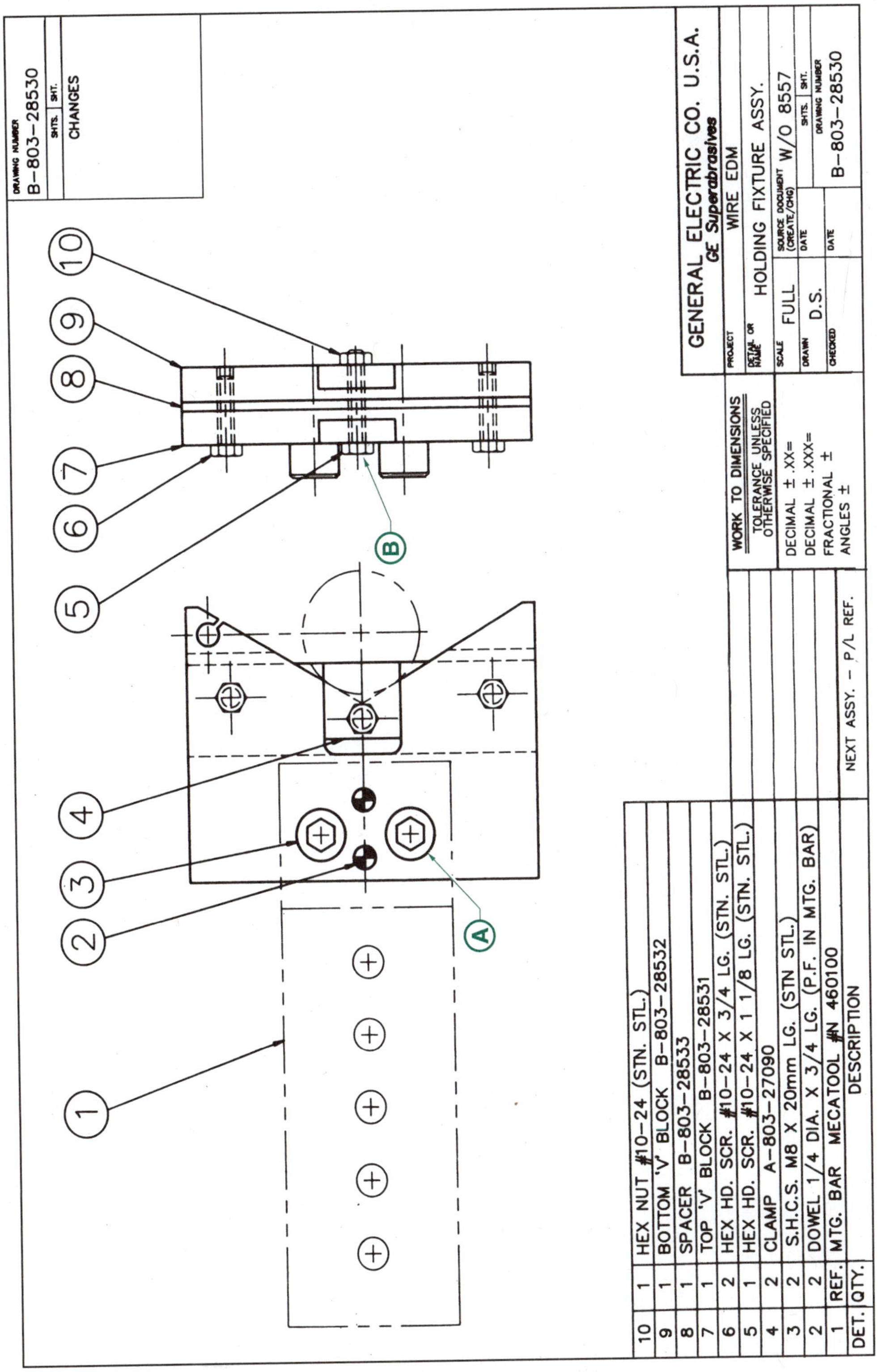

ASSEMBLY PRINT

Technical Sketching and Drawing

8 Specialized Drawings

INFORMATION REPRESENTATION

Specialized drawings convey information for specific industries or uses. In order to be clearly understood by a number of people, information should be represented in an organized and consistent way. Labels and data should convey their meaning while still being readable and conserving space. Symbols and graphs are two solutions for conveying large amounts of information while maintaining a concise drawing.

Symbols

A *symbol* is a graphic representation of an object, quantity, or unit. Symbols are often used to conserve space on a drawing. Symbols are not based on any specific language and can be easily recognized. When an option exists, the use of symbols is preferred over the use of abbreviations. Symbols are always used for equations and math operations.

Drawings use specialized symbols to represent architectural materials, plan components, electrical outlets and fixtures, plumbing fixtures and piping, and HVAC equipment. **See Figure 8-1** and **Appendix.** Conflicts may arise when symbols from multiple industries are used together because the same symbol may have different meanings depending on the context. Notes should be used to clarify the intent of a symbol that might have multiple meanings. A list of symbol meanings should also be included on a drawing or set of prints.

Graphs

Graphs provide a visual interpretation of data in a clear and concise manner. A *graph* is a diagram that shows the relationship between two or more variables. Most graphs contain a grid, scales, labels, plotted information, and a title. **See Figure 8-2.** Common graphs include line graphs, bar graphs, circle graphs, and pictographs.

Symbols are used on drawings such as prints as a way to clearly portray the required information while conserving space.

Symbols

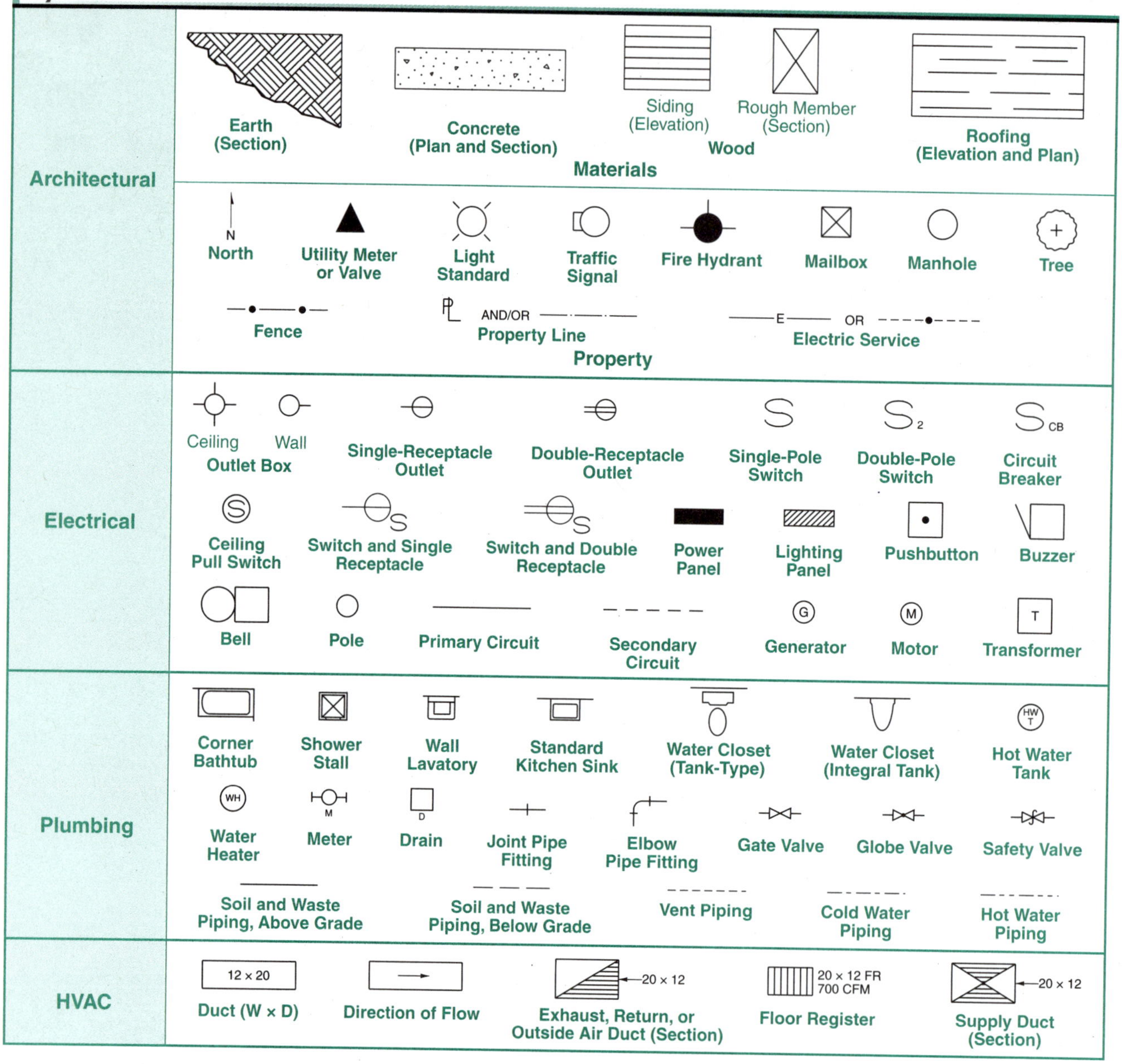

Figure 8-1. Specialized symbols are used on drawings to represent architectural materials, print components, electrical outlets and fixtures, plumbing fixtures and piping, and HVAC equipment.

Line Graphs. A *line graph* is a graph in which points representing variables are connected by a line. Line graphs show the relationship and change between variables. Line graphs have an x-axis (horizontal), a y-axis (vertical), and an origin point where the two axes meet. Points are plotted by determining the point in the graph space where the value on the x-axis corresponds to the value on the y-axis. **See Figure 8-3.**

Three types of line graphs are straight-line, curved-line, and broken-line. Straight-line and curved-line graphs have related variables so that an unknown value can be determined from a known value. Broken-line graphs contain variables with no set relationship to each other outside of the line graph.

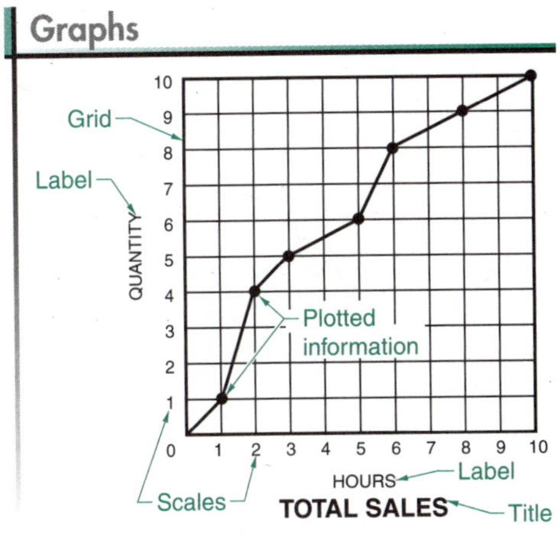

Figure 8-2. Graphs contain a grid, scales, labels, plotted information, and a title.

To draw a line graph, the given values should be arranged in a logical order, such as by size or sequence. The axes are assigned a variable, labeled, and given a scale of values. Information is plotted by determining the point in the graph space where the given x and y values meet and by marking each point with a small dot. The points are then connected with a straightedge or curve.

Bar Graphs. A *bar graph* is a graph in which values of variables are represented by bars. The bars may be drawn horizontally or vertically. The length of a bar indicates a value on the axis. **See Figure 8-4.** Bar graphs can also represent multiple values for a variable by using additional bars or shading methods for each value.

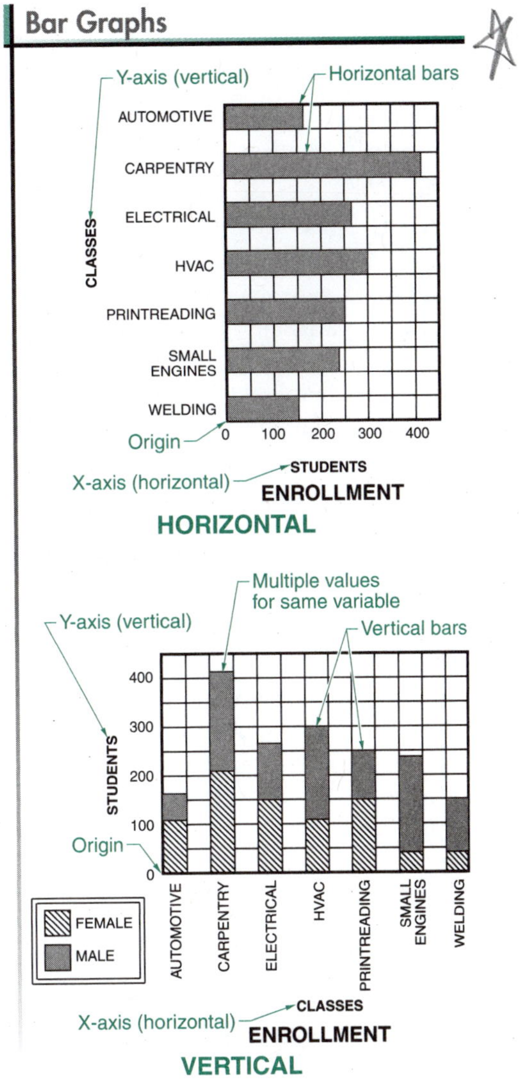

Figure 8-4. A bar graph uses horizontal or vertical bars to represent the values of variables.

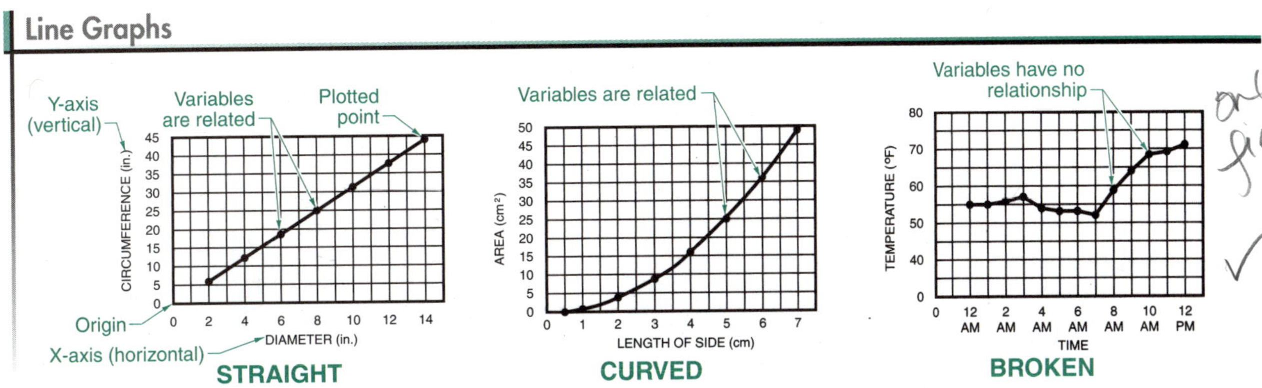

Figure 8-3. Line graphs show the relationship and change between variables through points connected by a line and are categorized as straight-line, curved-line, or broken-line.

To draw a bar graph, the given values should be arranged in a logical order. The axes are assigned a variable, labeled, and given a scale of values. The bars are then drawn to the length or height corresponding to the value.

Circle Graphs. A *circle graph (pie chart)* is a graph in which a circle represents the total of a variable, and sectors of the circle represent parts of the total. Values are represented by the sizes of the sectors. **See Figure 8-5.** Individual sector sizes can be found by determining the percentage of the sector value within the total value and then converting the percentage to an angle degree measurement. A circle contains 360°, which is 100% of the total. A 10% (0.10) sector value is equal to a 36° sector angle (360° × 0.10 = 36°). Each sector is generally labeled with the sector percentage. The corresponding sector value can be found by multiplying the decimal form of the sector percentage by the total value.

Circle Graphs

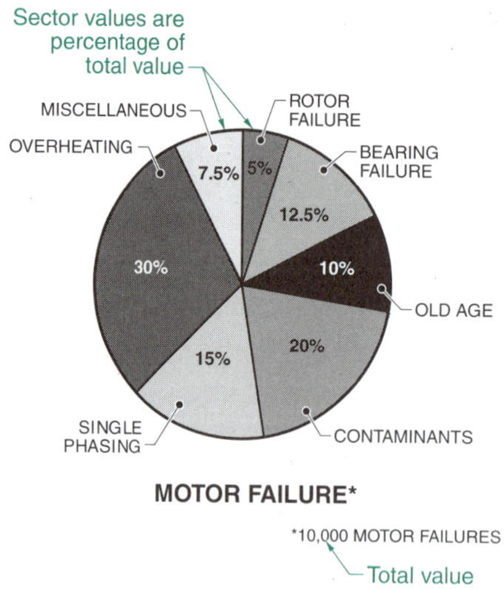

Figure 8-5. A circle graph (pie chart) shows the parts of a total as sectors in a circle.

To draw a circle graph, each sector value must be divided by the total to obtain sector percentages in decimal form. Each percentage is multiplied by 360° to get the sector angle. The angles are laid out within a circle, usually beginning with the smallest angle and working around the circle. Each consecutive angle shares a side with the previous one. The sum of all sector angles must equal 360°.

Pictographs. A *pictograph* is a graph in which symbols represent variables and their values. A pictograph shows approximations of values using whole and partial symbols. **See Figure 8-6.** To interpret pictographs, the value of each symbol must be known. The number of symbols are counted and then multiplied by the value of a symbol to obtain the variable.

Pictographs

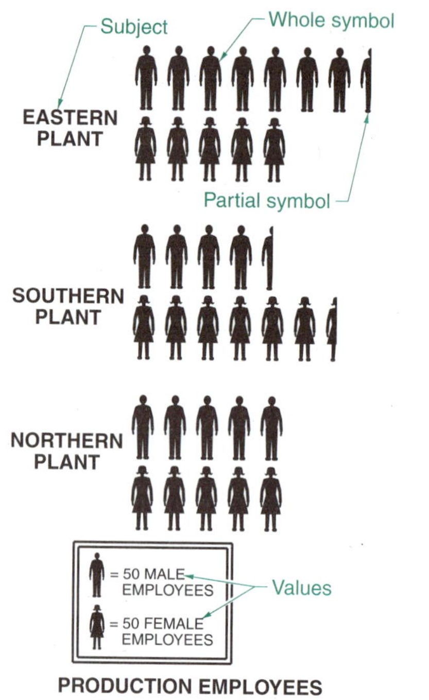

Figure 8-6. A pictograph uses whole or partial pictures or symbols to represent variables and their values.

To draw a pictograph, the variable is usually assigned to one axis and the values are assigned to the remaining axis. The symbol and value for each variable are chosen, and then symbols are drawn either whole or in parts as required for each value.

TOPOGRAPHY

The topography, or terrain, of the Earth is represented on a drawing using a topographic map. A *topographic map* is a map that depicts the surface features of an area. Topographic maps depict elevations, depressions, structures, water, and vegetation, and are generally included on an architectural plot plan or civil plans. **See Figure 8-7.** Contour lines and profiles are used to convey topographic information.

Contour Lines

Contour lines are the most commonly used method to portray the topography of a depicted area. A *contour line* is a line that shows points of the same elevation on a map of a land surface. Any point on the contour line is understood to be the same marked elevation. The increments in elevation for contour lines should remain constant per drawing.

Contour lines that appear close together on a map show a steeper terrain, while those farther apart show a shallower terrain. Smoother contour lines depict even terrain, while jagged contour lines depict rough terrain. Generally, every fifth contour line is marked heavier and its elevation is labeled for greater readability.

Profiles

A *profile* is a cross-section of a topographical map. A profile shows the terrain in an elevation view and is similar to a line graph. The elevations are plotted on a vertical scale. The horizontal scale is drawn to match the contour line map, with the elevation values carried down from the contour map cutting plane. **See Figure 8-8.**

TECH TIP

A profile depicts the relative elevations of a hill or depression that is cut by a cutting plane. Profiles depict what the top edge of the earth would look like if it were sectioned and cut into two parts. The line created appears like a plan view or silhouette of the earth at the place where it is cut.

Topographic Maps

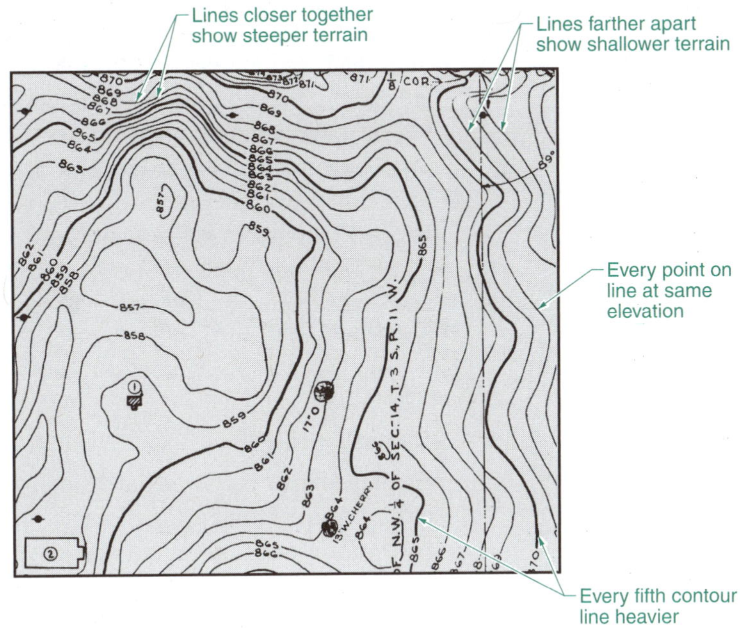

Figure 8-7. A topographic map depicts the surface features of an area and uses contour lines to show points of the same elevation.

Profiles

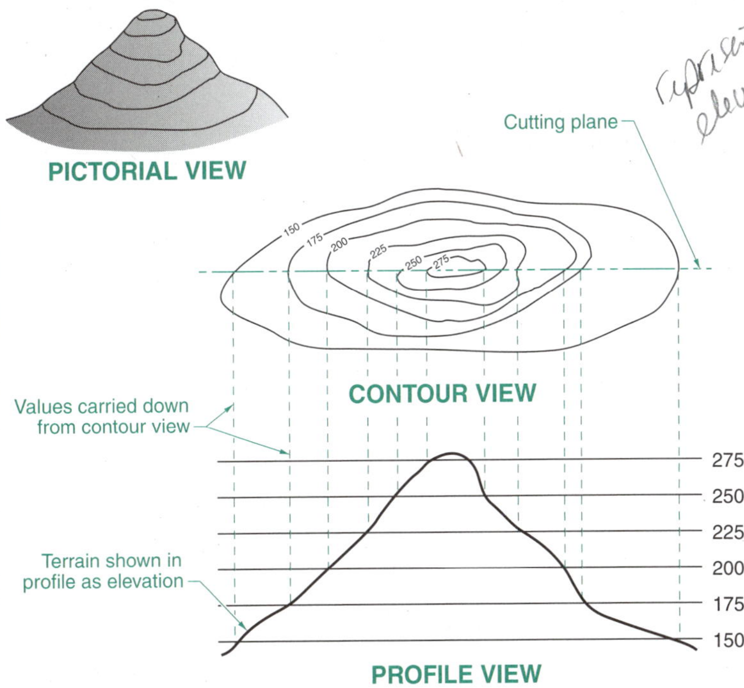

Figure 8-8. A profile is a cross-section of a topographical map and shows terrain in a similar manner to a line graph.

MANUFACTURING ELEMENTS

Certain manufacturing trades require specialized drawings and clear depictions of individual parts and elements. Developments are a way of determining two-dimensional layouts of three-dimensional objects. Sheet metal patterns for manufacturing metal parts are created from developments. Threaded fasteners, nonthreaded fasteners, springs, gears, and cams are also common parts that need to be represented according to specific guidelines.

Developments

A *development* is a two-dimensional pattern of a three-dimensional object that is laid out in a single flat plane with all views in true view. Developments are similar to multiview drawings, and they show all views side-by-side and in their true shape with no distortion. **See Figure 8-9.**

Layout is the method of developing the lines that form the outline for the finished shape. Developments are laid out along a baseline, or stretchout line. To assist in drawing irregularly shaped objects, points on the object can be labeled with number or letter labels. Projectors can be drawn from the labeled points to determine the exact form of the developed shape. The three most common methods used to develop a surface are parallel line development, radial line development, and triangulation. **See Figure 8-10.**

Parallel line development is a method of surface development used on objects having parallel fold lines, such as prisms and cylinders. Fold lines are parallel to each other and perpendicular to the stretchout line. To draw a development using parallel line development, the plan and elevation views of the object are used. The views are unfolded along the stretchout line, with each view drawn in true shape and sharing a side with any adjacent views.

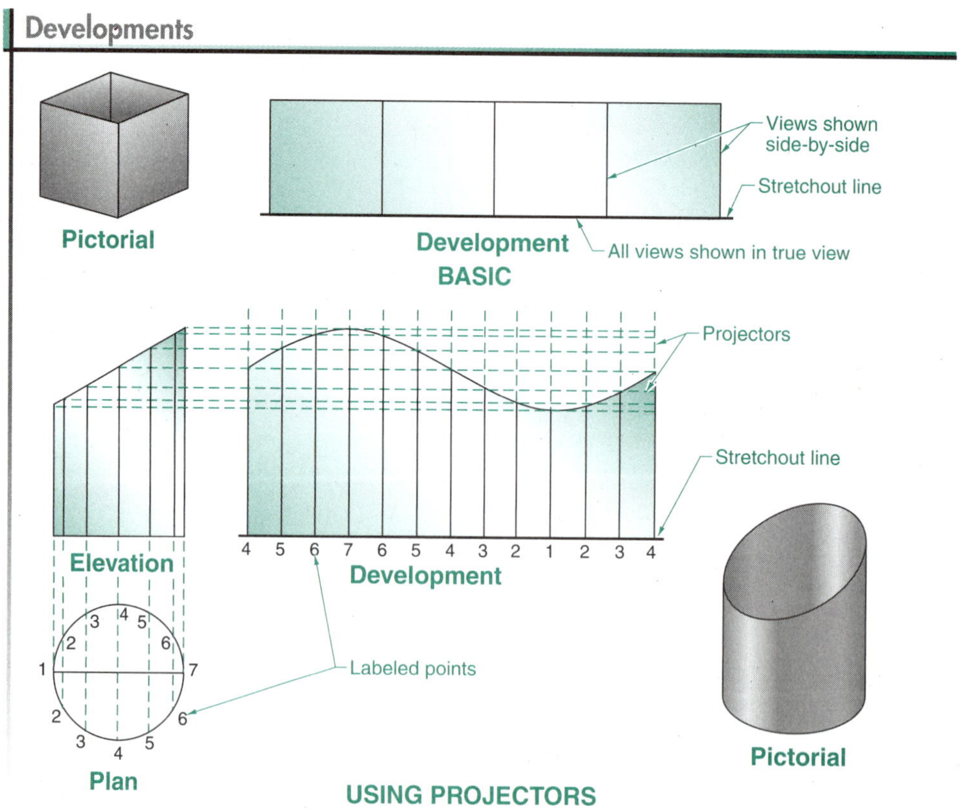

Figure 8-9. A development is a pattern of a three-dimensional object that is laid out in a single flat plane with all views in true view.

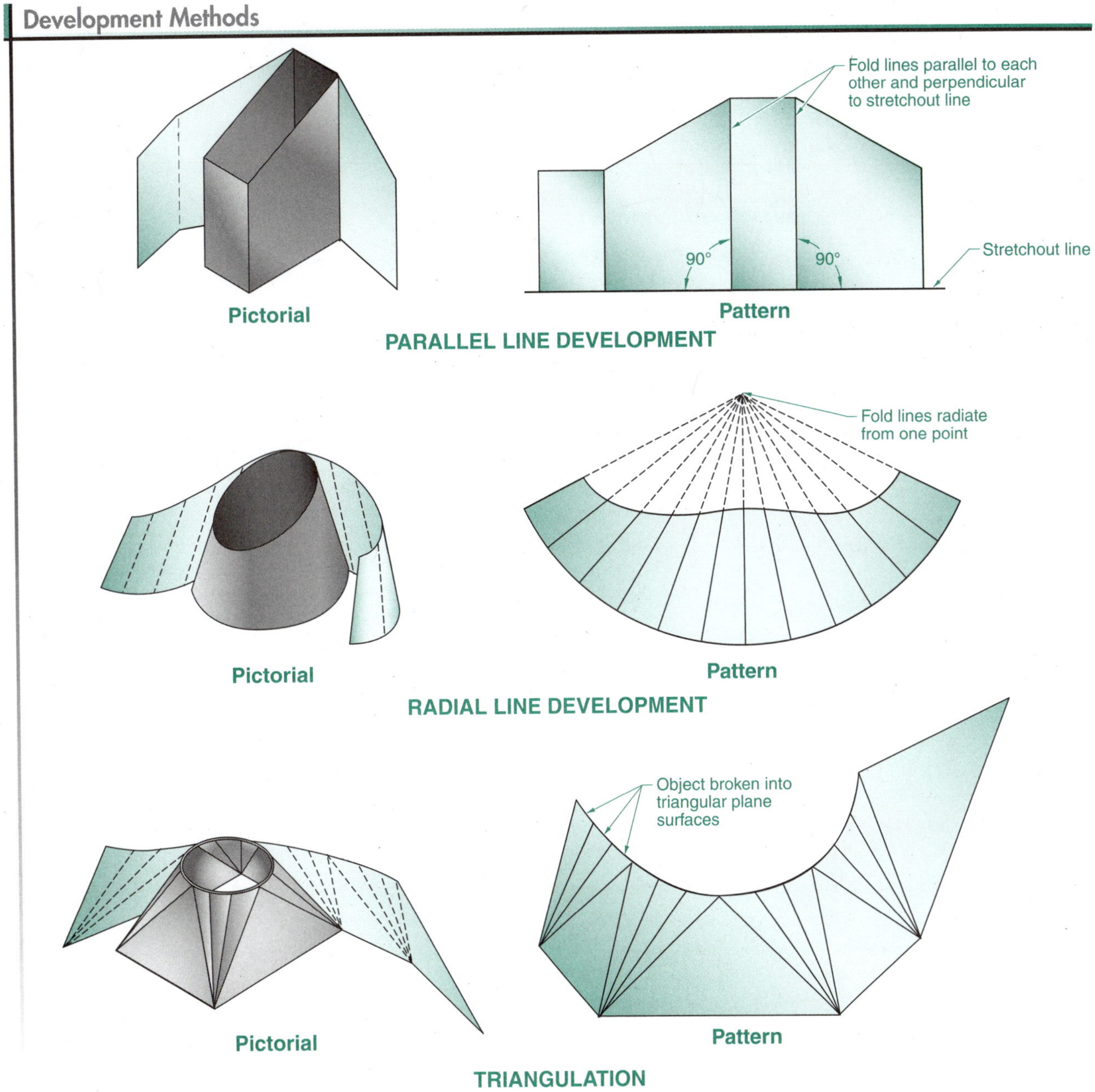

Figure 8-10. The three most common methods to develop a surface are parallel line development, radial line development, and triangulation.

Radial line development is a method of surface development used on objects that have fold lines that all radiate from one point. Radial line developments are used for shapes whose edges intersect at a vertex, such as pyramids and cones. To draw a radial line development, the plan and elevation views of the object are used. Lines are radiated from a common point to create a series of triangles in true view.

Triangulation is a method of surface development that breaks an object into triangular plane surfaces. Triangulation is a more complicated form of radial line development and is used for complex forms. Triangulation takes two known points to determine a third point, which forms a triangular plane on an object. To draw a development using triangulation, the surface is divided into triangles, the true lengths of all triangular sides are found, and each triangle is drawn one at a time and in true shape.

Sheet Metal Patterns

Sheet metal patterns are created from developments and are used to cut and fold flat sheet metal into the proper three-dimensional object. A *pattern* is a flat outline or stretchout of a developed surface. Sheet metal objects use pictorial drawings, multiview drawings, plans, and elevations to develop the shape of the pattern correctly. **See Figure 8-11.** Developed patterns may be drawn onto paper or metal and are then transferred to the specific piece of sheet metal to be used.

Threaded Fasteners

A *threaded fastener* is a device that joins two or more parts together with screw threads. A *screw thread* is a uniform ridge that forms a helix on the outside or inside surface of a cylinder or cone. **See Figure 8-12.** Threaded fasteners include bolts, nuts, studs, and screws.

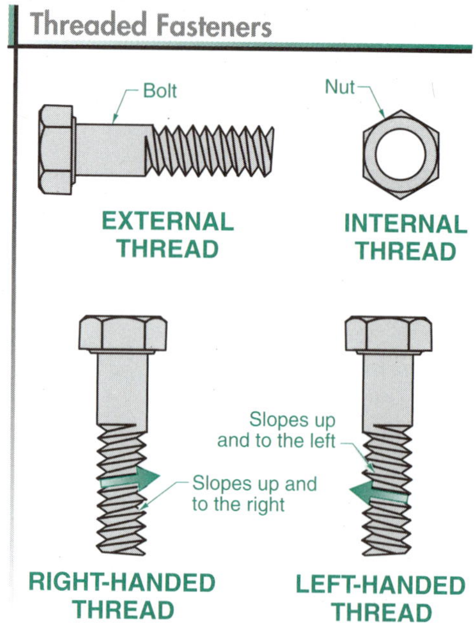

Figure 8-12. A threaded fastener joins two or more parts together with screw threads and can include bolts, nuts, studs, and screws.

> **TECH TIP**
> *Threaded fasteners are an advantageous method of joining parts together. They are easily installed and removed, can join either similar or dissimilar materials together, and are commercially available in many sizes, types, and materials.*

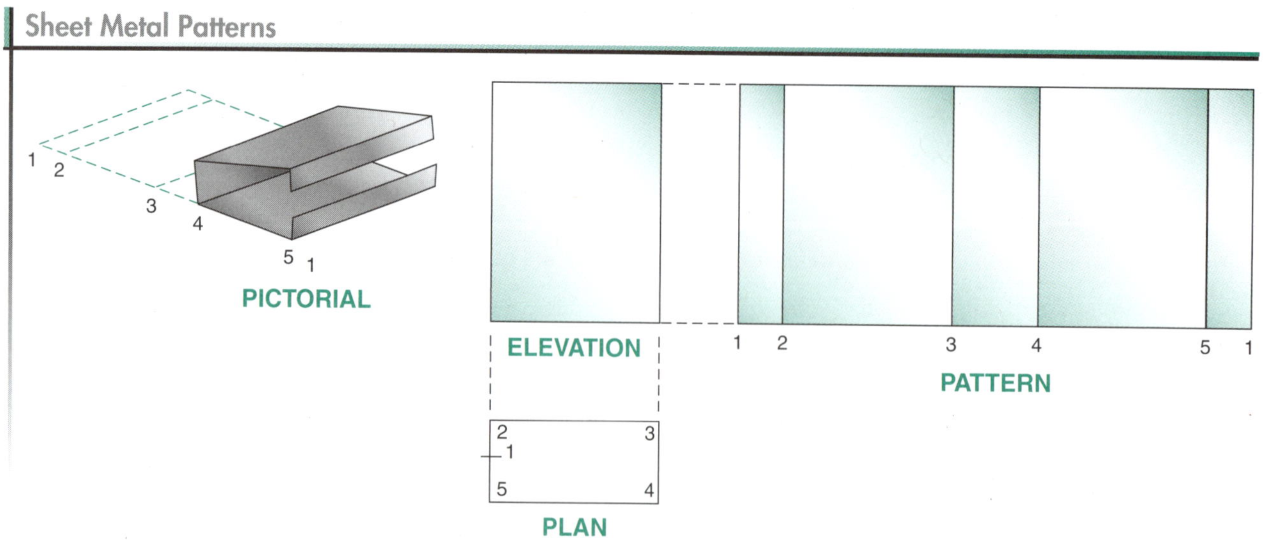

Figure 8-11. Sheet metal objects use pictorial drawings, multiview drawings, plans, and elevations to develop the shape of the pattern correctly.

Threads can be external and on shafts, such as bolts, or internal and on the inner surface of hollow cylinders, such as nuts. Threads can be right-handed or left-handed. External right-handed threads slope up and to the right. External left-handed threads slope up and to the left. Left-handed threads should be labeled LH on a drawing.

Thread Parts. Screw threads have crests, roots, and flanks. Crests and roots are comparable to the peaks and valleys of a mountain range, with flanks as the sloping sides. **See Figure 8-13.**

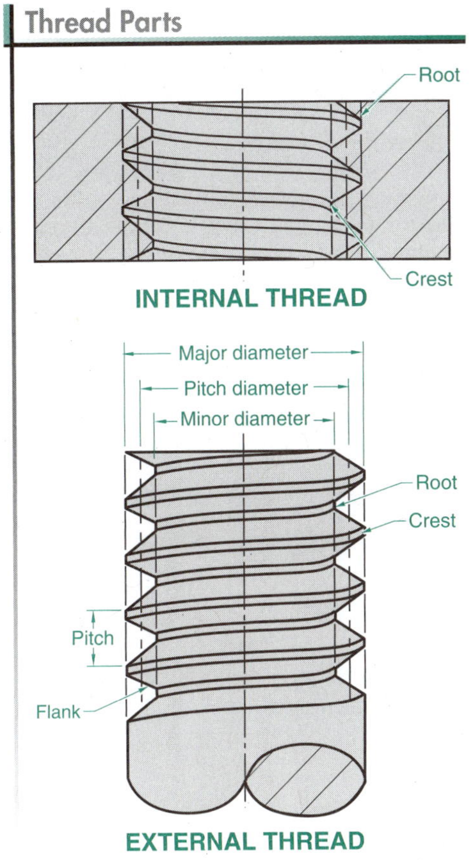

Figure 8-13. Screw threads have crests, roots, and flanks.

The dimensions of screw threads are based on the pitch and diameter. *Pitch* is the distance between corresponding points on adjacent threads measured parallel to the axis. Pitch determines the number of threads per inch. The diameter determines the width of the threaded form. The *major diameter* is the outer boundary of the crest of an external thread or the root of an internal thread. The *minor diameter* is the inner boundary of the root of an external thread or the crest of an internal thread.

Thread Representation. Threads should be represented on a drawing in a clear and simple way. The three most common methods used are simplified representation, schematic representation, and detailed representation. **See Figure 8-14.**

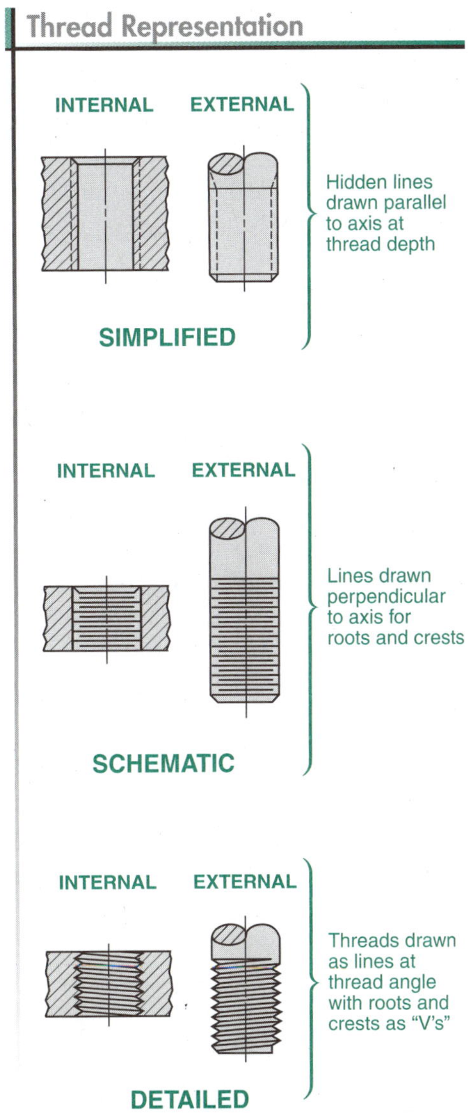

Figure 8-14. The three most common methods used to represent screw threads are simplified, schematic, and detailed representation.

Simplified thread representation is a method of thread representation in which hidden lines are drawn parallel to the axis of the cylinder at the approximate depth (minor diameter) of the thread. Simplified representation is the most commonly used method of thread representation. Pipe threads and other internal threads should be shown using the simplified method.

Schematic thread representation is a method of thread representation in which lines drawn perpendicular to the axis of the cylinder represent the roots and crests of the thread. This method is not used for hidden internal threads or for sectional views of external threads.

Detailed thread representation is a method of thread representation in which the threads are drawn as straight lines at the thread angle and the roots and crests are drawn as sharp "V's." Detailed representation provides a more accurate depiction of the thread, but is time-consuming to draw.

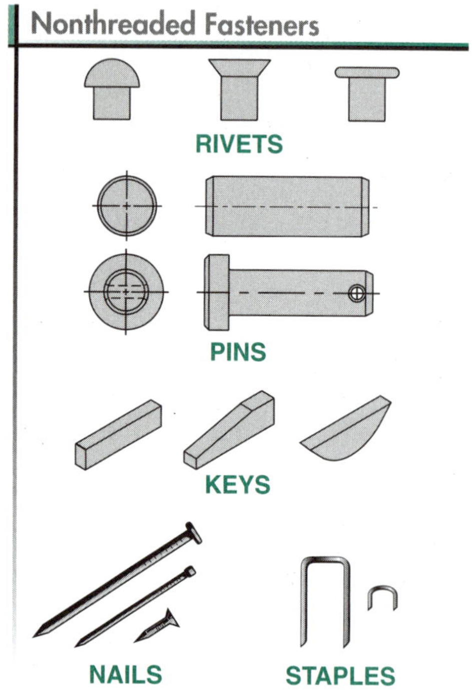

Figure 8-15. The most common types of nonthreaded fasteners are rivets, pins, keys, nails, and staples.

Nonthreaded Fasteners

A *nonthreaded fastener* is a device that joins or fastens parts together without threads. The most common types of nonthreaded fasteners are rivets, pins, keys, nails, and staples. **See Figure 8-15.** Nonthreaded fasteners should be designated on drawings by indicating their type, diameter, length, headshape, and metal or finish type as required. Nails are designated using their weight, length, and diameter.

Springs

A *spring* is a mechanical device that stores and applies energy and regains its original shape after being compressed or extended. Springs can be helical or flat. **See Figure 8-16.** Springs are rarely represented as true projections because they are very time-consuming to draw. Springs are drawn using schematic or detailed representation, which replaces curves with straight lines. To save time, springs can be partially drawn with phantom lines indicating the remainder of the spring.

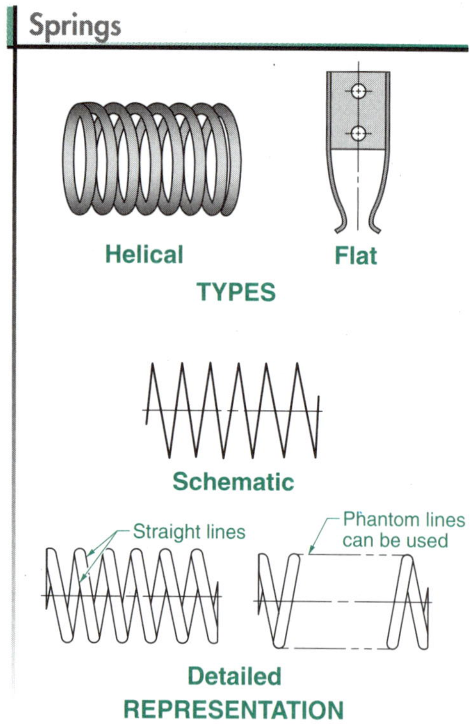

Figure 8-16. Springs can be helical or flat and are drawn using schematic or detailed representation.

Gears

A *gear* is a toothed wheel that is used in pairs to transmit power or motion from one part to another. Gears are represented on a drawing in their true shape. Specifications for the gear diameter, pitch, and number of teeth should be included in a gear drawing. **See Figure 8-17.**

Gears

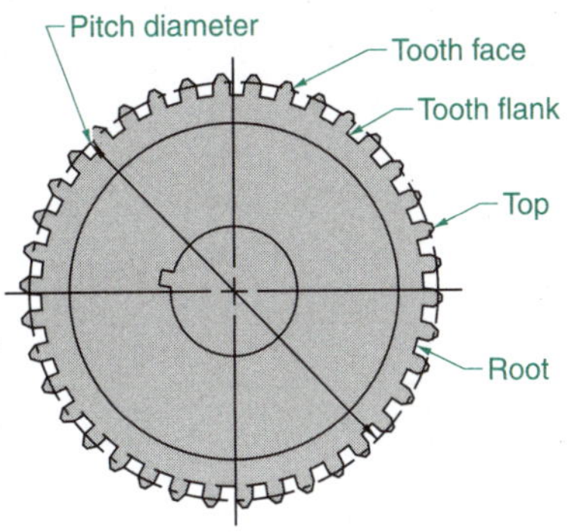

Figure 8-17. Gears are toothed wheels and are represented on a drawing in their true shape.

TECH TIP

Common gear types include spur, helical, herringbone, bevel, and worm gears. Gears are grouped into categories based on shaft orientation, including parallel shafts, intersecting shafts in the same plane, and nonintersecting shafts in the same plane.

Cams

A *cam* is a machine part that transmits motion using an irregular surface. A *cam follower* is a machine part that is in contact with the cam and moves in a designated path. **See Figure 8-18.** The shape of a cam determines the direction and motion of a cam follower. Because cams are custom designed for each application, a cam's features and motion are represented in a drawing in detail. Phantom lines are used to show the line of movement of one complete revolution of the cam (360°).

Cams

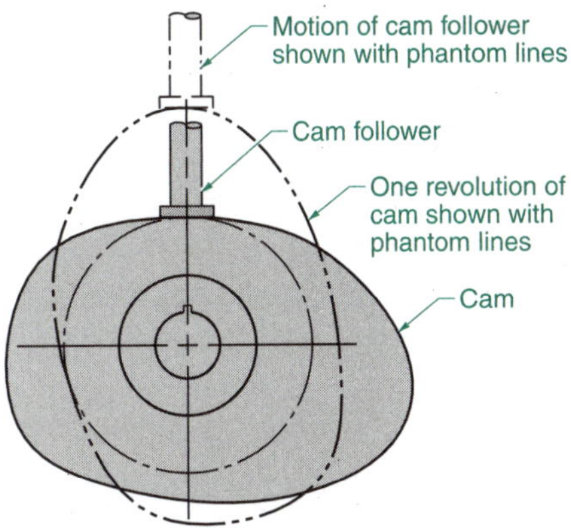

Figure 8-18. Cams are drawn showing one complete revolution of the cam using phantom lines.

Chapter 8 Quick Quiz®

Specialized Drawings

Review 8

Name ________________________________ Date ______________

REVIEW QUESTIONS

T F 1. Symbols are not based on any specific language and are preferred over the use of abbreviations.

__________ 2. A(n) ___ graph is a graph in which points representing variables are connected by a line.

__________ 3. The ___ point is where the x-axis and the y-axis meet.

__________ 4. ___ graphs contain variables with no set relationship to each other outside of the line graph.

T F 5. A bar graph is a graph in which values of variables are represented by bars.

__________ 6. A circle contains 360°, which is ___ of the total for a variable in a circle graph.

__________ 7. A(n) ___ is a graph in which whole and partial symbols represent variables and their values.

__________ 8. A(n) ___ map depicts the surface features of an area, including elevations, depressions, structures, water, and vegetation.

T F 9. Points on a contour line vary in marked elevation.

T F 10. Every fifth contour line is generally marked heavier and its elevation is labeled for greater readability.

__________ 11. A(n) ___ is a cross-section of a topographical map and shows the terrain in an elevation view.

__________ 12. A(n) ___ is a pattern of a three-dimensional object that is laid out in a single flat plane with all views in true view.

T F 13. Radial line development is a method of surface development used on objects having parallel fold lines, such as prisms and cylinders.

__________ 14. ___ is a method of surface development that breaks an object into triangular plane surfaces.

__________ 15. A(n) ___ is a flat outline or stretchout of a developed surface.

T F 16. Types of threaded fasteners include bolts, nuts, studs, and screws.

T F 17. The minor diameter is the outer boundary of the crest of an external thread or the root of an internal thread.

______________________ **18.** ___ thread representation is a method of thread representation in which hidden lines are drawn parallel to the axis of the cylinder at the approximate depth (minor diameter) of the thread.

______________________ **19.** A(n) ___ is a mechanical device that stores and applies energy and regains its original shape after being compressed or extended.

T F **20.** Gears are represented in their true shape on a drawing.

IDENTIFICATION—Graphs

______________________ **1.** Pictograph

______________________ **2.** Circle graph

______________________ **3.** Straight-line graph

______________________ **4.** Bar graph

______________________ **5.** Broken-line graph

______________________ **6.** Curved-line graph

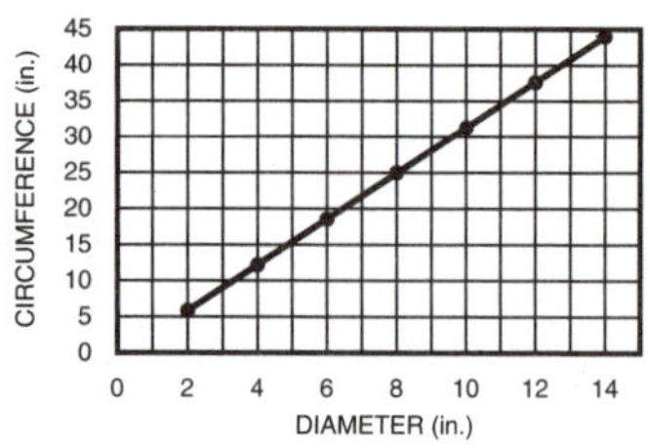

Ⓐ

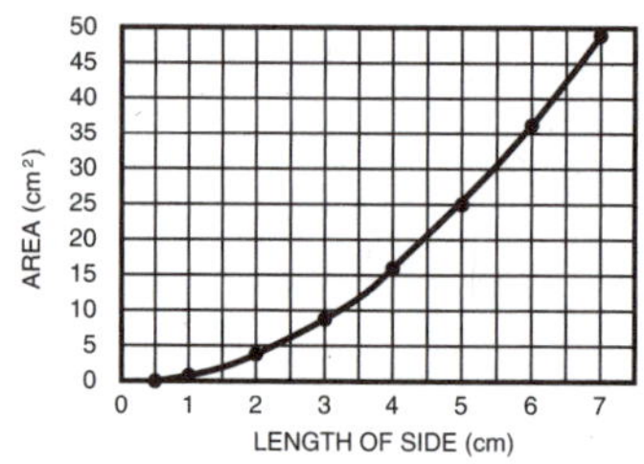

Ⓑ

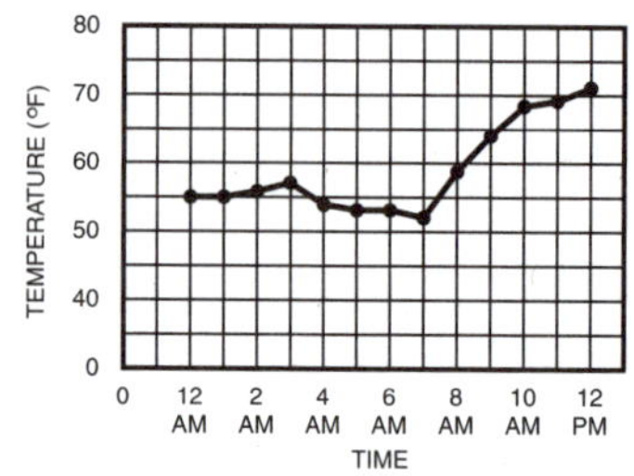

Ⓒ

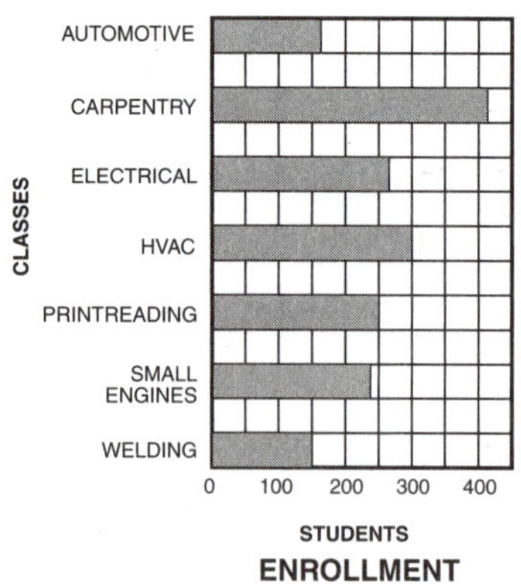

Ⓓ

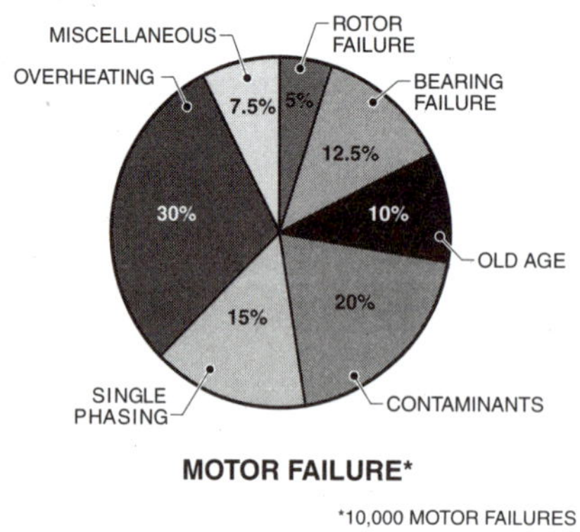

Ⓔ

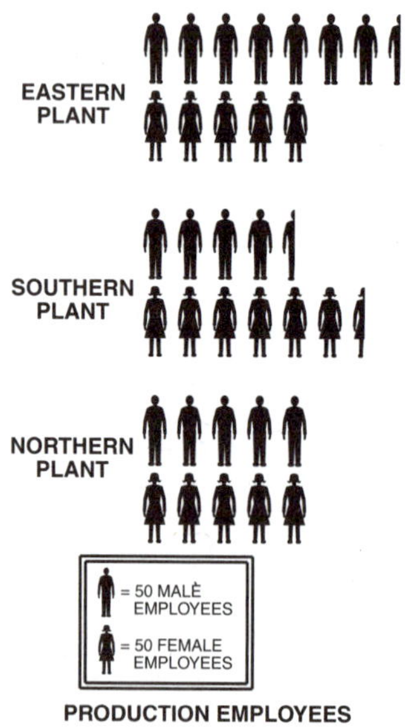

Ⓕ

IDENTIFICATION—Developments

________________ **1.** Radial line development

________________ **2.** Triangulation

________________ **3.** Parallel line development

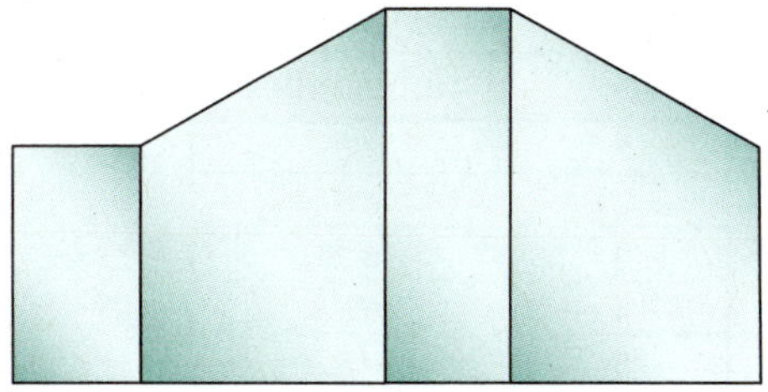

Ⓐ

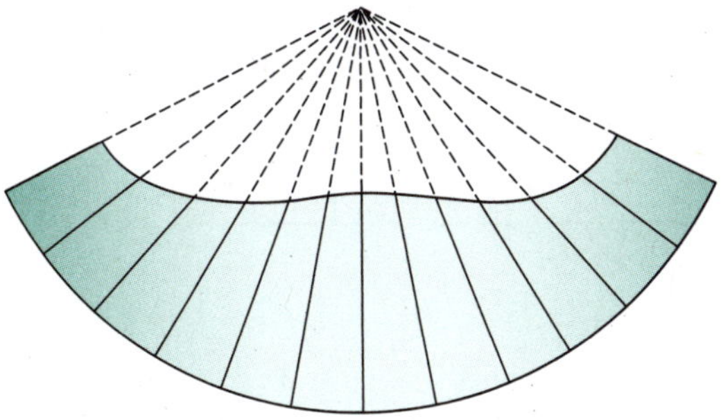

Ⓑ

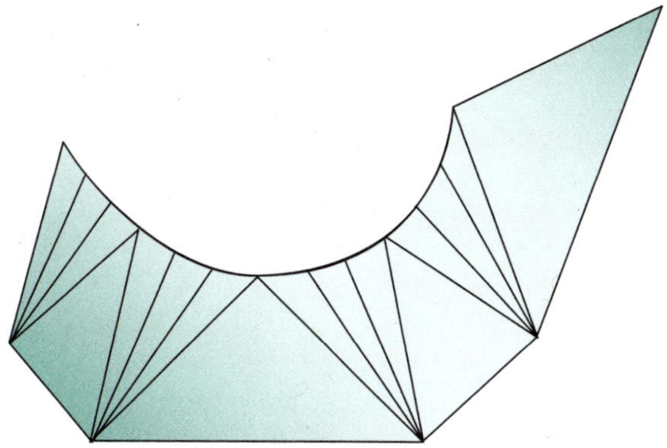

Ⓒ

EXERCISES

1. Create a line graph with the following Project-Cost information.

Projects (x-axis): 10, 20, 30, 40, 50, 60, 70

Cost (y-axis): 20, 35, 50, 65, 80, 95, 110

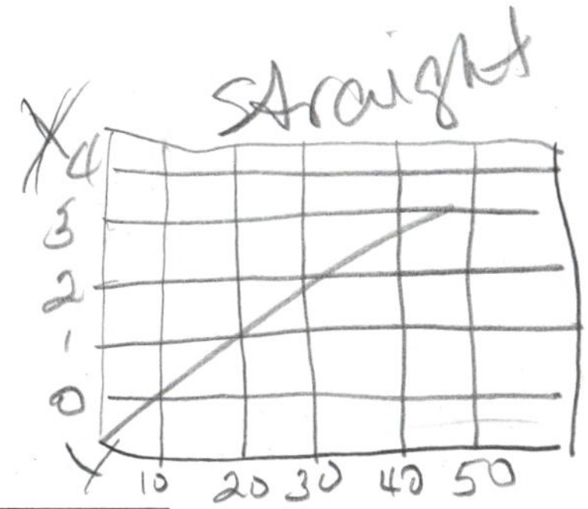

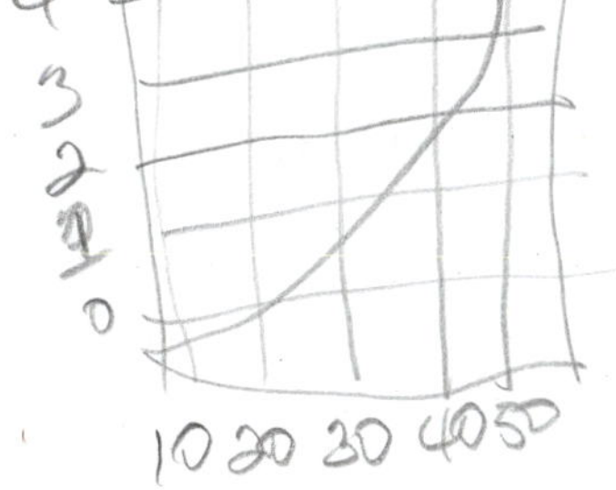

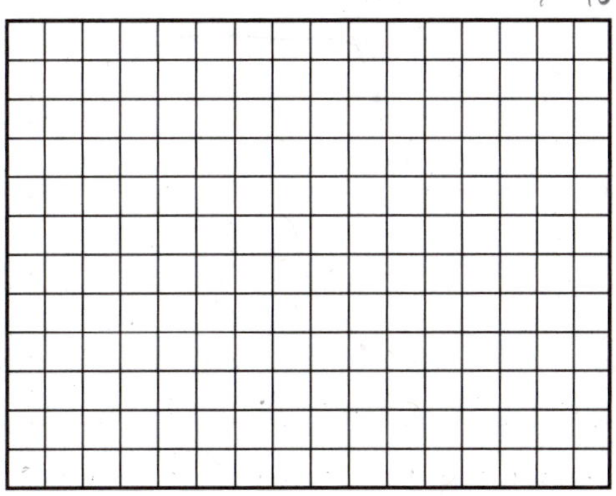

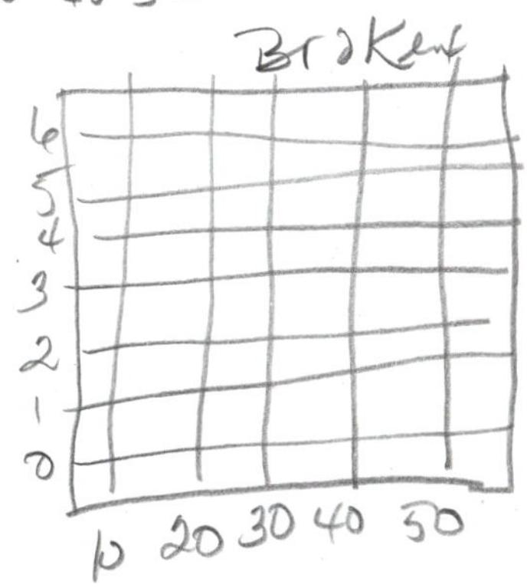

2. Create a bar graph titled Automobile Production Data based on the given data.

Year	Automobiles*
1	7.5
2	8.0
3	8.25
4	10.0
5	9.5
6	8.25
7	10.5

* in millions

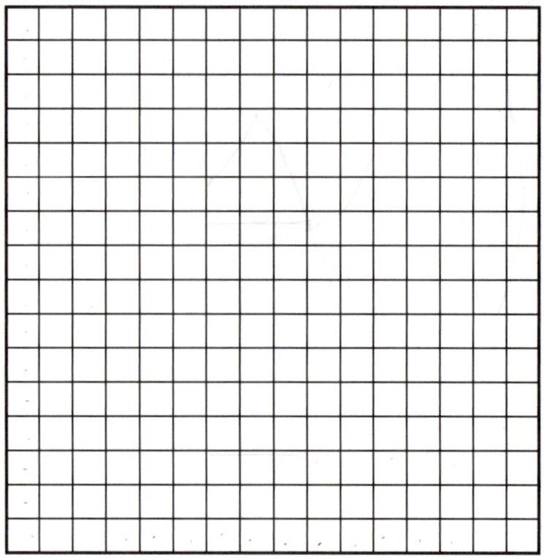

3. Label the scale and draw a profile view of the given contour lines and marked elevations.

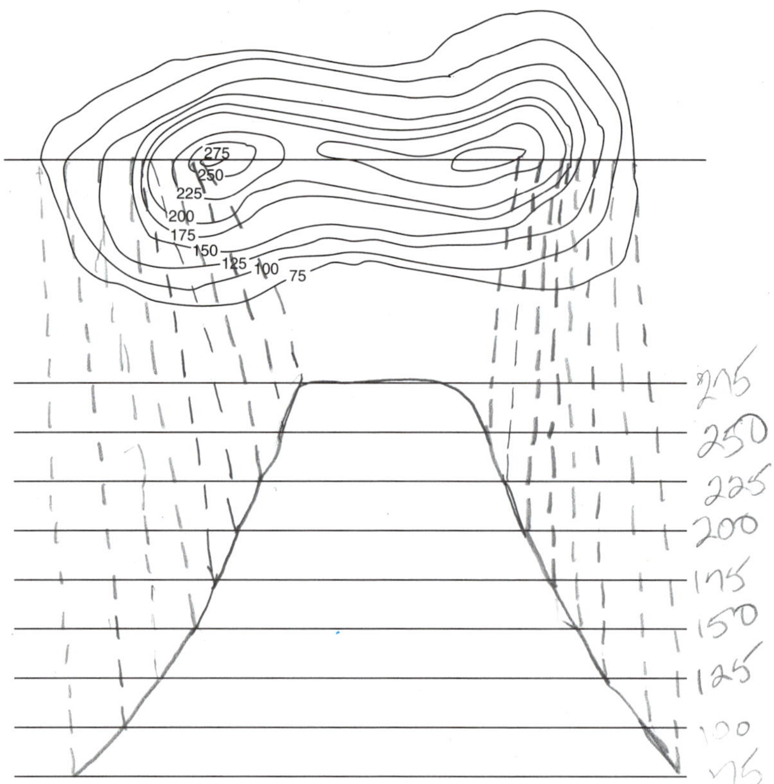

Appendix

Area—Plane Figures 192

Volume—Solid Figures 192

Drawing Limits 192

Decimal Equivalents of an Inch 193

Millimeter to Inch Equivalents 193

Pencil Lead Guidelines 193

Symbols 194

Manufacturing Printreading Symbols 195

Manufacturing Abbreviations 196

Architectural Abbreviations 197

Drawing Grid—Isometric 198

Drawing Grid—Oblique Cabinet 199

Isometric Ellipse Protractor 200

Area—Plane Figures

$A = l \times w$

where
A = area
l = length
w = width

SQUARE OR RECTANGLE

$A = \frac{1}{2} \times b \times h$

where
A = area
b = base
h = height

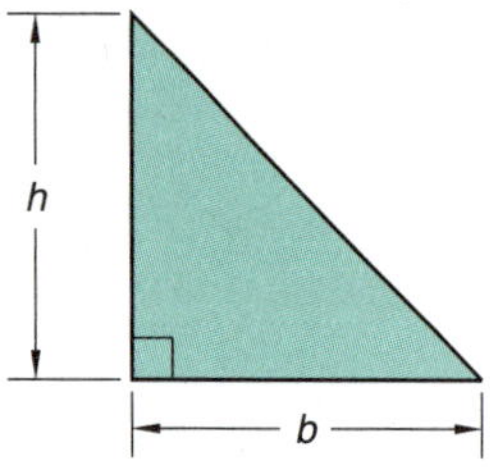

RIGHT TRIANGLE

$A = \pi r^2$

where
A = area
π = 3.1416
r = radius

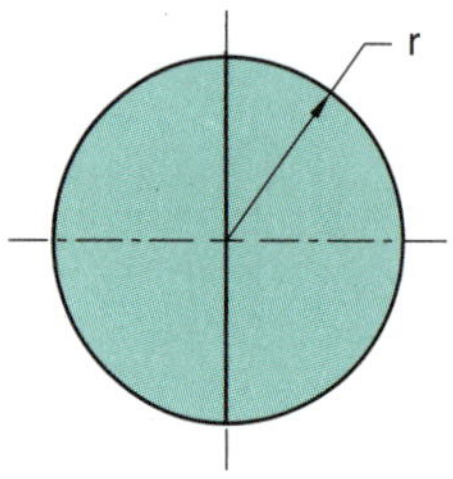

CIRCLE

Volume—Solid Figures

$V = l \times w \times h$

where
V = volume
l = length
w = width
h = height

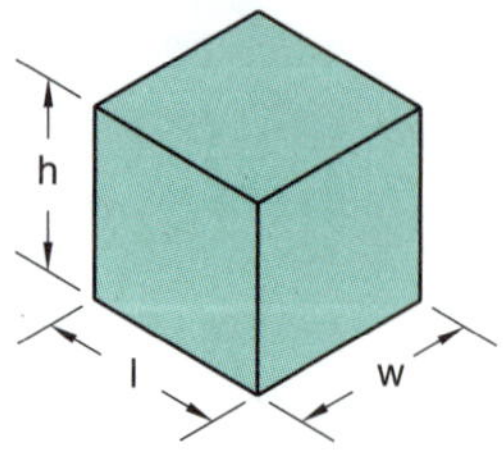

RIGHT RECTANGULAR PRISM

$$V = \frac{h\pi\,[(d_1)^2 + (d_2)^2 + (d_1 \times d_2)]}{12}$$

where
V = volume
h = height
π = 3.1416
d_1 = diameter of top
d_2 = diameter of base

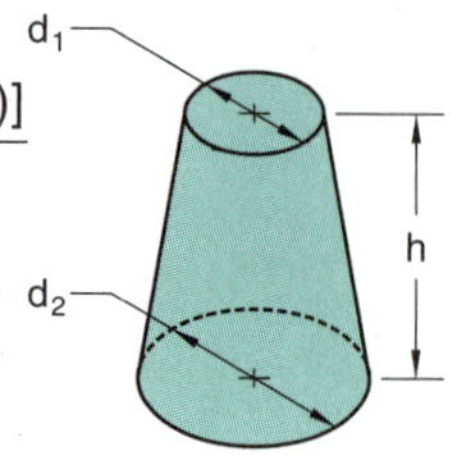

FRUSTRUM OF A CONE

$V = \pi r^2 \times h$

where
V = volume
π = 3.1416
r = radius
h = height

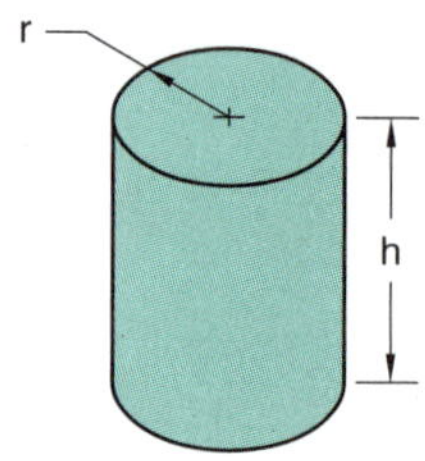

CYLINDER

Drawing Limits

Metric Units*					
Drawing Scale	210 × 297 (A4)	297 × 420 (A3)	420 × 594 (A2)	594 × 841 (A1)	894 × 1189 (A0)
1:200	42,000 × 59,400	59,400 × 84,000	84,000 × 118,800	118,800 × 168,200	168,200 × 237,800
1:100	21,000 × 29,700	29,700 × 42,000	42,000 × 59,400	59,400 × 84,100	84,100 × 118,900
1:50	10,500 × 14,850	14,850 × 21,000	21,000 × 29,700	29,700 × 42,050	42,050 × 59,450
1:20	4200 × 5940	5940 × 8400	8400 × 11,880	11,880 × 16,820	16,820 × 23,780
1:10	2100 × 2970	2970 × 4200	4200 × 5940	5940 × 8410	8410 × 11,890

Architectural Units			
Drawing Scale	8½″ × 11″ (Letter size)	11″ × 17″ (Ledger size)	24″ × 36″ (Architectural size D)
1/8″ = 1′-0″	68′ × 88′	88′ × 136′	192′ × 288′
1/4″ = 1′-0″	34′ × 44′	44′ × 68′	96′ × 144′
1/2″ = 1′-0″	17′ × 22′	22′ × 34′	48′ × 72′
3/4″ = 1′-0″	11′-4″ × 14′-8″	14′-8″ × 22′-8″	32′ × 48′
1″ = 1′-0″	8′-6″ × 11′	11′ × 17′	24′ × 36′

* in mm

Decimal Equivalents of an Inch

Fraction	Decimal	Fraction	Decimal	Fraction	Decimal	Fraction	Decimal	Fraction	Decimal
1/64	0.015625	7/32	0.21875	27/64	0.421875	5/8	0.625	53/64	0.828125
1/32	0.03125	15/64	0.234375	7/16	0.4375	41/64	0.640625	27/32	0.84375
3/64	0.046875	1/4	0.250	29/64	0.453125	21/32	0.65625	55/64	0.859375
1/16	0.0625	17/64	0.265625	15/32	0.46875	43/64	0.671875	7/8	0.875
5/64	0.078125	9/32	0.28125	31/64	0.484375	11/16	0.6875	57/64	0.890625
3/32	0.09375	19/64	0.296875	1/2	0.500	45/64	0.703125	29/32	0.90625
7/64	0.109375	5/16	0.3125	33/64	0.515625	23/32	0.71875	59/64	0.921875
1/8	0.125	21/64	0.328125	17/32	0.53125	47/64	0.734375	15/16	0.9375
9/64	0.140625	11/32	0.34375	35/64	0.546875	3/4	0.750	61/64	0.953125
5/32	0.15625	23/64	0.359375	9/16	0.5625	49/64	0.765625	31/32	0.96875
11/64	0.171875	3/8	0.375	37/64	0.578125	25/32	0.78125	63/64	0.984375
3/16	0.1875	25/64	0.390625	19/32	0.59375	51/64	0.796875	1	1.000
13/64	0.203125	13/32	0.40625	39/64	0.609375	13/16	0.8125		

Millimeter to Inch Equivalents*

mm	in.	mm	in.	mm	in.	mm	in.	mm	in.	mm	in.
½	= 0.01969	17	= 0.66929	34	= 1.33858	51	= 2.00787	68	= 2.67716	85	= 3.34645
1	= 0.03937	18	= 0.70866	35	= 1.37795	52	= 2.04724	69	= 2.71653	86	= 3.38582
2	= 0.07874	19	= 0.74803	36	= 1.41732	53	= 2.08661	70	= 2.75590	87	= 3.42519
3	= 0.11811	20	= 0.78740	37	= 1.45669	54	= 2.12598	71	= 2.79527	88	= 3.46456
4	= 0.15748	21	= 0.82677	38	= 1.49606	55	= 2.16535	72	= 2.83464	89	= 3.50393
5	= 0.19685	22	= 0.86614	39	= 1.53543	56	= 2.20472	73	= 2.87401	90	= 3.54330
6	= 0.23622	23	= 0.90551	40	= 1.57480	57	= 2.24409	74	= 2.91338	91	= 3.58267
7	= 0.27559	24	= 0.94488	41	= 1.61417	58	= 2.28346	75	= 2.95275	92	= 3.62204
8	= 0.31496	25	= 0.98425	42	= 1.65354	59	= 2.32283	76	= 2.99212	93	= 3.66141
9	= 0.35433	26	= 1.02362	43	= 1.69291	60	= 2.36220	77	= 3.03149	94	= 3.70078
10	= 0.39370	27	= 1.06299	44	= 1.73228	61	= 2.40157	78	= 3.07086	95	= 3.74015
11	= 0.43307	28	= 1.10236	45	= 1.77165	62	= 2.44094	79	= 3.11023	96	= 3.77952
12	= 0.47244	29	= 1.14173	46	= 1.81102	63	= 2.48031	80	= 3.14960	97	= 3.81889
13	= 0.51181	30	= 1.18110	47	= 1.85039	64	= 2.51968	81	= 3.18897	98	= 3.85826
14	= 0.55118	31	= 1.22047	48	= 1.88976	65	= 2.55905	82	= 3.22834	99	= 3.89763
15	= 0.59055	32	= 1.25984	49	= 1.92913	66	= 2.59842	83	= 3.26771	100	= 3.93700
16	= 0.62992	33	= 1.29921	50	= 1.96850	67	= 2.63779	84	= 3.30708		

* 10 mm = 1 centimeter = 0.3937″   25.4 mm = 1″

Pencil Lead Guidelines

Task	Lead Grade	Task	Lead Grade
Construction lines	3H, 4H	Phantom lines	2H, H
Guide lines	3H, 4H	Stitch lines	2H, H
Lettering	H, F, HB	Long break lines	2H, H
Dimension lines	2H, H	Visible lines	H, F, HB
Leader lines	2H, H	Cutting-plane lines	H, F, HB
Hidden lines	2H, H	Extension lines	2H, H
Cross-hatching lines	2H, H	Freehand break lines	H, F, HB
Center lines	2H, H		

Symbols . . .

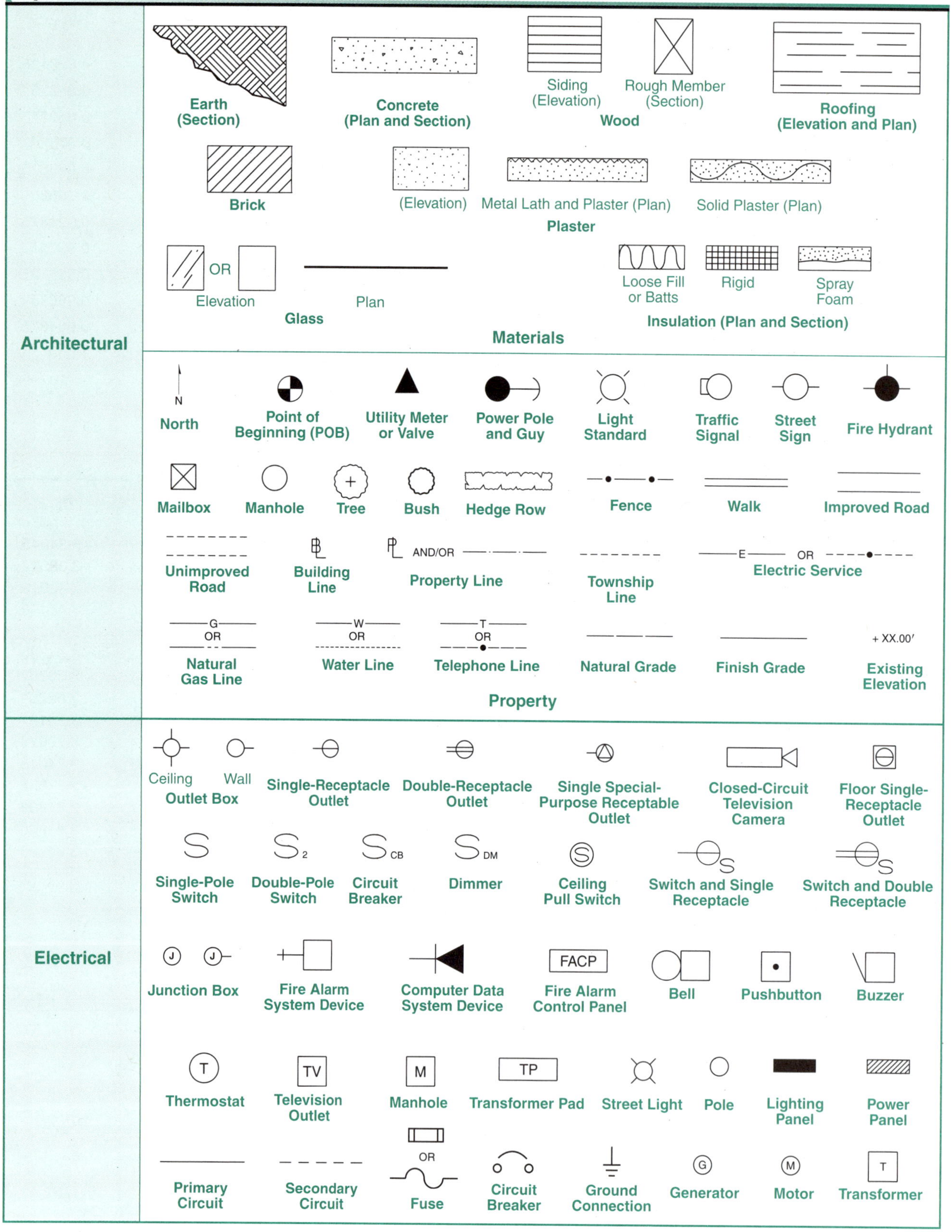
Architectural
Earth (Section)
Concrete (Plan and Section)
Siding (Elevation)
Rough Member (Section)
Wood
Roofing (Elevation and Plan)
Brick
(Elevation)
Metal Lath and Plaster (Plan)
Solid Plaster (Plan)
Plaster
OR
Elevation
Plan
Glass
Loose Fill or Batts
Rigid
Spray Foam
Insulation (Plan and Section)
Materials
N
North
Point of Beginning (POB)
Utility Meter or Valve
Power Pole and Guy
Light Standard
Traffic Signal
Street Sign
Fire Hydrant
Mailbox
Manhole
Tree
Bush
Hedge Row
Fence
Walk
Improved Road
Unimproved Road
Building Line
AND/OR
Property Line
Township Line
E
OR
Electric Service
G
OR
Natural Gas Line
W
OR
Water Line
T
OR
Telephone Line
Natural Grade
Finish Grade
+ XX.00′
Existing Elevation
Property
Electrical
Ceiling
Wall
Outlet Box
Single-Receptacle Outlet
Double-Receptacle Outlet
Single Special-Purpose Receptable Outlet
Closed-Circuit Television Camera
Floor Single-Receptacle Outlet
S
S_2
S_{CB}
S_{DM}
Single-Pole Switch
Double-Pole Switch
Circuit Breaker
Dimmer
Ceiling Pull Switch
Switch and Single Receptacle
Switch and Double Receptacle
J
Junction Box
Fire Alarm System Device
Computer Data System Device
FACP
Fire Alarm Control Panel
Bell
Pushbutton
Buzzer
T
Thermostat
TV
Television Outlet
M
Manhole
TP
Transformer Pad
Street Light
Pole
Lighting Panel
Power Panel
Primary Circuit
Secondary Circuit
OR
Fuse
Circuit Breaker
Ground Connection
G
Generator
M
Motor
T
Transformer

. . . Symbols

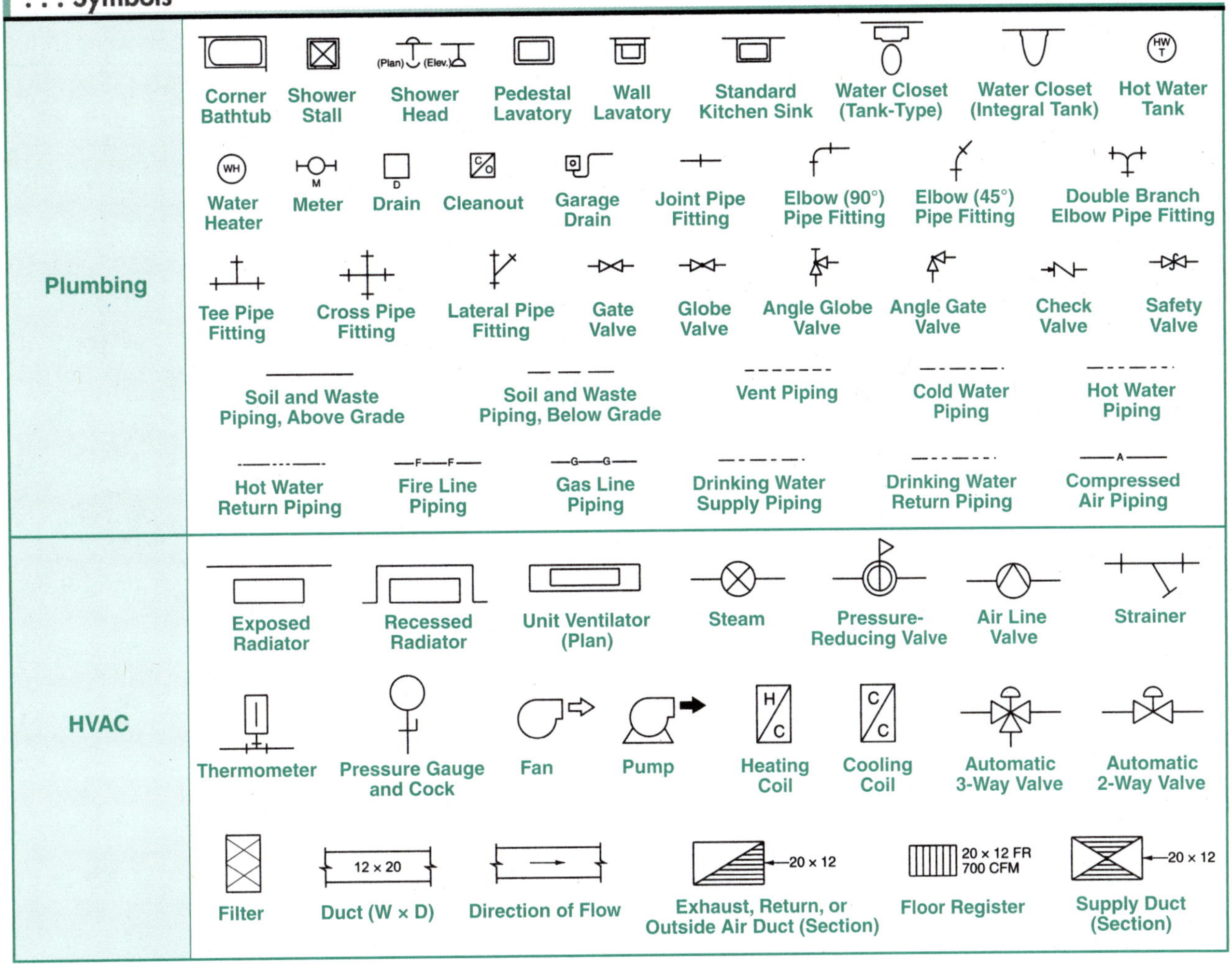

Manufacturing Printreading Symbols

Meaning	Symbol	Meaning	Symbol	Meaning	Symbol
Straightness	—	Concentricity	◎	Taper	⌳
Flatness	▱	Symmetry	⌯	Counterbore/Spotface	⌴
Circularity	○	Circular runout	*↗	Countersink	⌵
Cylindricity	⌭	Total runout	*⌰	Depth/deep	↧
Profile of a line	⌒	Maximum Material Condition	Ⓜ	Square	□
Profile of a surface	⌓	Least Material Condition	Ⓛ	Dimension not to scale	15 (underlined)
All around	⦵ (with leader)	Projected Tolerance Zone	Ⓟ	Arc length	⌒105
Angularity	∠	Diameter	Ø	Spherical radius	SR
Perpendicularity	⊥	Basic dimension	[50]	Spherical diameter	SØ
Parallelism	//	Reference dimension	(50)	Between	*↔
Position	⌖	Conical taper	⌲	Statistical tolerance	⟨ST⟩

* may be filled or not filled

Manufacturing Abbreviations

Term	Abbreviation	Term	Abbreviation	Term	Abbreviation
Actual	ACT	Fitting	FTG	Position	POS
Aluminum	AL	Fixture	FIX	Pound	LB
American wire gauge	AWG	Flat	F	Pounds per square inch	PSI
Amount	AMT	Flat head	FH	Primary	PRIM
Approximate	APPROX	Flexible	FLEX	Profile	PF
Arc weld	ARC/W	Forward	FWD	Project	PROJ
Area	A	Fractional	FRAC	Punch	PCH
Arrangement	ARR	Gauge	GA	Quadrant	QUAD
Assemble	ASSEM	General	GEN	Quality	QUAL
Assembly	ASSY	Glass	GL	Quantity	QTY
Auxiliary	AUX	Grade	GR	Radial	RAD
Base line	BL	Grind	GRD or GND	Radius	R
Bearing	BRG	Groove	GRV	Reassemble	REASM
Benchmark	BM	Head	HD	Rectangle	RECT
Between	BET	Headless	HDLS	Reference	REF
Bevel	BEV	Heat	HT	Reference line	REF L
Both faces	BF	Heavy	HVY	Reinforce	REINF
Both sides	BS	Height	HGT	Remove	REM
Both ways	BW	Hexagon	HEX	Required	REQD
Bottom	BOT	Horizontal	HORIZ	Return	RET
Bracket	BRKT	Inch	IN.	Reverse	REV
Brass	BRS	Inches per minute	IPM	Revolutions per minute	RPM
Break	BRK	Inside diameter	ID	Right hand	RH
Bronze	BRZ	Internal	INT	Rivet	RIV
Cabinet	CAB	Intersect	INT	Root diameter	RD
Capacity	CAP	Iron	I	Round	RD
Cap screw	CAP SCR	Irregular	IRREG	Schedule	SCH
Cast	C	Joint	JT	Schematic	SCHEM
Cast Iron	CI	Junction	JCT	Screw	SCR
Casting	CSTG	Lateral	LAT	Secondary	SEC
Center	CTR	Left hand	LH	Section	SECT
Center line	CL	Length	LG	Shaft	SFT
Chamfer	CH or CHAM	Limit	LIM	Sheet	SH
Circle	CIR	Linear	LIN	Side	S
Circular pitch	CP	Lubricate	LUB	Similar	SIM
Circumference	CIRC	Manual	MAN	Space	SP
Clearance	CL	Manufactured	MFD	Specification	SPEC
Clockwise	CW	Material	MAT or MATL	Speed	SP
Concentric	CONC	Material list	ML	Spherical	SPHER
Counterclockwise	CCW	Maximum	MAX	Spring	SPG
Cross section	XSECT	Mechanical	MECH	Square	SQ
Cubic foot	CU FT	Metal	MET	Stainless steel	SST or SS
Cubic inch	CU IN	Millimeter	MM	Standard	STD
Cylinder	CYL	Minimum	MIN	Steel	STL
Depth	DP or DEEP	Minute	MIN	Straight	STR
Degree	DEG	Miscellaneous	MISC	Structural	STR
Density	D	Mounted	MTD	Surface	SURF
Design	DSGN	Mounting	MTG	Symbol	SYM
Detail	DET	Multiple	MULT	Symmetrical	SYM
Diagonal	DIAG	Near face	NF	Tangent	TAN
Diagram	DIAG	Near side	NS	Taper	TPR
Diameter	DIA	Normal	NORM	Technical	TECH
Diametral pitch	DP	Not to scale	NTS	Teeth	T
Dimension	DIM	Number	NO.	Teeth per inch	TPI
Drafting	DFTG	Octagon	OCT	Temperature	TEMP
Drawing	DWG	On center	OC	Template	TEMP
Drill	DR	Opening	OPNG	Tension	TENS
Duplicate	DUP	Opposite	OPP	Thick	THK
Each	EA	Original	ORIG	Thread	THD
Enclose	ENCL	Outside diameter	OD	Threads per inch	TPI
Equipment	EQUIP	Pair	PR	Through	THRU
Equivalent	EQUIV	Parallel	PAR	Tolerance	TOL
Existing	EXIST	Part	PT	Tool steel	TS
Extension	EXT	Permanent	PERM	Typical	TYP
Fabricate	FAB	Perpendicular	PERP	Vertical	VERT
Far side	FS	Pitch	P	Volume	VOL
Feet	FT	Pitch circle	PC	Washer	WASH
Feet per minute	FPM	Pitch diameter	PD	Weight	WT
Figure	FIG	Plate	PL	Width	W
Finish	FIN	Point	PT	Wire	W

Architectural Abbreviations

Abbreviation	Term	Abbreviation	Term
AC	Air conditioner	LAU	Laundry
ACS	Access	LAV	Lavatory
ACSP or AP	Access panel	LCL or LCLO	Linen closet
ADD or ADH	Adhesive	LDG	Landing
B	Bathroom	LINO or LINOL	Linoleum
BC	Between centers	LR	Living room
BCL	Broom closet	LT	Light
BLK	Block	MC	Medicine cabinet
BLKG	Blocking	MLD or MLDG	Molding
BR	Bedroom	OP	Operator
BT	Bathtub	P	Porch
BYP	Bypass	PAN.	Pantry
CA	Cold air	PASS.	Passage
CAB.	Cabinet	PL	Plate
CFM	Cubic feet per minute	PLAS	Plaster
CHM	Chimney	PLAT	Platform
CLG	Cooling or ceiling	PREFAB or PFB	Prefabricated
CL GL	Clear glass	PTN	Partition
CLO	Closet	PW	Projected window
CNCL	Concealed	QT	Quarry tile
CTR	Counter	R	Riser
CW	Cold water	R or RM	Room
DET or DTL	Detail	R or RNG	Range
DK	Deck	RAD	Radiator
DMPR	Damper	REC	Recessed
DMR	Dimmer	REF or REFR	Refrigerator
DNG R or DR	Dining room	REG	Register
DR	Door (elevation views)	RF	Roof
DW	Dishwasher	RTN	Return
EMT	Electric metallic tubing	SCH or SCHED	Schedule
ENTR	Entrance	SD or SL DR	Sliding door
EQPT	Equipment	SDG	Siding
EXT	Extinguisher or exterior	SEC or SECT.	Section
FL	Flush	SH	Shower
FLMT	Flush mount	SHELV	Shelving
FLR or FLG	Flooring	SK or S	Sink
FLUR or FLUOR	Fluorescent	SM	Sheet metal
FNSH	Finish	ST	Stairs
FP	Fireplace	STOR	Storage
GB	Gypsum board	STWY	Stairway
GFCI	Ground-fault circuit interrupter	THERMO or T	Thermostat
GRB	Garbage	U RM	Utility room
GYP	Gypsum	V	Vent
HA	Hot air	VENT	Ventilation
HD	Head	VP	Vent pipe
HDR	Header	VS	Vent stack
HTD	Heated	WA	Warm air
HVAC	Heating, ventilating, and air conditioning	WC	Water closet
HW	Hot water	WIND	Window
INTR or INT	Interior	WTHPRF	Weatherproof
K or KIT	Kitchen	WV	Wall vent

Drawing Grid—Isometric

Drawing Grid—Oblique Cabinet

Isometric Ellipse Protractor

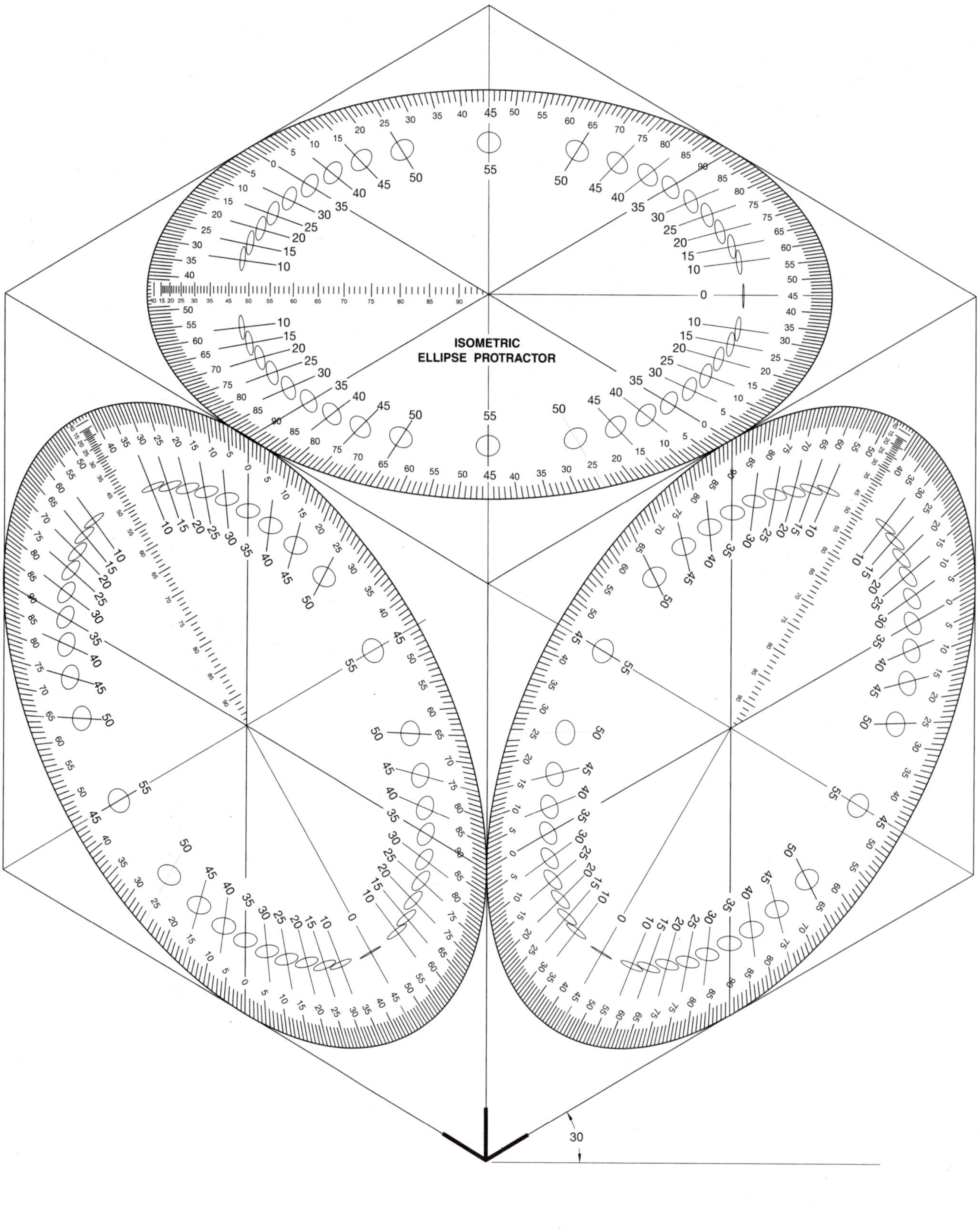

Glossary

A

abbreviation: A letter or group of letters that represents a term or phrase.

actual texture: A texture that is a physically tactile surface that can be felt or touched.

acute angle: An angle measuring less than 90°.

airbrush: A small spray gun that uses air to spray paint or other colored liquids onto a surface.

aligned section: A sectional view in which the cutting plane is bent to pass through angled features.

angle: The intersection of two lines.

angled plane: A plane that is not parallel to any of the axonometric axes.

angular dimension: A dimension that measures the degree of separation between two adjoining lines or points.

arc: A curved line that is part of the circumference of a circle.

arrowhead: A symbol that indicates the extent of a dimension.

ascender: The portion of a lowercase letter that rises above the height of the lowercase letter "x."

assembly print: A print that shows how two or more parts fit together.

auxiliary section: A sectional view that is not one of the principal planes.

auxiliary view: A view that shows the true shape of a plane or feature on an object that is not parallel to one of the six planes of projection.

axonometric drawing: A pictorial drawing that shows three sides of an object and contains no true view of any side.

B

background: The subjects or areas of a sketch that appear to recede toward the back of the image.

bar graph: A graph in which values of variables are represented by bars. The bars may be drawn horizontally or vertically.

base: The face of a polyhedron that is situated at the top or bottom.

basic dimension: A dimension that is used to describe a theoretical exact size, shape, or location of a feature on an object.

blocking in: The method of quickly marking the structure of a subject on a sketch by breaking the subject into common shapes and lines.

block shading: A method of shading that uses flat areas of black.

break line: A line that is used to indicate a partial view of a larger object or part.

broken-out section: A sectional view in which a small portion is removed and is designated by a freehand break line.

C

cabinet drawing: An oblique drawing with receding lines foreshortened to one-half the scale of lines in the true view.

cam: A machine part that transmits motion using an irregular surface.

cam follower: A machine part that is in contact with the cam and moves in a designated path.

cast shadow: The dark silhouette created by an object that blocks the light source and is cast onto another surface.

cavalier drawing: An oblique drawing with receding lines drawn to the same scale as the lines in the true view.

center line: A thin line that locates axes or center points of arcs and circles.

chord: A straight line with two end points on the circumference of a circle.

circle: A plane figure that is a regular, closed curve with all points on the circle an equal distance from a center.

circle graph: A graph in which a circle represents the total of a variable, and sectors of the circle represent parts of the total. Also called *pie chart.*

compass: An instrument used to draw arcs and circles.

composition: The arrangement and relationship of shapes and elements on the picture plane.

cone: A polyhedron with a circular or elliptical base and one face that wraps around the form, meeting at a vertex.

conic section: A curve produced by an intersecting plane on a three-dimensional cone form.

contour: The identifying outline that separates all or part of a subject from the background.

contour line: A line that shows points of the same elevation on a map of a land surface.

core shadow: The area of an object that is directly opposite the light source.

curved line: A line that bends and has no angles.

cutaway drawing: A pictorial drawing of an object or assembly that is cut into sections to show interior parts.

cutting-plane line: A thick line that shows where an object would have an imaginary cut in order to view internal features.

cylinder: A type of prism created by a straight axis rotating one full revolution between two parallel circular or elliptical bases.

D

datum: A theoretical exact point, axis, or plane that serves as the origin for the location or geometric characteristics of object features.

datum feature: A datum that is an actual feature of an object that is used to establish a datum.

datum target: A datum established by a point, line, or area designated on an object.

descender: The portion of a lowercase letter that drops below the main baseline of the line of letters.

detail: A portion of a drawing that is enlarged to show features more clearly.

detailed thread representation: A method of thread representation in which the threads are drawn as straight lines at the thread angle and the roots and crests are drawn as sharp "V's."

development: A pattern of a three-dimensional object that is laid out in a single flat plane with all views in true view.

dimension: A numerical value that gives the size, form, or location of objects on a drawing.

dimensioning: A method of adding dimensions to a drawing to indicate the geometrical characteristics of an object.

dimension line: A line that is used with dimensions to show the size or location.

dimetric drawing: An axonometric drawing with two axes drawn on equal angles to the third, but with an unequal angle between them.

direct tolerancing: The practice of specifying a dimension's permissible range directly within the dimensioning lines.

divider: An instrument that is used to transfer and compare measurements. Also called *hinge divider.*

dual dimension: A dimension that shows both English and metric dimensions.

E

elevation: An orthographic view of an object with the viewing plane parallel to one side of the object.

ellipse: An elongated circle that is formed when the sum of the distances from a point on the curve to two fixed focus points remains the same.

ellipsoid: A three-dimensional form that is created by rotating an ellipse on a single axis.

equilateral triangle: A triangle with three equal sides and three equal angles.

eraser: A handheld piece of rubber or rubber-like material that removes unwanted marks from paper.

erasing shield: A template that has cutouts in different sizes and shapes.

exploded view drawing: A pictorial drawing that shows an assembly pulled apart and its parts aligned to show their location within the whole assembly.

extension line: A line that extends from a surface feature of an object and ends a dimension line.

F

face: The surface of a side in a polyhedron.

feature: Any surface, angle, hole, round, or other characteristic of an object that can be dimensioned and controlled.

feature control: A method of identifying and controlling features on an object.

feature control frame: A rectangular frame divided into sections that encloses characteristic symbols, numerical tolerances, datum references, and modifier symbols.

figure: The subject of a sketch being drawn and emphasized.

first-angle projection: A projection method that places the object between the plane of projection and the observer.

flexible curve: An instrument made of flexible metal or plastic that can be bent by hand to form the necessary curve. Also called *adjustable curve.*

focus: A point on an ellipse that is found by drawing arcs with a radius equal to half the major axis and with a center at one end of the minor axis.

foreground: The subjects or areas in a sketch that appear to come forward to the front of an image.

foreshortening: A technique used to create an illusion that a surface or distance on an object not seen in true view is shorter than actual size.

form: The shape and structure of all or part of a subject and includes a sense of mass and volume.

french curve: A template shaped for different curves and used to draw smooth curves of varying radii. Also called *irregular curve.*

frontal plane: A plane that is parallel to the height and width axes.

front auxiliary view: A primary auxiliary view that has the inclined surface perpendicular to the front principal plane.

full section: A sectional view in which the cutting plane passes entirely through an object.

G

gear: A toothed wheel that is used in pairs to transmit power or motion from one part to another.

general note: A note that applies to the entire drawing.

general oblique drawing: A drawing with receding lines foreshortened at any scale between half and full.

geometric dimensioning and tolerancing: A method of specifying the size, shape, and location of features and their allowable tolerances.

graph: A diagram that shows the relationship between two or more variables.

ground: The area surrounding the figure.

H

half section: A sectional view in which two cutting planes are passed at right angles to each other along center lines or symmetrical axes.

helix: The curve traced on a cylinder or cone by a point rotating at a right angle to the axis.

hidden line: A thin line that represents the shape and edges of an object that cannot be seen.

highlight: A bright place on the illuminated surface of an object that is directly in front of the light source.

horizon: The eye level of the viewer.

horizontal line: A line that is parallel to the horizon.

horizontal plane: A plane that is parallel to the width and depth axes.

hyperbola: A curve that is formed when the difference of the distances from the foci remains the same.

hypotenuse: The longest side of a right triangle that is opposite the right angle.

I

illuminated surface: The area of an object that is turned toward the light source.

inclined line: A slanted line that is neither horizontal nor vertical.

individual feature: A feature that is an independent characteristic of a part.

inking: An illustration process where pencil lines are drawn over in ink.

interior arrangement drawing: A pictorial drawing where the exterior of an object or assembly is drawn transparent or lightly shaded in order to reveal interior parts. Also called *phantom drawing.*

isometric drawing: An axonometric drawing with the axes equally spaced 120° apart.

isometric piping diagram: An isometric diagram of a plumbing system.

L

layout: The method of developing the lines that form the outline for the finished shape.

lead: The distance along the axis for one revolution around the cylinder or cone.

leader: An inclined (slanted) line that connects a dimension, note, or specification with a particular feature of an object. Also called *leader line.*

limit dimensioning: A direct tolerancing method that includes only the maximum and minimum values of a dimension on a drawing.)

line: A mark that has a continuous length that connects one or more points.

linear dimension: A dimension that measures a straight distance and is commonly expressed in English or metric units of measure.

line graph: A graph in which points representing variables are connected by a line.

local note: A note that applies only to one or more specific items in a drawing.

location dimension: A dimension that uses the components of an object to locate an angle or feature on an object.

M

major axis: The axis in an ellipse that is longer.

major diameter: The outer boundary of the crest of an external thread or the root of an internal thread.

mechanical pencil: A pencil that uses an internal mechanism to push a thin lead through a hole at the tip of a lead holder.

middle ground: The subjects or areas in a sketch that exist between the foreground and the background.

minor axis: The axis in an ellipse that is shorter.

minor diameter: The inner boundary of the root of an external thread or the crest of an internal thread.

modifier symbol: A symbol that represents a specific condition of the part or feature when the tolerance is applied.

multiview drawing: A two-dimensional representation of a three-dimensional object shown with no distortion of lines or angles.

N

negative space: The space that both surrounds and fits between a form or part of a form.

nonisometric: Any feature that is not parallel to the height, width, or depth of the object, and therefore is not parallel to one of the three isometric axes.

nonthreaded fastener: A device that joins or fastens parts together without threads.

note: A line (or lines) of explanatory text that is added to a drawing.

O

oblique drawing: A pictorial drawing that shows one side of an object in true view with the remaining sides receding back in space.

oblique polyhedron: A polyhedron with an axis at an oblique angle to the base(s).

obtuse angle: An angle measuring more than 90°.

offset section: A sectional view in which the cutting-plane line is offset from the main axis to pass through features that are not located in a straight line.

one-point perspective: A drawing with one vanishing point on the horizon where parallel lines appear to recede.

orthographic projection: A method of representing the true shape of one view of an object onto a single plane.

P

parabola: A curve that is formed by all points being equidistant from a given point, focus, and line.

parallel: When two lines or planes remain the same distance apart and never intersect.

parallel line development: A method of surface development used on objects having parallel fold lines, such as prisms and cylinders.

parallelogram: A type of quadrilateral that has opposite sides parallel to each other.

pattern: A flat outline or stretchout of a developed surface.

perpendicular: When two lines or planes form a right angle.

perspective drawing: A pictorial drawing where all receding lines converge at one or more vanishing points.

phantom line: A thin line that shows an object's alternate position or a repeated detail.

pictograph: A graph in which symbols represent variables and their values.

pictorial drawing: A three-dimensional representation of an object.

picture plane: The two-dimensional space on paper where the flattened image of a three-dimensional subject is depicted.

pitch: The distance between corresponding points on adjacent threads measured parallel to the axis.

plan: An orthographic drawing of an object viewed from above and contains a horizontal cutting plane that passes through the object.

plane: A two-dimensional surface that has no depth.

plane figure: A two-dimensional shape bound by lines that exist on one plane.

plane of projection: The plane upon which the object is projected.

plus and minus tolerancing: A direct tolerancing method that provides an ideal dimension and includes the allowable deviations in the positive and negative directions.

point: An exact location in space that has no dimension.

point-to-point dimensioning: A dimensioning method that specifies sizes along an object from one feature to another.

point of view: The viewing position of the observer when viewing an object.

polar coordinate dimensioning: A dimensioning method that determines location using angle and radius dimensions.

polygon: A plane figure that has at least three straight sides.

polyhedron: A three-dimensional form that is constructed of plane figures.

positive space: The space that forms a shape that can be clearly seen against a background.

primary auxiliary view: A view that is projected to an auxiliary plane that is perpendicular to one of the principal planes but inclined to the remaining planes.

print: A drawing that shows the structure and related characteristics of an object, building, or site.

prism: A polyhedron with bases that are parallel equal polygons and three or more side faces that are parallelograms.

profile: A cross-section of a topographical map.

profile plane: A plane that is parallel to the height and depth axes.

projector: A line extending from an edge or contour of an object that passes through the plane of projection and the point of view.

proportion: The comparative size relationship between one part of a sketch or object to another.

protractor: A flat, semicircular tool that is used to measure angles.

pyramid: A polyhedron with a polygon for a bottom base and three or more triangular faces that connect in common vertex.

Q

quadrilateral: A polygonal plane figure that has four straight sides.

R

radial line development: A method of surface development used on objects that have fold lines that all radiate from one point.

radial lines: A series of lines that radiate from the same point.

rectangular coordinate dimensioning: A dimensioning method that specifies all dimensions from a common baseline.

reference dimension: A dimension derived from other dimensions and is indicated by the placement of parentheses around the dimension.

reflected light: Light that shines on an object after bouncing off other objects and therefore does not come directly from the light source.

regular polygon: A polygon that has all angles of equal size and all sides of equal length.

regular polyhedron: A polyhedron that has regular polygons for all of its faces.

related feature: A feature that is associated with a specific datum or datums.

removed section: A sectional view that is detached from the projected view and located elsewhere on the sheet.

rendering: An illustration technique that shades an image using blacks, whites, and intermediate grays to create a three-dimensional, photo-realistic appearance.

revolved section: A sectional view that shows the cross-sectional shape of elongated parts.

right polyhedron: A polyhedron with an axis perpendicular to the base(s).

right triangle: is a triangle with one right angle that measures 90°.

S

scale: An instrument used to measure and proportionally enlarge or reduce lines.

schedule: A chart used to conserve space and display information in a concise and organized format.

schematic diagram: A drawing that shows the electrical connections and functions of a circuit using graphic symbols.

schematic piping diagram: An orthographic diagram of a plumbing system without regard to scale and exact location of the fixtures shown.

schematic thread representation: A method of thread representation in which lines drawn perpendicular to the axis of the cylinder represent the roots and crests of the thread.

screw thread: A uniform ridge that forms a helix on the outside or inside surface of a cylinder or cone.

secondary (double) auxiliary view: A view that is projected onto a plane that is oblique to all of the principal planes.

sectional view or section: A drawing of a cross section of an object.

section lines: A grouping of thin, parallel lines that identify the internal features of an object cut by a cutting-plane line.

shaded surface: The area or areas of an object that are turned away from the light source.

shape: The extent of all or part of a subject that is often contained within a contour.

sheet note: A note that provides information for a specific item on the print on which it appears.

side auxiliary view: A primary auxiliary view that has the inclined surface perpendicular to the side principal plane.

sighting: A method of viewing and measuring a three-dimensional space in order to transfer components to their appropriate place and proportion on a sketch.

simplified thread representation: A method of thread representation in which hidden lines are drawn parallel to the axis of the cylinder at the approximate depth (minor diameter) of the thread.

size dimension: A dimension that gives the overall size of an angle or feature.

sketching: A two-dimensional visual representation of a subject.

specification: A supplemental document with written instructions and information that is not necessarily shown on the print.

sphere: A three-dimensional form that is created when a circle is rotated on a single axis.

spring: A mechanical device that stores and applies energy and regains its original shape after being compressed or extended.

straight-line shading: A method of shading that uses a series of straight or curved parallel lines.

surveying: The process of measuring land by recording elevations, noting directions and lengths of property lines, and noting sizes of structures.

symbol: A graphic representation of an object, quantity, or unit.

symmetrical object: An object in which one half is the mirror image of the other half.

symmetry line: A thin line that is used on a partial view of an object to indicate the remaining portion is symmetrical.

T

tabular dimensioning: A dimensioning method that assembles all numerical feature dimensions in a table.

tangent: A line or arc that shares a point and direction with a given curve in at least one location.

tangent point: The point where a line and an arc or two arcs touch.

technical pen: A specialized writing instrument that is used to make lines of consistent width on a technical drawing.

template: A flat piece of plastic or metal that has cutouts of commonly used shapes.

texture: The combination of tone and line that creates the illusion of a physical tactile surface.

thin section: A sectional view that is too thin to be shown by the ordinary cross-sectioning convention.

third-angle projection: A projection method that places the plane of projection between the object and the observer.

threaded fastener: A device that joins two or more parts together with screw threads.

three-point perspective: A drawing with two vanishing points on the horizon and a third vanishing point above or below the horizon where parallel lines appear to recede.

title block: The area of a print that contains identifying information about the contents of the print.

tolerance: The amount of measured variation allowed in a given dimension.

tone: A tint or shade of black, white, or color.

top auxiliary view: A primary auxiliary view that has the inclined surface perpendicular to the top principal plane.

topographic map: A map that depicts the surface features of an area.

triangle: 1. An instrument used to draw vertical and inclined lines. **2.** A polygonal plane figure with three straight sides that intersect to form three angles.

triangulation: A method of surface development that breaks an object into triangular plane surfaces.

trimetric drawing: An axonometric drawing with all axes drawn at different angles.

true view: A view where the face of the object being shown is parallel to the picture plane and has no distortion.

T-square: An instrument used to draw horizontal lines and to position triangles.

two-point perspective: A drawing with two vanishing points on the horizon where parallel lines appear to recede.

V

value: The relative lightness or darkness of an object as perceived by the viewer.

value contrast: The relationship between two areas of value.

vanishing point: A point on the horizon where parallel lines appear to meet.

vertex: The point opposite the base where the face(s) meet at a point.

vertical line: A line that is perpendicular to the horizon.

viewfinder: A viewing tool with a proportional window that can be used to sight the composition and accurately place objects in a sketch.

viewing-plane line: A thick line that shows the viewpoint of an additional drawing showing another view or removed portion of an object.

visible line: A thick line that defines the visible shape of an object. Also called *object line.*

visual movement: The visual path or paths the eye follows as it moves across the page.

visual texture: The illusion of texture as shown on a sketch.

volume: The three-dimensional space of all or part of a subject.

Z

zone: An area on a print designated by horizontal and vertical labels along the margin.

Index

A

abbreviation, 151, *151*
actual texture, 13, *14*
acute angle, 28
airbrush, 35, *35*
aligned section, 116, *117*
angled plane, 82, *82*
angles, 9, *9,* 28, 57, *57*
  acute, 28
  bisecting, 57, *57*
  obtuse, 28
angular dimension, 130, *130*, *133*
angular line, 28–29
arc, 24, 133, *133*
  bisecting, 56, *57*
  tangent, *61,* 61–63, *62, 63*
architect's scale, 24, *26*
architectural plans, 154, *155*
  floor, 154, *155*
  foundation, 154, *155*
  plat, 154, *155*
  plot, 154, *155*
  structural, 154, *155*
  utility, 154, *155*
architectural prints, 149, *150,* 154, *155*. *See also* architectural plans
  elevations
    exterior, 154, *156*
    interior, 154, *156*
arrowhead, *127,* 128, *128*
ascender, 37
assembly print, 161
  orthographic, 161, *162*
  pictorial, 161, *162*
atmospheric effects, 10, *11*
auxiliary section, 118, *118*
auxiliary view, 106, *106*
auxiliary view drawings, 108–112, *109*
  creating, 111–112, *112, 113*
  primary, 110–111
    front, 110, *110*
    partial, *110,* 111
    side, 110, *110*
    symmetrical, *110,* 111
    top, 110, *110*
  secondary, *111,* 111–112
axonometric drawing, 82–85
  dimetric, 84, *84*
  isometric, *82,* 82–84, *85*
  trimetric, 84, *84*

B

background, 4, *5*
bar graph, 173–174
  horizontal, 173, *173*
  vertical, 173, *173*
base, 51
basic dimension, 137, *137*
bird's eye view, 88, *89*
bisecting components, 56–57, *57*
  angle, 57
  arc, 56, *57*
  line, 56, *57*
blind contour drawing, 3, *3*
blocking in, 3–4
block shading, 34, *34*
bow compass, 24, *25*
bow divider, 24, *25*
break line, *31,* 32
broken-out section, 116, *117*

C

cabinet drawing, 87, *88*
cam, 181, *181*
cam follower, 181, *181*
cast shadow, 13, *13*
cavalier drawing, 87
center line, 30, *31*, 132
chord, 49, *50, 58*
circle, *3,* 49, *50, 58*
  chord, 49, *50, 58*
  circumference, 49, *50*
  diameter, 49, *50*
  geometric construction, 57–63, *58, 59, 60, 61, 62, 63*
  radius, 49, *50*
  tangent, 59, *59*
circle graph, 174, *174*
circular cone, 51, *52*
circular construction, 57–63
  circles, 58, *59, 60*
  tangents, 58–63, *59, 60, 61, 62, 63*
    arc, *61,* 61–63, *62, 63*
    circle, 59, *60*
    line to circle, 59–60, *60*
circumference, 49, *50*
civil prints, 150, *150,* 159, 161, *161*
  surveying, 159
compass, 24, *25*
  bow, 24, *25*
  drop-bow, 24, *25*
composition, 4, 5, *5*
  background, 3–4, *5*
  foreground, 3–4, *5*
  horizontal, *5*
  middle ground, 3–4, *5*
  vertical, *5*
  visual movement, 4–5, *6*
cone, 51, *52*
  circular, 51, *52*
  elliptical, 51, *52*
  oblique, 51, *52*

right, 51, *52*
vertex, 51, *52*
conic section, 69, *70*
circle, 69, *70*
ellipse, 69, *70*
hyperbola, 69, *70*
parabola, 69, *70*
contour, 1, *1*
contour line, 175, *175*
contrast, 11, *12*
high, 11, *12*
low, 11, *12*
medium, 11, *12*
core shadow, 13, *13*
crest, 179, *179*
cross-hatching, 11, *12*
curved line, 29, *30*
cutaway drawing, 93, *93*
cutting-plane lines, *31*, 31–32, 114, *114*
cylinder, 51, *52*
oblique, 51, *52*
right, 51, *52*

D

datum, 138
datum feature, 138, *139*
datum target, 138, *139*
deltoid, *50,* 51
depth illusion, 9–10
atmospheric effects, 10, *11*
overlapping shapes, 9, *10*
relative size, 9, *10*
vertical positioning, 10, *10*
descender, 37
detail, 106
detail drawings, 154, *154*
detailed thread representation, *179,* 180
detail prints, 161, *161*
developments, *176,* 176–178
layout, 176
parallel line, 176, *177*
radial line, 177, *177*
triangulation, *177,* 177–178
diameter, 49, *50*
dimension, 127
dimensioning, 127–133
indicating, 27–128
arrowhead, *127,* 128, *128*
dimension line, 127, *127*, *128*
extension line, *127,* 128, *129*
leader, *127,* 128, *129*
limit, 134, *134*
methods, 135–137
point-to-point, 135, *135*
polar coordinate, 137, *137*
rectangular coordinate, 135, *136*
tabular, 135, *136*
notes, 130–131
general, 130–131, *131*
local, 131, *131*
placement. *See* dimension placement
types. *See* dimension types
units of measure, 130, *130*
dimension line, 127, *127*, *128*
dimension placement, *131,* 131–133
angle, 133, *133*
arc, 133, *133*
circles, 132, *132*
reference, 131
dimension types, 129–130
angular, 130, *130*
dual, 129, *129*
linear, 129–130, *129*
dimension units of measure, 130, *130*
dimetric drawing, 84, *84*
direct tolerancing, 134, *134*
divider, 24, *25*
bow, 24, *25*
proportional, 24, *25*
dividing lines, 55, *56*
double auxiliary view. *See* secondary auxiliary view
drawing lines, 28–29, 68–70
angular, 28–29
curved, 29, *30*
dividing, 55, *56*
horizontal, 28
inclined, *28,* 28–29, *29*
parallel, 53, *54*
perpendicular, 53–54, *55*
tangent, 53
tangent point, 53
vertical, 28
drop-bow compass, 24, *25*
dual dimension, 129, *129*

E

electrical diagrams, 159, *160*
ladder (line), 159, *160*
one-line, 159, *160*
wiring, 159, *160*
electrical prints, 150, *150, 159,* 159–162
electrical diagrams, 159, *160*
schematic diagrams, 159, *160*
elevation, 150, *150,* 154
exterior, 154, *156*
interior, 154, *156*
ellipses, *3,* 69–71, *71*
foci method, 70, *71*
focus, 69
major axis, 69–71
minor axis, 69–71
parallelogram method, 70, *71*
ellipsoid, 51, *52*
elliptical cone, 51, *52*
engineer's scale, 24, *26*
enlarging shapes, 68, *69*
using radial lines, 68, *69*
equilateral triangle, 50, *64*
eraser, 23
erasing shield, 23, *23*
exploded view drawing, 93, *94*
extension line, *127,* 128, *129*
exterior elevation, 154, *156*
eye-level view, 88, *89*

F

face, 51
feature, 138
feature control, 138–140
frame, 138, 140, *140*
individual, 138–140
related, 138–140
figure, 5, *6*
fire protection system prints, 158, *158*
first-angle projection, 105, *105, 106*
flank, 179, *179*
flexible curve, 26, *27*
floor plans, 154, *155*
foci method, 70
focus, 69–71
foreground, 4, *5*
foreshortening, 82
form, 1, *1*
foundation plans, 154, *155*
french curve, 26, *27*
frontal plane, 82, *82, 83*
front auxiliary view, 110, *110*
full section, 115, *115*

G

gear, 181, *181*
general note, 130–131, *131,* 152, *153*
general oblique drawing, 87

geometric construction, 53–74
bisecting, 56–57, *57*
angle, 57
arc, 56, *57*
line, 56, *57*
circular construction, 57–63, *58, 59, 60, 62, 63*
circles, 58, *59, 60*
tangents, 58–63, *59, 62, 63*
complex construction, 69–74, *70*
conic section, 69, *70*
ellipse, 69–70, *71*
helix, 73–74, *74*
hyperbola, 72–73, *73*
parabola, 72
drawing lines, 53–55
dividing, 55, *56*
parallel, 53, *54*
perpendicular, 53–54, *55*
enlarging shapes, 68, *69*
helix, 73–74, *74*
plane figures, 63–68, *64, 65, 66, 67, 68, 69*
polygon, *66,* 66–68, *67, 68, 69*
quadrilateral, 64–65, *65*
triangle, 63–64, *64*
reducing shapes, 68, *69*
geometric dimensioning and tolerancing, 137–140
datum, 138
features, 138, *139*
targets, 138, *139*
defined, 137
feature control, 138–140
frame, 140, *140*
individual, 138–140
related, 138–140
geometric dimensioning, 137–138
basic, 137, *137*
reference, 138, *138*
geometric shapes, 49–52, *50, 52*
three-dimensional solids, 51–52
ellipsoid, 51, *52*
polyhedron, 51–52, *52*
sphere, 51, *52*
two-dimensional shapes, 49–51, *50*
circle, 49, *50*
polygon, 50, *50*
quadrilateral, *50,* 51
triangle, *50,* 50–51
geometric terminology, 49
parallel, 49
perpendicular, 49
plane, 49
point, 49
graph, 171–174, *173*
bar, 173–174
horizontal, 173, *173*
vertical, 173, *173*
circle, 174, *174*
line
broken, 172, *173*
curved, 172, *173*
straight, 172, *173*
pictograph, 174, *174*
ground, 5, *6*

H

half section, 116, *116*
hatching, 11, *12*
helix, 73–74, *74*
lead, 73, *74*
left-handed, 73–74, *74*
right-handed, 73–74, *74*
hexagon, 50, *66, 68*
hidden line, 30, *31*
highlight, 13, *13*
hinge divider, 24
horizon, 88, *89*
horizontal line, *3,* 28, *28*
horizontal plane, 82, *82, 83*
HVAC prints, 158, *158*
hyperbola, *72,* 72–73, *73*
hypotenuse, *50,* 51, *64*

I

illuminated surface, 13, *13*
illustrating technical drawings, 32–35
inking, 33, *33*
rendering, 35, *35*
shading, 33–34, *34*
block, 34, *34*
straight-line, 34, 34
inclined line, *28,* 28–29, *29*
angular line, 28–29
indicating dimensions, 127–128
arrowhead, *127,* 128, *128*
dimension line, 127, *128, 127*
extension line, *127,* 128, *129*
leader, *127,* 128, *129*
individual feature, 138–140
information representation, 171–174
graph, 171–174, *173*
bar, *173,* 173–174
circle, 174, *174*
line, 172–173, *173*
pictograph, 174, *174*
symbols, 171, *172*
inking, 33, *33*
instrument drawings, *21,* 21–27
standard tools, 21–23
eraser, 23
erasing shield, 23, *23*
paper, 21, *21*
pencil, 22, *22*
pen, 22, *23*
technical drawing instruments, 23–26
compass, 24, *25*
divider, 24, *25*
flexible curve, 26, *27*
french curve, 26, *27*
protractor, 26, *27, 29*
scale, 24–26, *26*
template, 26, *27*
triangle, 23, *24, 29*
T-square, 23, *24*
interior arrangement drawings, 92, *93*
interior elevation, 154, *156*
isometric drawings, *82,* 82–84, *83, 85*
circles, 85–86
isometric piping diagrams, 157

L

ladder (line) diagrams, 159, *160*
layout, 176–178
lead, 73
leader, *127,* 128, *129*
leader line, *127,* 128, *129*
letter formation, *36,* 37
lettering technical drawings, 35–37
formation, *36,* 37
sizing, 35–37, *36*
ascender, 37
descender, 37
lowercase, *36,* 37
uppercase, 36, *36*
spacing, 37, *37*
letter sizing, 35–37
lowercase, *36,* 37
uppercase, 36, *36*
letter spacing, 37, *37*
letter width, *36,* 37
lighting effects
cast shadow, 13, *13*
core shadow, 13, *13*
highlight, 13, *13*

illuminated surface, 13, *13*
reflected light, 13, *13*
shaded surface, 13, *13*
limit dimensioning, 134, *134*
line
bisecting, 56, *57*
angle, 57
arc, 56, *57*
breaks, *32*
defined, 1
types, 30–32, *31*
break, *31,* 32
center, 30, *31*
cutting-plane, *31,* 31–32
hidden, 30, *31*
parallel, 53, *55*
perpendicular, 53–54, *55*
phantom, *31,* 32
section, *31,* 32
symmetry, 31, *31*
viewing-plane, *31,* 31–32
visible, 30, *31*
weights, 32, *32*
linear dimension, 129, *129*
line diagram, 159, *160*
line drawing, 27–32
angular, 28–29
curved, 29, *30*
horizontal, 28, *28*
inclined, *28,* 28–29, *29*
vertical, 28, *28*
line graphs
broken, 172, *173*
curved, 172, *173*
straight, 172, *173*
line weights and breaks, 32, *32*
local note, 131, *131*
location dimension, 127
lowercase lettering, 36, *36*

M

major axis, 69–71
major diameter, 179, *179*
manufacturing elements, 176–181
cams, 181, *181*
cam follower, 181, *181*
development, *176,* 176–178
layout, 176
parallel line, 176, *177*
radial line, 177, *177*
triangulation, *177,* 177–178
gears, 181, *181*
nonthreaded fasteners, 180
keys, 180, *180*
nails, 180, *180*
pins, 180, *180*
rivets, 180, *180*
staples, 180, *180*
sheet metal patterns, 178, *178*
springs
flat, 180, *180*
helical, 180, *180*
threaded fasteners, *178,* 178–180
parts, 179, *179*
representation, *179,* 179–180
manufacturing prints
assembly, 161
orthographic, 161, *162*
pictorial, 161, *162*
detail, 161, *161*
materials
paper, 2, 21, *21*
pencils, *2,* 22, *22*
mechanical engineer's scale, 25, *26*
mechanical pencil, 22, *22*
mechanical prints, 149, *150,* 156–157
fire protection system, 158, *158*
HVAC, 158, *158*
plumbing, 157, *157*
piping diagrams, 157, *157*
middle ground, 4, *5*
minor axis, 69–71
minor diameter, 179, *179*
modifier symbol, 140
multiview drawing, 81, *82,* 103–108, *108*
creating, 107–108, *109*
orthographic projection, 103, *104*
plane of projection, 103
point of view, 103
projector, 103
projection methods, 103–106
first-angle, 105, *105, 106*
third-angle, *105, 106,* 106
view placement, 106–107
arrangement, 106, *106*
reference arrows, 107, *107*

N

negative space, *6,* 7
nonisometric, 83
nonthreaded fastener
keys, 180, *180*
nails, 180, *180*
pins, 180, *180*
rivets, 180, *180*
staples, 180, *180*
note, 130–131, 152
general, 130–131, *131,* 152, *153*
local, 131, *131*
sheet, 152, *153*

O

oblique cone, 51, *52*
oblique cylinder, 51, *52*
oblique drawing, *86,* 86–87
cabinet, 87, *87, 88*
cavalier, 87, *87*
general, 87, *87*
oblique polyhedron, 51, *52*
oblique prism, 51, *52*
oblique pyramid, 52
obtuse angle, 28
octagon, 50, *66, 69*
offset section, 116, *117*
one-line diagrams, 159, *159–161, 160*
one-point perspective, 88–90, *89, 91*
orthographic print, 161, *162*
orthographic projection, 103, *104*
plane of projection, 103
point of view, 103
projector, 103
overlapping shapes, 9, *10*

P

paper
sizes, 21, *21*
types, 2
parabola, *72,* 72–73
parallel, 49
parallel line, 53, *54*
parallelogram, *50,* 51
parallelogram method, 70
parallel line development, 176, *177*
partial auxiliary view, *110,* 111
pattern, 178, *178*
pencil grades, 2, *2*
pencil types
mechanical, 22, *22*
wood, 22, *22*
pens, 22, *23*
technical, 22, *23*
pentagon, 50, *66, 67*
perpendicular, 49
perpendicular line, 53–54, *55*

perspective drawings, 88–91, *89*
  one-point, 88–90, *89, 91*
  three-point, *89,* 90–91, *92*
  two-point, *89,* 90, *91*
  views, 88, *89*
phantom drawing, 92, *93*
phantom line, *31,* 32
pictograph, 174, *174*
pictorial drawings, 81–93
  axonometric, *81,* 82–85
    dimetric, 84, *84*
    isometric, *82,* 82–84, *83, 85*
    trimetric, 84, *84*
  cutaway, 93, *93*
  exploded view, 93, *94*
  interior arrangement, 92, *93*
  oblique, *81, 86,* 86–87
    cabinet, 87, *87, 88*
    cavalier, 87, *87*
    general, 87, *87*
  perspective, *81,* 88–91, *89*
    one-point, 88–90, *89, 91*
    three-point, *89,* 90–91, *92*
    two-point, *89,* 90, *91*
pictorial print, 161, *162*
picture plane, 1
pie chart, 174, *174*
piping diagrams, 157
  isometric, 157
  schematic, 157
pitch, 179, *179*
plane, 49, 82
  angled, 82, *82*
  frontal, 82, *82, 83*
  horizontal, 82, *82, 83*
  profile, 82, *82, 83*
plane figures, 49–51, 63–68
  circle, 49, *50*
  polygon, 50, *66,* 66–68, *67, 68, 69*
  quadrilateral, *50,* 51, 64–65, *65*
  triangle, *50,* 50–51, 63–64, *64*
plane of projection, 103
plans, 150, *150*, 154, *155*
plat plans, 154, *155*
plot plans, 154, *155*
plumbing prints, 157, *157*
  piping diagrams, 157, *157*
    isometric, 157, *157*
    schematic, 157, *157*
plus and minus tolerancing, *134,* 134–135
point, 49
point-to-point dimensioning, 135, *135*
point of view, 103
polar coordinate dimensioning, 137, *137*
polygons, 50, *66,* 66–68, *67, 68, 69*
  hexagon, 50, *66, 68*
  octagon, 50, *66, 69*
  pentagon, 50, *66, 67*
  quadrilateral, *50,* 51, 64–65, *65*
  regular, 50, *50*
  triangle, *50,* 50–51
polyhedrons
  base, 51, *52*
  cone, 51, *52*
  cylinder 51, *52*
  face, 51
  oblique, 51, *52*
  prism, 51, *52*
  pyramid, 52
  regular, 51, *52*
  right, 51, *52*
positive space, *6,* 7
primary auxiliary view, 110–111
  creating, 111, *112*
  front, 110, *110*
  partial, *110,* 111
  side, 110, *110*
  symmetrical, *110,* 111
  top, 110, *110*
print format and features, 150–154
  abbreviations, 151, *151*
  detail drawings, 154, *154*
  elevation, 150, *150*
  notes
    general, 152, *153*
    sheet, 152, *153*
  plan, 150, *150*
  revision block, 151, *151*
  schedule, 151, *152*
  section drawings, 154, *154*
  specifications, 152, *153*
  symbols, 151, *151*
  title block, 151, *151*
  zone, 152, *152*
printreading, 149–150
prints, *149,* 149–150
  architectural, 149, *150*
  civil, 150, *150*
  electrical, 150, *150*
  mechanical, 149, *150*
  structural, 149, *150*
prism, 51, *52*
  oblique, 51, *52*
  right, 51, *52*
profile, 175, *175*
profile plane, 82, *82, 83*
projection methods
  first-angle, 105, *105, 106*
  third-angle, *105,* 106, *106*
projector, 103
proportion, 7
  estimating, *8*
proportional divider, 24, *25*
protractor, 26, *27, 29*
pyramid, 52, *52*

Q

quadrilateral, *50,* 51, 64–65, *65*
  deltoid, *50,* 51
  parallelogram, *50,* 51
  rectangle, *50,* 51
  rhomboid, *50,* 51
  rhombus, *50,* 51
  square, *50,* 51
  trapezium, *50,* 51
  trapezoid, *50,* 51

R

radial line, 68, *69*
radial line development, 177, *177*
radius, 49, *50*
rectangle, *50,* 51
rectangular coordinate dimensioning, 135, *136*
reducing shapes, 68, *69*
reference arrows, 107, *107*
reference dimension, 131, 138, *138*
reflected light, 13, *13*
regular polygon, 50, *50*
regular polyhedron, 51, *52*
related feature, 138–140
relative size, 9, *10*
removed section, 118, *118*
rendering, 35
revision block, 151, *151*
revolved section, 117, *118*
rhomboid, *50,* 51
rhombus, *50,* 51
right cone, 51, *52*
right cylinder, 51, *52*
right polyhedron, 51, *52*
right prism 51, *52*
right pyramid, 52, *52*
right triangle, *50,* 50–51, *64*
roots, 179, *179*

S

scale, 24–26, *26*
  architect's, 24, *26*
  engineer's, 24, *26*
  mechanical engineer's, 25, *26*
schedule, 151, *152*
schematic diagram, 159, *160*
schematic piping diagram, 157
schematic thread representation, *179,* 180
screw thread, 178–179
scribbling, 11, *12*
secondary auxiliary view, *111,* 111–112
  creating, 112
section, 112
sectional view, 112
sectional view drawings, 112–119, *113*
  cutting-plane lines, 114, *114*
  section lines, 114–115, *115*
  types, 115–119
    aligned section, 116, *117*
    auxiliary section, 118, *118*
    broken-out section, 116, *117*
    full section, 115, *115*
    half section, 116, *116*
    offset section, 116, *117*
    removed section, 118, *118*
    revolved section, 117, *118*
    thin section, 119, *119*
  section drawings, 154, *154*
  section lines, *31,* 32, 114–115, *115*
  shaded surface, 13
  shading, 33–34, *34*
    block, 34, *34*
    straight-line, 34, *34*
  shapes, 1, *1,* 68, *69*
    enlarging, 68, *69*
    reducing, 68, *69*
  sheet metal pattern, 178, *178*
  sheet note, 152, *153*
  side auxiliary view, 110, *110*
  sighting, 7
    viewfinder, *7,* 7–8
  simplified thread representation, *179,* 180
  size dimension, 127
  sizing
    letters, 35–37, *36*
      ascender, 37
      descender, 37
      lowercase, *36,* 36–37
      uppercase, 36, *36*
  paper, 21, *21*
sketching
  basics, 2–4
    blocking in, 3–4
    contour drawings, *3*
    warming up, 2-3, *3*
  composition, 4, *5*
    horizontal, *5*
    vertical, *5*
  defined, 1
  depth illusion, 9–10
  figure, 13
  ground, 13
  materials, *2,* 13
  picture plane, 1
  sighting, 13–14
  texture, 1, 11–13, *14*
  tone, 11
space
  negative, *6,* 7
  positive, *6,* 7
spacing
  letter, 37, *37*
specifications, 152, *153*
sphere, 51, *52*
springs, 180, *180*
square, *50,* 51
standard tools, 21–23
  eraser, 23
  erasing shield, 23, *23*
  paper, 21
  pencil, 22, *22*
  pen, 22, *23*
stippling, 11, *12*
straight-line shading, 34, *34*
structural plans, 154, *155*
structural prints, 149, *150, 156,* 156
surface treatment
  actual texture, 13, *14*
  visual texture, 13–14, *14*
surveying, 159
symbol, 151, *151,* 171, *172*
symmetrical auxiliary view, *110,* 111
symmetrical object, 116
symmetry line, 31, *31*

T

tabular dimensioning, 135, *136*
tangent, 53, 58–63, *59, 60,* 61, *62, 63*
  arc, *61,* 61–63, *62, 63*
  circle, *59, 60,* 61
  line to circle, *59,* 59–60, *60*
tangent point, 53, *53,* 58–63
technical drawing instruments, 23–26
  compass, 24, *25*
    bow, 24, *25*
    drop-bow, 24, *25*
  divider, 24, *25*
    bow, 24, *25*
    proportional, 24, *25*
  flexible curve, 26, *27*
  french curve, 26, *27*
  protractor, 26, *27, 29*
  scale, 24–26, *26*
    architect's, 24, *26*
    engineer, 24, *26*
    mechanical engineer's, 25, *26*
  template, 26, *27*
    flexible curve, 26, *27*
    french curve, 26, *27*
  triangle, 23, *24, 29*
  T-square, 23, *24*
technical pen, 22, *23*
template, 26, *27*
texture, 1, 11–13, *14*
thin section, 119, *119*
third-angle projection, *105,* 106, *106*
threaded fastener, 178–180
  bolts, 178, *178*
  nuts, 178, *178*
  parts, 179
    crests, 179, *179*
    flanks, 179, *179*
    major diameter, 179, *179*
    minor diameter, 179, *179*
    pitch, 179, *179*
    roots, 179, *179*
  representation, 179–180
    detailed, *179,* 180
    schematic, *179,* 180
    simplified, *179,* 180
  screws, 178, *178*
  studs, 178, *178*
thread parts, 179
  crests, 179, *179*
  flanks, 179, *179*
  major diameter, 179, *179*
  minor diameter, 179, *179*
  pitch, 179, *179*
  roots, 179, *179*
thread representation, 179–180
  detailed, *179,* 180
  schematic, *179,* 180
  simplified, *179,* 180
three-dimensional solids, 51–52, *52*

ellipsoid, 51, *52*
polyhedron, 51–52, *52*
cone, 51, *52*
cylinder, 51, *52*
oblique, 51, *52*
prism, 51, *52*
pyramid, 52
regular, 51, *52*
right, 51, *52*
sphere, 51, *52*
three-point perspective, *89,* 90–91, *92*
title block, 151, *151*
tolerance, 134
tolerancing, 134–137
dimensioning methods, 135–137
point-to-point, 135, *135*
polar coordinate, 137, *137*
rectangular coordinate, 135, *136*
tabular, 135, *136*
direct, 134, *134*
limit dimensioning, 134, *134*
plus and minus, 134–135, *134*
tone, 1
top auxiliary view, 110, *110*
topographic map, 175, *175*
topography
contour lines, 175, *175*
profile, 175, *175*
trapezium, *50,* 51
trapezoid, *50,* 51
triangle, 23, *24, 29, 50,* 50–51, 63–64, *64*
equilateral, 50, *64*
hypotenuse, *50,* 51, *64*
right, 50, *64*
triangulation, *177,* 177–178
trimetric drawing, 84, *84*
true view, 81, *83*
T-square, 23, *24*
two-dimensional shapes, 49–51, *50*
circle, 49, *50*
chord, 49, *50*
circumference, 49, *50*
diameter, 49, *50*
radius, 49, *50*
polygon, 50, *50*
hexagon, 50, *50*
octagon, 50, *50*
pentagon, 50, *50*
quadrilateral, 50, *50*
regular, *50,* *50*
triangle, 50, *50*
quadrilateral, *50,* 51
deltoid, *50,* 51
parallelogram, *50,* 51
rectangle, *50,* 51
rhomboid, *50,* 51
rhombus, *50,* 51
square, *50,* 51
trapezium, *50,* 51
trapezoid, *50,* 51
triangle, *50,* 50–51
equilateral, 50
hypotenuse, *50,* 51
right, *50,* 50–51
two-point perspective, *89,* 90, *91*

U

uppercase lettering, 36, *36*
utility plans, 154, *155*

V

value, 11
cross-hatching, 11, *12*
hatching, 11, *12*
scribbling, 11, *12*
stippling, 11, *12*
value contrast, 11, *12*
vanishing point, 88, *89*
vertex, 51, *52*
vertical line, *3,* 28, *28*
vertical positioning, 10, *10*
view arrangement, 106, *106*
viewfinder, *7,* 7–8
viewing-plane line, *31,* 31–32
view placement, 106–107
arrangement, 106, *106*
reference arrows, 107, *107*
view relationships, 103
glass box, 103, *104*
views, 88, *89*
visible line, 30, *31*
visual movement, 4–5, *6*
circular, *6*
diagonal, *6*
horizontal, *6*
vertical, *6*
visual texture, 13–14, *14*
volume, 1, *1*

W

warming up, 2-3, *3*
circles and ellipses, *3*
coils, *3*
horizontal lines, *3*
vertical lines, *3*
wiring diagrams, 159, *160*
wood pencil, 22
worm's eye view, 88, *89*

Z

zone, 152, *152*

USING THE TECHNICAL SKETCHING AND DRAWING CD-ROM

Before removing the CD-ROM from the protective sleeve, please note that the book cannot be returned for refund or credit if the CD-ROM sleeve seal is broken.

System Requirements

To use this Windows®-compatible CD-ROM, your computer must meet the following minimum system requirements:

- Microsoft® Windows Vista™, Windows XP®, Windows 2000®, or Windows NT® operating system
- Intel® Pentium® III (or equivalent) processor
- 256 MB of available RAM
- 90 MB of available hard-disk space
- 800 × 600 monitor resolution
- CD-ROM drive
- Sound output capability and speakers
- Microsoft® Internet Explorer 5.5, Firefox® 1.0, or Netscape® 7.1 web browser and Internet connection required for Internet links

Opening Files

Insert the CD-ROM into the computer CD-ROM drive. Within a few seconds, the home screen will be displayed allowing access to all features of the CD-ROM. Information about the usage of the CD-ROM can be accessed by clicking on USING THIS CD-ROM. The Quick Quizzes®, Illustrated Glossary, Flash Cards, Interactive Tutorials, Interactive Exercises, Print and Drawing Library, Media Clips, and ATPeResources.com can be accessed by clicking on the appropriate button on the home screen. Clicking on the American Tech web site button (www.go2atp.com) accesses information on related educational products. Unauthorized reproduction of the material on this CD-ROM is strictly prohibited.

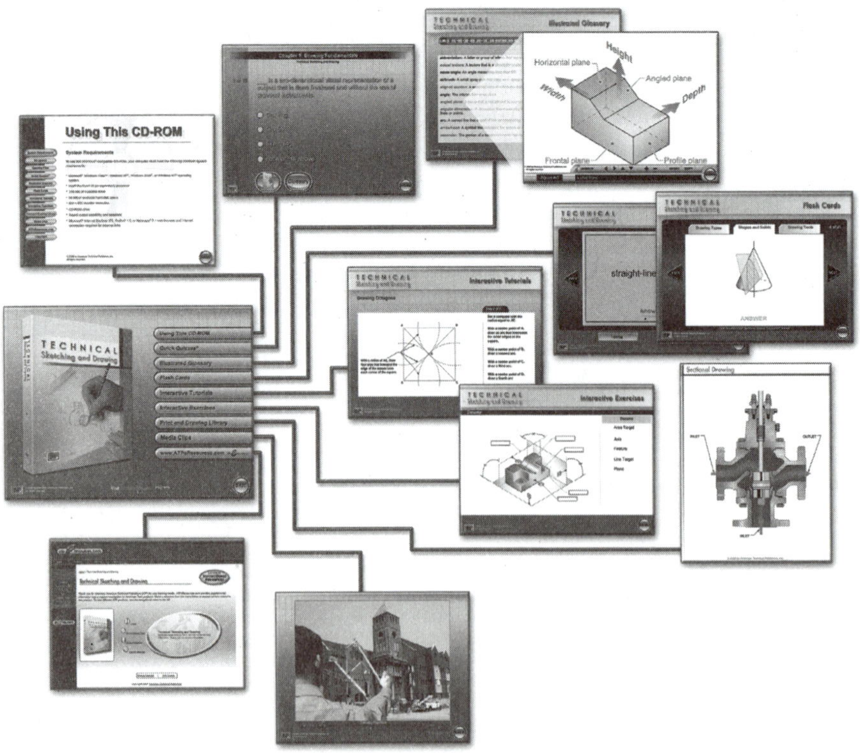

Intel and Pentium are registered trademarks of Intel Corporation or its subsidiaries in the United States and other countries. Microsoft, Windows Vista, Windows XP, Windows 2000, Windows NT, and Internet Explorer are either registered trademarks or trademarks of Microsoft Corporation in the United States and/or other countries. Netscape is a registered trademark of Netscape Communications Corporation in the United States and other countries. Firefox is a registered trademark of Mozilla Corporation in the United States and other countries. Quick Quiz and Quick Quizzes are registered trademarks of American Technical Publishers, Inc. All other trademarks are the properties of their respective owners.